MODERN METHODS OF SPEECH PROCESSING

THE KLUWER INTERNATIONAL SERIES IN ENGINEERING AND COMPUTER SCIENCE

VLSI, COMPUTER ARCHITECTURE AND DIGITAL SIGNAL PROCESSING

Consulting Editor
Jonathan Allen

Other books in the series:

FORMAL SEMANTICS FOR VHDL, Carlos Delgado Kloos
ISBN: 0-7923-9552-2
ON OPTIMAL INTERCONNECTIONS FOR VLSI, Andrew B. Kahng, Gabriel Robins
ISBN: 0-7923-9483-6
SIMULATION TECHNIQUES AND SOLUTIONS FOR MIXED-SIGNAL COUPLING IN INTEGRATED CIRCUITS, Nishath K. Verghese, Timothy J. Schmerbeck, David J. Allstot
ISBN: 0-7923-9544-1
MIXED-MODE SIMULATION AND ANALOG MULTILEVEL SIMULATION, Resve Saleh, Shyh-Jye, A. Richard Newton
ISBN: 0-7923-9473-9
CAD FRAMEWORKS: Principles and Architecutres, Pieter van der Wolf
ISBN: 0-7923-9501-8
PIPELINED ADAPTIVE DIGITAL FILTERS, Naresh R. Shanbhag, Keshab K. Parhi
ISBN: 0-7923-9463-1
TIMED BOOLEAN FUNCTIONS: A UNIFIED FORMALISM FOR EXACT TIMING ANALYSIS, William K. C. Lam, Robert K. Brayton
ISBN: 0-7923-9454-2
AN ANALOG VLSI SYSTEM FOR STEREOSCIPIC VISION, Misha Mahowald
ISBN: 0-7923-9444-5
ANALOG DEVICE-LEVEL LAYOUT AUTOMATION, John M. Cohn, David J. Garrod, Rob A. Rutenbar, L. Richard Carley
ISBN: 0-7923-9431-3
VLSI DESIGN METHODOLOGIES FOR DIGITAL SIGNAL PROCESSING ARCHITECTURES, Magdy A. Bayoumi
ISBN: 0-7923-9428-3
CIRCUIT SYNTHESIS WITH VHDL, Roland Airiau, Jean-Michel Berge, Vincent Olive
ISBN: 0-7923-9429-1
ASYMPOTIC WAVEFORM EVALUATION, Eli Chiprout, Michel s. Nakhla
ISBN: 0-7923-9413-5
WAVE PIPELINING: THEORY AND CMOS IMPLEMENTATION, C. Thomas Gray, Wentai Liu, Ralph K. Cavin, III
ISBN: 0-7923-9398-8
CONNECTIONIST SPEECH RECOGNITION: A Hybrid Appoach, H. Bourlard, N. Morgan
ISBN: 0-7923-9396-1
BiCMOS TECHNOLOGY AND APPLICATIONS, SECOND EDITION, A.R. Alvarez
ISBN: 0-7923-9384-8
TECHNOLOGY CAD-COMPUTER SIMULATION OF IC PROCESSES AND DEVICES, R. Dutton, Z. Yu
ISBN: 0-7923-9379
VHDL '92, THE NEW FEATURES OF THE VHDL HARDWARE DESCRIPTION LANGUAGE, J. Bergé, A. Fonkoua, S. Maginot, J. Rouillard
ISBN: 0-7923-9356-2
APPLICATION DRIVEN SYNTHESIS, F. Catthoor, L. Svenson
ISBN:0-7923-9355-4

MODERN METHODS OF SPEECH PROCESSING

edited by

Ravi P. Ramachandran
Richard J. Mammone
CAIP Center, Rutgers University

KLUWER ACADEMIC PUBLISHERS
Boston / Dordrecht / London

Distributors for North America:
Kluwer Academic Publishers
101 Philip Drive
Assinippi Park
Norwell, Massachusetts 02061 USA

Distributors for all other countries:
Kluwer Academic Publishers Group
Distribution Centre
Post Office Box 322
3300 AH Dordrecht, THE NETHERLANDS

Library of Congress Cataloging-in-Publication Data

A C.I.P. Catalogue record for this book is available
from the Library of Congress.

Printed on acid-free paper.

Printed in the United States of America

To my wife, parents and grandparents.

R.P.R.

To my wife and sons.

R.J.M.

CONTENTS

CONTRIBUTORS

Jont B. Allen
AT&T Bell Laboratories
Murray Hill, New Jersey

John S. Collura
Department of Defense
Ft. Meade, Maryland

Vladimir Cuperman
Simon Fraser University
Burnaby, B. C., Canada

Kevin R. Farrell
Dictaphone Corporation
Stratford, Connecticut

Oded Ghitza
AT&T Bell Laboratories
Murray Hill, New Jersey

P. S. Gopalakrishnan
IBM T. J. Watson Research Center
Yorktown Heights, NY

Vishwa Gupta
Bell Northern Research
Montreal, Canada

Jesper Haagen
Tele Denmark Research
Horsholm, Denmark

B. H. Juang
AT&T Bell Laboratories
Murray Hill, New Jersey

Yu-Hung Kao
Texas Instruments
Dallas Texas

W. Bastiaan Kleijn
AT&T Bell Laboratories
Murray Hill, New Jersey

Stephen V. Kosonocky
IBM T. J. Watson Research Center
Yorktown Heights, NY

Peter Kroon
AT&T Bell Laboratories
Murray Hill, New Jersey

Matthew Lennig
Bell Northern Research
Montreal, Canada

Peter Lupini
Simon Fraser University
Burnaby, B. C., Canada

Richard J. Mammone
Rutgers University
Piscataway, New Jersey

Lorin Netsch
Texas Instruments
Dallas Texas

Stephen Oh
Texas Instruments
Dallas Texas

James Ooi
Massachusetts Institute of Technology
Cambridge, Massachusetts

Douglas O'Shaughnessy
INRS Telecommunications
Montreal, Canada

P. K. Rajasekaran
Texas Instruments
Dallas Texas

Ravi P. Ramachandran
Rutgers University
Piscataway, New Jersey

Michael A. Ramalho
Bell Communications Research
Red Bank, New Jersey

Jan Robin Rohlicek
BBN HARK Systems Corporation
Cambridge, Massachusetts

Vishu Viswanathan
Texas Instruments
Dallas Texas

PREFACE

The term speech processing refers to the scientific discipline concerned with the analysis and processing of speech signals for getting the best benefit in various practical scenarios. These different practical scenarios correspond to a large variety of applications of speech processing research. Examples of some applications include enhancement, coding, synthesis, recognition and speaker recognition. A very rapid growth, particularly during the past ten years, has resulted due to the efforts of many leading scientists. The ideal aim is to develop algorithms for a certain task that maximize performance, are computationally feasible and are robust to a wide class of conditions.

The purpose of this book is to provide a cohesive collection of articles that describe recent advances in various branches of speech processing. The main focus is in describing specific research directions through a detailed analysis and review of both the theoretical and practical settings. The intended audience includes graduate students who are embarking on speech research as well as the experienced researcher already working in the field. For graduate students taking a course, this book serves as a supplement to the course material. As the student focuses on a particular topic, the corresponding set of articles in this book will serve as an initiation through exposure to research issues and by providing an extensive reference list to commence a literature survey. Experienced researchers can utilize this book as a reference guide and can expand their horizons in this rather broad area.

With the above thoughts, we now expand on the various topics covered bearing in mind that as is the case with any book, the areas covered are by no means exhaustive. Although we have tried to partition this vast field into sections for the purposes of effective book organization, we realize that there are no strict boundaries of knowledge. Each part of the book is devoted to a goal of speech research. Part 1 deals with effectively communicating speech from one point to another through coding. Part 2 covers the issue of recognizing a word or other speech unit independently of the speaker it came from. The topic of Part 3 is the task of successfully recognizing a speaker from his or her speech utterance. Part 4 concentrates on the transformation of text into speech. In Part 5, specific applications of signal processing concepts and modeling phenomena to speech are illustrated.

We now further describe the contents of the book by expanding on each part separately. First, consider Part 1. The article by Ramachandran on pitch prediction focuses on a specific component of predictive speech coders that is used to regenerate the periodicity in the signal. The second article by Collura describes quantization strategies for coding another component of predictive coders, namely, the parameters of the near-sample predictor which reinserts the formant structure. The article by Kroon and Kleijn on the analysis-by-synthesis paradigm describes a particular technique used in low bit rate predictive coding. A recent concept, waveform interpolation, for producing high quality speech at low bit rates is discussed in the fourth article by Haagen and Kleijn. Part 1 ends with an exposition of speech coding at variable bit rates in the article by Cuperman and Lupini.

Part 2 commences with an article on word spotting by Rohlicek. The following article by Kosonocky describes the use of neural networks for speech recognition. The next two articles deal with techniques to recognize a word or sequence of words which form part of a large vocabulary. The first of these articles by Gopalakrishnan further focuses on the concepts of feature extraction and modeling in continuous speech recognition. The second article by Gupta and Lennig looks at isolated word recognition. The fifth article by Juang addresses the very important issue of getting high recognition accuracy under different environmental conditions. This will result in automatic systems that perform well even under unexpected or adverse conditions thereby enhancing robustness. The final article by Allen discusses human speech recognition. A deep understanding of how humans recognize speech will play a vital role in improving automatic machine based recognition systems.

Both articles in Part 3 examine the speaker recognition problem. The use of data fusion to augment performance for both text-independent speaker identification and text-dependent speaker verification is illustrated in the first article by Farrell and Mammone. The second article by Kao, Netsch and Rajasekaran concentrates on speech transmitted over long distance telephone channels. The theme of robustness to channel effects is central to this article.

Part 4 has one article by O'Shaughnessy devoted to getting natural sounding speech from text. Current approaches that attempt to alleviate the inadequacy in modeling human speech production are examined.

In Part 5, four different areas having applications to speech processing are depicted. First, the use of a beamforming algorithm with a microphone array for hands-free voice communication in a car is described in the article by Oh and Viswanathan. The next article by Ramalho and Mammone discusses a new speech model (known as the Pitch Mode Modulation Model) and its applications in speech enhancement, speaker identification and speech synthesis. The area of auditory modeling and its use in coding and recognition is the topic of the third article by Ghitza. This part ends with an article by Ooi and Viswanathan that discusses the use of wavelets in speech with particular emphasis on a type of analysis-by-synthesis coder. The above constitutes a synopsis of the material contained in this book.

Ravi P. Ramachandran

Richard J. Mammone

ACKNOWLEDGEMENTS

At the outset, we thank all the authors for their contributions. It is worthwhile to note that some have assisted in contributing more than one chapter. We particularly are happy to have Vishu Viswanathan take the initiative in offering his second chapter on wavelets. Our gratitude goes to the research support provided by the CAIP center at Rutgers. For timely help in giving suggestions on the use of LATEX for typesetting this book and for supplying the macros for certain fonts, we thank the computer support staff at CAIP, the staff at Kluwer and Peter Kroon of Bell Laboratories. The secretarial assistance given by Kathyrn Bryan is gratefully acknowledged. Portions of the manuscript were proofread by John Collura, Jesper Haagen, Vidhya Ramanujam and Roopashri Ramachandran. Their suggestions improved the quality of certain chapters. We appreciate the assistance of Peter Kroon and Kevin Farrell in helping coordinate the material for Part 1 and Part 3 respectively. Sincere thanks goes to our respective families for their constant encouragement in this very rewarding effort.

PART 1

SPEECH CODING

1

THE USE OF PITCH PREDICTION IN SPEECH CODING

Ravi P. Ramachandran

Caip Center, Rutgers University
Piscataway, New Jersey, 08855-1390, USA

1 INTRODUCTION

Two major types of correlations are present in a speech signal. These are known as near-sample redundancies and distant-sample redundancies. Near-sample redundancies are those which are present among speech samples that are close together. Distant-sample redundancies are due to the inherent periodicity of voiced speech. Predictive speech coders make use of these correlations in the speech signal to enhance coding efficiency. In predictive speech coders, the cascade of two nonrecursive prediction error filters process the original speech signal. The formant filter removes near sample redundancies. The pitch filter acts on distant-sample waveform similarities. The result is a residual signal with little sample to sample correlations. The parameters that are quantized and coded for transmission include the filter coefficients and the residual signal. From the coded parameters, the receiver decodes the speech by passing the quantized residual through a pitch synthesis filter and a formant synthesis filter. The filtering steps at the receiver can be viewed in the frequency domain as first inserting the fine pitch structure and then, shaping the spectral envelope to insert the formant structure. The formant and pitch filters are adaptive in that the analysis to determine the coefficients is carried out frame by frame. Also, the bits representing the quantized parameters are transmitted on a frame by frame basis. The bit rate of the coder is the total number of bits transmitted in one frame divided by the time duration of the analysis frame. The frame duration is chosen to be long enough to keep the bit rate required small, yet short enough to allow the speech signal under consideration to be adequately described by a set of constant parameters.

The focus of this chapter is on the description of the pitch predictor and its role in producing the periodicity of the decoded speech in predictive coders. The basic issues in the analysis and design of pitch predictors include:

- Filter transfer function
- Parameter computation
- Stability
- Use in speech coders from the following points of view:
  - Generating the correct periodicity to enhance quality
  - Coding strategies

The fundamental objective in using pitch predictors is to generate the speech of the highest possible perceptual quality at the lowest possible bit rate.

Pitch predictors are used in different predictive speech coders. One example is the adaptive predictive coder [1][2] in which the pitch predictor is placed in a feedback loop around the quantizer. Coders based on the analysis by synthesis paradigm [3] include the regular pulse coder, multipulse coder [4] and Code-Excited Linear Prediction (CELP) [5]. The difference among these coders is in the method of generating the quantized residual. In the particular case of CELP (most emphasized in this article), the residual is vector quantized by a stochastic codebook containing a repertoire of waveforms consisting of Gaussian random numbers with unit variance. The analysis by synthesis strategy involves an exhaustive search in which each waveform in the codebook is passed through the synthesis filters to allow for a comparison with the original speech. The waveform that leads to the closest resemblance to the original speech is chosen.

The aim of this chapter is to present a review of the research done in the area of pitch prediction and discuss its impact on the efficiency of speech coders. The outline of the chapter is as follows. Section 2 discusses the pitch filter transfer function. The stability issue is described in Section 3. The concept of fractional delays is introduced in Section 4. The use of pitch predictors in CELP coders is described in Section 5. The aspects of periodicity and coding efficiency are the topics of Sections 6 and 7 respectively. The chapter is summarized in Section 8.

2 FILTER TRANSFER FUNCTION

A formant predictor has a transfer function $F(z) = \sum_{i=1}^{Q} a_i z^{-i}$ where Q is between 8 and 16 for 8 kHz sampled speech. The speech signal $s(n)$ is passed through the filter $1-F(z)$ to generate a residual $r(n)$ that is free of near-sample correlations. The corresponding synthesis filter is $H_F(z) = 1/(1-F(z))$. There exist methods to find the a_i based on minimizing the weighted mean-square error of $r(n)$ over a frame of N samples [6]. The simplest form of the pitch predictor has one tap whose transfer function is given by

$$P(z) = \beta_1 z^{-M} \tag{1.1}$$

where the integral delay M represents the pitch period. Since the sampling frequency is unrelated to the pitch period, the individual samples do not show a high period to period or distant sample correlation [2]. Therefore, a 3 tap predictor serves like an interpolation filter and provides for interpolated estimates that show higher period to period correlations. The transfer function is

$$P(z) = \beta_1 z^{-M+1} + \beta_2 z^{-M} + \beta_3 z^{-M-1} \tag{1.2}$$

The pitch synthesis filter has a system function $H_P(z) = 1/(1 - P(z))$.

In computing the predictor coefficients and M, consider the situation of a signal that is passed through the prediction error filter $1-P(z)$ to generate a residual $e(n)$. The signal can either be the input speech $s(n)$ or the residual $r(n)$ formed after formant prediction. This is known as an open-loop analysis in that the parameters are determined by considering only the clean input signals and not imposing any dependence on any decoded signals. The method of using the decoded or synthesized signals to compute the parameters is known as a closed-loop analysis and will be discussed later. Assuming a given value of M, the coefficients of $P(z)$ are chosen to minimize the mean-squared residual

$$E_{\text{mse}} = \sum_{n=1}^{N} e^2(n) \tag{1.3}$$

where (if $r(n)$ is the input signal)

$$e(n) = r(n) - \beta_1 r(n-M+1) - \beta_2 r(n-M) - \beta_3 r(n-M-1) \tag{1.4}$$

and N is the number of samples in one frame. The minimization of E_{mse} leads to a system of equations which can be written in matrix form as $\mathbf{Ac} = \mathbf{d}$. For a 3 tap predictor, the entries of the matrix $\mathbf{A}$ are

$$A(i,j) = \phi(M+i, M+j) = \sum_{n=1}^{N} r(n-M-i)r(n-M-j) \tag{1.5}$$

for $-1 \leq i, j \leq 1$. The vector

$$\mathbf{c} = [\beta_1 \ \ \beta_2 \ \ \beta_3]^T \tag{1.6}$$

and the vector

$$\mathbf{d} = [\phi(0, M-1) \ \ \phi(0, M) \ \ \phi(0, M+1)]^T \tag{1.7}$$

Note that for the 1 tap case, the predictor coefficient is determined as $\beta_1 = \phi(0, M)/\phi(M, M)$.

Methods to determine M are described in [7]. The methods are based on analyzing an expression for the resulting minimum mean-squared error E_{res} given that the coefficients are obtained by solving the system of equations given above. The resulting error E_{res} is

$$E_{\mathrm{res}} = \phi(0,0) - \mathbf{c}^T\mathbf{d} \tag{1.8}$$

in which the second term is a function of M. The optimal value of M is that which maximizes $\mathbf{c}^T\mathbf{d}$. The procedure is to do an exhaustive search of all integral values of M within an allowable range (usually between 20 and 147 samples for 8 kHz sampled speech) to find the optimal value. For the 1 tap case, the expression $\phi^2(0, M)/\phi(M, M)$ is maximized. However, for 3 tap predictors, it is computationally expensive to do an exhaustive search. Two suboptimal approaches as outlined in [7] are as follows. First, the value found for the one tap case can be used for 3 tap filters. The second approach assumes that the input signal is $r(n)$. In this case, the off-diagonal terms of $\mathbf{A}$ which represent near-sample redundancies can be neglected. Then, an approximation to $\mathbf{c}^T\mathbf{d}$ which facilitates an exhaustive search is given by

$$\mathbf{c}^T\mathbf{d} \approx \sum_{m=M-1}^{M+1} \frac{\phi^2(0, m)}{\phi(m, m)} \tag{1.9}$$

The prediction gain measures the extent to which a predictor removes redundancies in a signal. It is the ratio of the average energy of the input signal to the average energy of the residual. Experiments have shown that processing speech first through the formant filter and then through the pitch filter results in a higher overall prediction gain than the reverse arrangement [7]. Given that the formant filter coefficients are updated on a frame by frame basis, the pitch filter parameters (coefficients and delay) are updated over segments of smaller duration known as subframes. Within a certain frame, there are an integral number of subframes. This more frequent update of the pitch filter is needed to get a higher prediction gain and to obtain decoded speech of higher quality. In practice, the formant filter is updated every 20 ms and the pitch filter every 5 ms.

3 PITCH FILTER STABILITY

Consider the canonical case of the prediction error filters $1-F(z)$ and $1-P(z)$ in cascade with a quantizer followed by the synthesis filters $H_P(z)$ and $H_F(z)$. The quantizer can be modeled as an additive noise source. The quantized residual has two components, namely, the residual formed by filtering the original speech signal and the additive noise. When the quantized residual is passed through the synthesis filters, the first component due to the signal will reproduce the original speech. For this component, the stability aspect of the synthesis filters is not a problem. However, the quantization noise passes only through the synthesis filters. Therefore, instability can cause the noise to accentuate thereby diminishing speech quality. Since the filters are updated regularly, periods of instability may exist only over finite (or possibly very small) time durations. However, if these periods of instability correspond to high energy voiced segments, the quality of the decoded speech will degrade as confirmed in [8].

The autocorrelation method assures that $H_F(z)$ is stable. The method of determining the pitch predictor coefficients and delay as described above does not assure a stable $H_P(z)$. Given the predictor coefficients and M, the stability in the 1 tap case is easy to check ($|\beta_1| < 1$). However, for 3 tap filters, conventional stability tests involve the checking of about M conditions which can be quite high for pitch predictors. Furthermore, known tests do not lend themselves to a technique that can stabilize an unstable $H_P(z)$. To overcome these problems, a stability test for 3 tap filters based on a tight sufficient condition was formulated in [8] and is as follows. Let $a = \beta_1 + \beta_3$ and $b = \beta_1 - \beta_3$.

1. If $|a| \geq |b|$, then satisfying the condition $|\beta_1| + |\beta_2| + \beta_3| < 1$ is sufficient for stability.
2. If $|a| < |b|$, two conditions must be satisfied.
   (a) $|\beta_2| + |a| < 1$.
   (b) Either $b^2 \leq |a|$ or $b^2\beta^2 - (1-b^2)(b^2-a^2) < 0$.

The proposed test is independent of the order M, is computationally much simpler than known tests based on necessary and sufficient conditions, and naturally leads to a stabilization technique. It is also shown in [8] that the set of test conditions, although only sufficient, is tight for finite M and comprises a set of necessary and sufficient conditions in the limit of large M.

When stabilizing a pitch filter, we start from an optimal filter $1 - P(z)$ that minimizes the mean-squared residual (or equivalently maximizes the prediction gain) but is not minimum phase. The coefficients are modified to get a minimum phase pitch filter. However, this minimum phase pitch filter is suboptimal in that the prediction gain will be less than the maximum value achieved by the optimal filter. The stabilization technique is formulated such that $H_P(z)$ becomes stable and the reduction in prediction gain offered by modifying $1 - P(z)$ is minimized. The technique is based on scaling β_1, β_2 and β_3 by a common factor t [8]. If $|a| \geq |b|$, the value of t is

$$t = \frac{1}{|\beta_1| + |\beta_2| + |\beta_3|} \tag{1.10}$$

If $|a| < |b|$ and $b^2 \leq |a|$, then

$$t = \frac{1}{|a| + |\beta_2|} \tag{1.11}$$

If $|a| < |b|$ and $b^2 > |a|$, then

$$t = \sqrt{\frac{b^2 - a^2}{b^4 + b^2\beta_2^2 - b^2a^2}} \tag{1.12}$$

The method is computationally simple and noniterative. Experiments have shown that stabilization results in a negligible reduction in prediction gain and leads to speech of better perceptual quality.

4 FRACTIONAL DELAY PITCH FILTERS

The original motivation of going from a 1 tap to a 3 tap pitch predictor was to provide interpolation between the samples since the pitch period is unrelated to the sampling frequency. Also, 3 tap predictors provide a higher prediction gain than 1 tap filters. Another way of increasing the distant sample correlations is to use a fractional delay pitch predictor as proposed in [9]. A fractional delay predictor has 1 tap but provides for better temporal resolution by allowing the delay M to be expressed as an integer plus a fraction l/D where $0 \leq l < D$ and l and D are integers. By using this fractional delay, the sampling frequency effectively increases by a factor D thereby providing a better match between the pitch delay M and the pitch period of the underlying continuous time speech signal. The implementation of a fractional delay is done by using a nonrecursive interpolation filter with linear phase. An efficient implementation results by using a polyphase structure. Although many design approaches for interpolation filters exist, a $\sin x/x$ function weighted by a Hamming window was used in [9]. With this implementation, the method for finding the best fractional delay M via an exhaustive search and the coefficient β_1 is as described earlier.

A comparison of 1 tap fractional delay predictors and 3 tap predictors using integer delays reveals the following. When using the formant predicted residual as the input to $1 - P(z)$, the prediction gain improves with the degree of resolution D but virtually saturates for $D \geq 8$. A one tap filter with $D = 4$ results in about the same prediction gain as a 3 tap filter with integer delays [9]. However, the exhaustive search for the best fractional delay M requires more computation. Fast search procedures that start with a judicious sampling of the possible values of M leads to a negligible decrease in the prediction gain [9]. Although the stability check for 3 tap filters and any consequent stabilization is easily performed, a 1 tap filter is still more convenient from this viewpoint. Regarding the coding effort, first consider the pitch delay. For a 3 tap pitch filter using integer delays from 20 to 147, 7 bits are required. The additional fractional delays for a delay resolution D requires an extra $\log_2 D$ bits. If scalar quantization is used for the coefficients, the additional bits needed to code three coefficients is usually more than the extra bits needed to allow for fractional delays. This implies that the parameters for a fractional delay predictor can be coded more efficiently than for a 3 tap pitch predictor with integer delays. The number of bits for coding 3 pitch coefficients can be decreased by using a vector quantizer [10]. However, only scalar quantizers are needed for the 1 tap fractional delay predictor. Subjective tests in a CELP coder reveal that the use of a 1 tap fractional delay predictor shows noticeable improvements in the speech quality [9].

5 USE IN CELP SYSTEMS

The pitch predictor is a significant component of a CELP system in that it regenerates the periodicity in the decoded speech. The operation of the CELP coder is based on the analysis by synthesis paradigm. This approach starts by configuring a model to synthesize speech. Then, a mathematical analysis determines the model parameters to minimize the weighted mean-square error between the synthesized speech and the original speech. This type of analysis which considers the synthesized speech is known as a closed-loop analysis which was first introduced in [4]. The synthesis model is an excitation signal (having little or no redundancies) filtered by $H_P(z)$ and $H_F(z)$. The different methods of generating the excitation signal leads to different coders [3]. In CELP, the excitation signal is chosen from a stochastic codebook populated by white Gaussian sequences of unit variance and appropriately scaled to get a desired variance. The closed-loop analysis must be generally formulated to find the best entry of the stochastic codebook, the stochastic codebook gain and the parameters of $H_P(z)$ and $H_F(z)$. Since including the formant filter as part of the closed-loop search leads to a nonlinear problem, an open-loop analysis that assures a stable $H_F(z)$ is first performed. This leads to an analysis by synthesis structure as shown in Fig. 1. Note that in Fig. 1, $s(n)$ is the original speech signal, $r(n)$ is the residual generated after formant prediction, $q(n)$ is the synthesized residual, $c(n)$ is the stochastic codebook entry and $e(n)$ is the error or difference between the residuals $r(n)$ and $q(n)$. Since the aim is to minimize the error between the speech signals and not the residuals, a weighted error signal $e_w(n)$ is considered by filtering $e(n)$ by $W(z)$. Although the weighting filter $W(z)$ would be expected to be given by $H_F(z)$, a modified form $H_F(z/\gamma)$ $(0.7 \leq \gamma \leq 0.8)$ is used due to perceptual considerations [3][5].

We continue our discussion on CELP assuming that a one tap pitch predictor with fractional delays is used. In this case, the pitch predictor operation is nearly equivalent to an adaptive codebook formulation [11]. The corresponding structure is shown in Fig. 2. The adaptive codebook entries are past versions of the synthesized residual $q(n)$. The signal $q(n)$ is a linear combination of two components, a scaled adaptive codebook entry $\beta_1 q(n-M)$ and a scaled stochastic codebook entry $gc(n)$.

Each frame of speech (consisting of N samples usually corresponding to 20 ms) is divided into subframes each consisting of L samples. For a 20 ms frame, there are usually 4 subframes of 5 ms duration. Each frame is processed as follows:

- Find $F(z)$ by an open-loop analysis over the entire frame and interpolate parameters such that they are held constant for each subframe.
- For each subframe, do the following:
  - Generate the residual $r(n)$.
  - Parameterize $H_F(z/\gamma)$ by an impulse response $h(n)$.
  - Find the zero input response of $W(z)$, call it $w_0(n)$ and subtract it from $r(n) \star h(n)$ to get $f_0(n)$.
  - Form the initial weighted error signal.

- Find the optimal adaptive codebook entry and gain.
- Form a new weighted error signal by subtracting the optimal adaptive codebook contribution.
- Find the optimal stochastic codebook entry and gain.

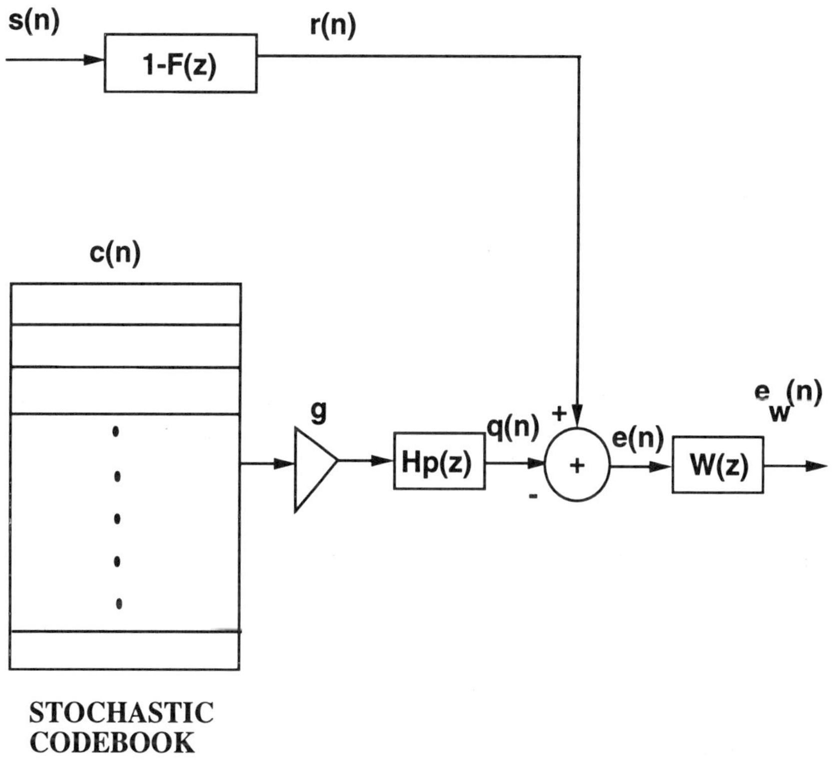

Figure 1 CELP system with pitch predictor

For the purposes of this article, we will focus on the determination of the adaptive codebook parameters. For more details, the reader is referred to [3]. The initial weighted error signal is

$$e_w(n) = f_0(n) - \beta_1 q(n - M) \star h(n) \tag{1.13}$$

The minimization of

$$E_{\text{mse}} = \sum_{n=1}^{L} e_w^2(n) \tag{1.14}$$

with respect to β_1 results in an optimal β_1 as given by

$$\beta_1 = \frac{\theta_1(M)}{\theta_2(M)} \tag{1.15}$$

where

$$\theta_1(M) = \sum_{n=1}^{L} f_0(n)[q(n-M) \star h(n)] \tag{1.16}$$

and

$$\theta_2(M) = \sum_{n=1}^{L} [q(n-M) \star h(n)]^2 \tag{1.17}$$

The resulting error E_{res} is

$$E_{\text{res}} = \sum_{n=1}^{L} f_0^2(n) - \frac{\theta_1^2(M)}{\theta_2(M)} \tag{1.18}$$

The optimal adaptive codebook entry (or equivalently the best choice of M) is that which maximizes $\theta_1^2(M)/\theta_2(M)$. As in the open-loop case, M can be found by an exhaustive search of a predetermined set of permissible fractional values. Then, β_1 is computed. The procedure continues by updating the weighted error and determining the best stochastic codebook parameters in a similar manner.

The information transmitted to the decoder includes the formant filter parameters, the adaptive codebook parameters and the stochastic codebook parameters. From this information, the decoded speech is generated by first forming $q(n)$ exactly as in Fig. 2 and passing $q(n)$ through $H_F(z)$. The number of permissible values of M (equivalent to the number of adaptive codebook entries) is dictated by the number of bits used. Moreover, the permissible values of M need not be uniformly distributed between the minimum (20 samples) and maximum (147 samples) values. In [12], a 9 bit adaptive codebook with 512 nonuniformly distributed fractional delay values is described and used in CELP coders operating at different bit rates. The gain β_1 is coded by a 3 bit nonuniform scalar quantizer. In the presence of gain quantization, the best delay is determined as described above. However, since only a small finite number of gain values are permitted, an exhaustive search to find the quantized gain that leads to the smallest error is performed. Note that forcing the quantized values to be between -1 and 1 guarantees the stability of $H_P(z)$.

An algorithmic description of the role of a pitch predictor in CELP coders has been given. Two outstanding issues still remain. First, ways to improve the performance of the pitch predictor by examining methods to generate the correct periodicity in the decoded speech are discussed. Second, a rapid update interval of 5 ms for the pitch delay leads to a high bit rate for the pitch predictor parameters. The question of why this is necessary given that the pitch period varies very slowly over time should be addressed so that the bit rate can be lowered without compromising the quality of the decoded speech.

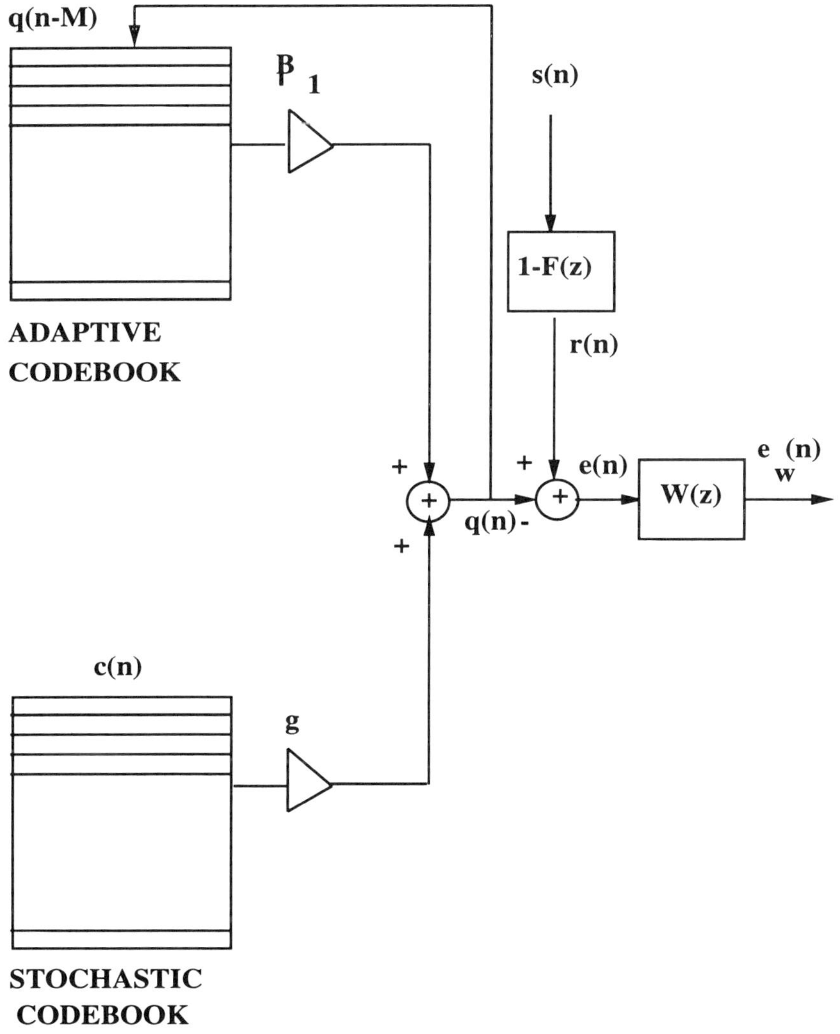

Figure 2 CELP system with adaptive codebook

6 PERIODICITY CONSIDERATIONS

In addition to enhancing the coding efficiency, the pitch predictor must generate the correct level of periodicity in the decoded speech. The perceived level of periodicity diminishes as the bit rate is lowered [13]. This causes a degradation in the speech quality in that the speech becomes noisy. If the level of periodicity is too high, a buzziness type of artifact emerges [13]. In analysis by synthesis coders like CELP, the level of periodicity of the decoded speech is usually lower than that of the original. We will review different procedures to correct for any disparity in the levels of periodicity of the original and decoded speech.

An early method of improving speech quality is the use of postfiltering [3][10]. Two postfilters are commonly used, one based on the formant filter and one based on the pitch filter. The postfilter based on $H_F(z)$ sharpens the formant peaks and deepens the spectral valleys [14]. This results in a tradeoff between noise suppression and signal distortion. The postfilter based on $H_P(z)$ is given by $1/(1-\epsilon\beta_1 z^{-M})$ [3][10] where $0 < \epsilon < 1$. The frequency response is that of a comb filter. The noise energy between the pitch harmonics is lowered. The value of ϵ is experimentally determined to be around 0.3 [10] so as to achieve an optimum balance between noise suppression and speech distortion. Note that no frame by frame adaptation of ϵ is performed. At the receiver, the decoded speech is passed through the pitch postfilter and the formant postfilter for the purpose of enhancement.

In CELP, the excitation to $H_F(z)$ consists of a linear combination of an adaptive codebook and a stochastic codebook contribution. Although the role of the stochastic codebook is significant in refining the description of the excitation and modeling its random component, it injects a noisy character that brings down the level of periodicity particularly at low bit rates of 5 kb/s or less. The approach in [15] adaptively controls the contribution of the stochastic codebook depending on the level of the adaptive codebook contribution. The adaptive codebook contribution is dominant during the high energy voiced portions of the speech signal. For these portions, the amount of noise injected by the stochastic codebook is reduced in order to better match the periodicity of the original speech signal. However, in unvoiced and transitional regions, there is less control over the stochastic codebook since the adaptive codebook is not as dominant [15].

The contribution of the adaptive codebook is measured by

$$S_P = \frac{||s(n)||}{||s(n) - w_0(n) - \beta_1 q(n-M) \star h(n)||} \tag{1.19}$$

where $||\cdot||$ denotes the L_2 norm. Note that β_1 and M are optimal parameters. The requirement for controlling the stochastic codebook excitation is that

$$\frac{||s(n)||}{||gc(n) \star h(n)||} \geq T(S_P) \tag{1.20}$$

where $T(S_P)$ is linear for lower values of S_P and then flat after a certain cutoff value. The slope in the linear region and the cutoff value are determined experimentally [15]. In each subframe, S_P and $T(S_P)$ are first calculated. Then, the best stochastic codebook entry and the quantized gain are found subject to the above requirement (details in [15]). Note that when S_P is high, the stochastic codebook contribution is small. If S_P is relatively low, the role of the stochastic codebook is more apparent. It is shown that this method improves speech quality for a bit rate of 4.8 kb/s [15]. In a related technique [16], two parallel blocks are used at the transmitter. The best quantized stochastic codebook gain is computed and transmitted. However, the signal that is fed back to the adaptive codebook has a component reflecting a modified stochastic codebook gain which allows for sharper pitch pulses in voiced regions. As the analysis proceeds from one subframe to the next, the best stochastic codebook gain is automatically altered to allow for greater dominance of the adaptive codebook in voiced regions.

The concept of deviating from the optimal stochastic codebook gain shows that the mean-square error criterion although mathematically tractable, is inadequate from a perceptual point of view. The use of weighting attempts to improve the situation by exploiting the noise masking capability of speech. The filter $H_F(z/\gamma)$ (mentioned earlier) utilizes the fact that regions corresponding to formants can tolerate more noise. An additional weighting filter, $1-\epsilon\beta_1 z^{-M}$ placed in cascade with $H_F(z/\gamma)$, deemphasizes the part of the error having the same harmonic structure as the speech [17]. This weighting filter is related to the pitch postfilter. Unlike the postfilter, the weighting filter is part of the analysis by synthesis mechanism. Note that the problem of finding an error criterion that is perceptually very meaningful and yet mathematically tractable still remains.

An actual means of quantifying periodicity is proposed in [13]. For a signal $x(n)$ with a pitch period M, the periodicity is the expectation value of the measure $Y(x,k)$ as defined by [13]

$$Y(x,k) = \frac{\sum_{n=k}^{k+M-1} x(n)x(n-M)}{\sum_{n=k}^{k+M-1} x^2(n-M)} \tag{1.21}$$

Let $x(n)$ be quasi-periodic. Then, it can be decomposed as $x(n) = \lambda x(n-M) + u(n)$ where $u(n)$ is uncorrelated with $x(n-M)$. For this case, $E[Y(x,k)] = \lambda$ [13] where E is the expectation operator. If $v(n)$ is the decoded speech, then (assuming no gain quantization) [13]

$$E[Y(v,k)] = \lambda - E[\frac{\sum_{n=k}^{k+M-1} u(n)(x(n-M) - v(n-M))}{\sum_{n=k}^{k+M-1} v^2(n-M)}] \tag{1.22}$$

The interpretation of the above equation is as follows [13]. If $u(n)$ in the current pitch cycle is independent of $x(n)$ and $v(n)$ in previous pitch cycles, the periodicity remains unchanged independently of the bit rate. This occurs when the speech is generated with an autoregressive model. Generally, there is a change in the periodicity which grows with the reconstruction error or equivalently with decreasing bit rate. For periodic signals with additive noise, the periodicity after coding increases. For signals having slowly evolving pitch cycles with the change from one pitch cycle to the next being about the same over a few pitch cycles, the periodicity of the decoded signal diminishes. This is characteristic of voicing onsets and transitions thereby suggesting a reason as to why CELP systems lead to a reduction in the periodicity. One remedy to this problem is to increase the periodicity of the original speech prior to encoding [18]. This is done by taking a current speech segment and averaging it with highly correlated segments that are about one pitch period ahead and behind the current segment [18]. This type of smoothing can be made frequency selective by picking only certain frequency bands.

7 CODING EFFICIENCY

For coding the pitch predictor parameters, 12 bits are typically used over an update interval of 5 ms [12]. This results in a rather high bit rate of 2.4 kb/s. As mentioned earlier, since the pitch period varies slowly, it is surprising why the pitch delay has to be updated so often. If the update interval is increased to 10 ms, the bit rate decreases by a factor of two but at the expense of speech quality. For female speakers whose pitch periods are low, two or more pitch pulses are present over these longer intervals. It is difficult for the adaptive codebook to accurately resolve multiple pitch pulses with only one delay factor. This section discusses techniques that attempt to lower the bit rate for transmitting the pitch parameters without compromising perceptual speech quality.

The restrictive pitch deviation coding technique [19] combines an open-loop and closed-loop search as follows. Consider the usual case when a frame is 20 ms in duration and is divided into 4 subframes of 5 ms duration. An open-loop analysis over the entire frame determines a candidate pitch delay. The full range of values from 20 to 147 samples is considered. The deviation or offset from this candidate delay in each subframe is restricted for the closed-loop analysis. Now, a much smaller subset of the full range of delay values is used thereby imposing much less of a computational burden and simultaneously requiring less number of bits. The transmitted parameters include the candidate pitch delay (one per frame) and the offset value (one per subframe). As an example, suppose 9 bits are used to code the candidate delay. A commonly used offset limit is 1 sample [19]. If 3 bits are used to represent the offset, the bit rate for the delay information is 1.05 kb/s. Given that β_1 is updated every subframe and quantized to 3 bits, the bit rate for the pitch parameters is 1.65 kb/s. This is about 70 percent of the bit rate needed for the standard approach.

Another method that combines open-loop and closed-loop analyses and is based on coding the delay offsets is described in [20]. The main drawback of coding the pitch delay offsets is that the overall delay trajectory for the entire frame (referred to as the frame delay trajectory in [20]) is suboptimal. A frame delay trajectory is a sequence of subframe delay values over the entire frame. When looked at on a sample by sample basis, the trajectory is that of a stepped delay contour. The conditions on the trajectory impose no restriction on the delay in the first subframe [20]. The delay in the next subframe must be within a maximum deviation from the delay in the first subframe. The same type of restriction holds for the delays in subsequent subframes. The transmitted information consists of the delay for the first subframe and the offsets for the remaining subframes. An open-loop global optimization of the frame delay trajectory is first performed. A search over all possible frame delay trajectories is impractical. For each subframe, a set of candidate delays are found based on the prediction gains they yield. With these candidate delays, different frame trajectories are formed subject to the constraints on the delay offsets. The trajectory giving the highest overall prediction gain is kept and refined by a closed-loop analysis.

The concept of pitch parameter interpolation eliminates the need to transmit delay offsets. Again, consider one frame of speech that is divided into four subframes. The pitch predictor is first modified such that a continuous delay contour that varies on a sample by sample basis is used instead of the conventional stepped delay contour [21] [22]. When a stepped delay contour is used, there are discontinuities in the delay from subframe to subframe. This leads to discontinuities in the synthesized residual $q(n)$ in that portions of the signal may be lost or repeated [22]. For a continuous delay contour, the delay varies linearly in each subframe and is continuous across subframe boundaries. This avoids the discontinuities in $q(n)$.

In a particular frame, the endpoint delay value serves as the initial condition. The endpoint delay of the current frame is determined in an open-loop fashion. The delay contour is linearly interpolated (on a sample by sample basis) between the initial and endpoint values. Due to interpolation, the delay value need only be transmitted once per frame. Compared to transmitting the delay once per subframe, this reduces the bit rate by a factor of four. Consider any subframe within the frame. Since the delay contour is already specified, the adaptive codebook contribution (denoted as $t(n)$) is known. However, the adaptive codebook contribution is not optimal for that subframe. Hence, there is a mismatch between the original and synthesized residuals particularly so between the pitch pulses of the two signals. This causes a degradation in the speech quality. The degradation is avoided by warping the original residual $r(n)$ to force an optimal mean-square match to the adaptive codebook contribution $t(n)$. An example of a family of time warping functions that transform $r(n)$ to $r_w(n)$ is given in [22]. A set of warping functions within this family is searched to find the best warped signal $r_w(n)$ that fits $t(n)$. For each considered warp, the zero input response of $W(z) = H_F(z/\gamma)$ is subtracted from $r_w(n) \star h(n)$ to get $f_0(n)$. The warp that is selected is what leads to the maximization of the quantity $\theta_1^2/(\theta_2\theta_3)$ [22] where

$$\theta_1 = \sum_n f_0(n)[t(n) \star h(n)] \tag{1.23}$$

$$\theta_2 = \sum_n f_0^2(n) \tag{1.24}$$

and

$$\theta_3 = \sum_n [t(n) \star h(n)]^2 \tag{1.25}$$

Note that the gain β_1 is calculated in the usual way and is transmitted once per subframe (a discussion of gain interpolation will come later). The adaptive codebook is best matched to a modified or warped original speech signal which has a slightly different pitch track than the original speech. This warping procedure will succeed as long as the perceptual quality of the original and modified speech is the same. This can be assured by only considering slight time warps that bring about slight changes in the pitch that serve to merely correct for the mismatch between $t(n)$ and $r(n)$. The modification of the original speech that makes delay interpolation successful is a generalization of the analysis by synthesis paradigm [21] [23]. Subjective tests confirm that the bit rate can be lowered without compromising perceptual quality [22].

Regarding pitch gain interpolation, note that the gain is held constant in a subframe and the interpolation is done in a stepwise fashion. The problem of interpolation errors propagating from subframe to subframe emerges. This causes large errors in the energy contour of the reconstructed signal [22]. A less sensitive parameter, namely, the root mean-square energy of the adaptive codebook contribution E_{LTP} is interpolated and converted to the pitch gain in each subframe. In a particular frame, the endpoint value of E_{LTP} serves as the initial condition and is determined in an open-loop fashion. The energy contour is linearly interpolated (on a stepwise basis) between the initial and endpoint values. Consider any subframe within the frame. The adaptive codebook contribution $t(n)$ is known and the best warped signal $r_w(n)$ has been determined. The interpolated energy is converted to the pitch gain β_1. The warped signal $r_w(n)$ is amplitude scaled to optimally match $\beta_1 t(n)$ [22]. The scaling is restricted so as not to affect the perceptual quality of the speech.

Pitch delay interpolation has been implemented as a part of fully quantized speech coders [24] [25]. To alleviate the computational burden, the modified residual is formed by time shifting (rather than time warping) the original residual to optimally match the adaptive codebook contribution. Coders operating between 4 and 6 kb/s are described in [25]. They perform better than a standard 4.4 kb/s coder and similar to a standard 13 kb/s coder as revealed by formal perceptual tests.

8 BRIEF SUMMARY

The aim of this chapter is to present a review of the research done in the area of pitch prediction and discuss its impact on the efficiency of speech coders.

Starting from the system function description of the pitch filter, various issues concerning the important role of pitch filters in speech coders have been discussed. Continued effort is put in to lower the bit rate of speech coders and maintain high perceptual quality. With this objective in mind, the implementation of interpolation schemes, quantization strategies and methods to generate the proper periodicity are extremely significant.

REFERENCES

[1] B. S. Atal and M. R. Schroeder, "Adaptive predictive coding of speech signals", *Bell System Technical Journal,* vol. 49, pp. 1973–1986, Oct. 1970.

[2] B. S. Atal and M. R. Schroeder, "Predictive coding of speech signals and subjective error criteria", *IEEE Trans. on Acoust., Speech and Signal Proc.,* vol. ASSP-27, pp. 247–254, June 1979.

[3] P. Kroon and E. F. Deprettere, "A class of analysis-by-synthesis predictive coders for high quality speech coding at rates between 4.8 and 16 kbits/s", *IEEE Jour. on Selec. Areas in Commun.,* vol. 6, pp. 353–363, Feb. 1988.

[4] S. Singhal and B. S. Atal, "Improving performance of multi-pulse LPC coders at low bit rates", *Proc. IEEE Int. Conf. on Acoust., Speech and Signal Proc.,* San Diego, California, pp. 1.3.1–1.3.4, March 1984.

[5] M. R. Schroeder and B. S. Atal, "Code-excited linear prediction (CELP): High-quality speech at low bit rates", *Proc. IEEE Int. Conf. on Acoust., Speech and Signal Proc.,* Tampa, Florida, pp. 25.1.1–25.1.4, March 1985.

[6] L. R. Rabiner and R. W. Schafer, *Digital Processing of Speech Signals,* Prentice-Hall, 1978.

[7] R. P. Ramachandran and P. Kabal, "Pitch prediction filters in speech coding", *IEEE Trans. on Acoust., Speech and Signal Proc.,* vol. 37, pp. 467–478, April 1989.

[8] R. P. Ramachandran and P. Kabal, "Stability and performance analysis of pitch filters in speech coders", *IEEE Trans. on Acoust., Speech and Signal Proc.,* vol. ASSP-35, pp. 937–946, July 1987.

[9] P. Kroon and B. S. Atal, "On improving the performance of pitch predictors in speech coding systems", in *Advances in Speech Coding,* edited by B. S. Atal, V. Cuperman and A. Gersho, Kluwer Academic Publishers, pp. 321–327, 1991.

[10] P. Kroon and B. S. Atal, "Quantization procedures for the excitation in CELP coders" *Proc. IEEE Int. Conf. on Acoust., Speech and Signal Proc.*, Dallas, Texas, pp. 36.8.1–36.8.4, April 1987.

[11] W. B. Kleijn, D. J. Krasinski and R. H. Ketchum, "An efficient stochastically excited linear predictive coding algorithm for high quality low bit rate transmission of speech", *Speech Communication*, vol. 7, pp. 305–316, 1988.

[12] P. Kroon and K. Swaminathan, "A high-quality multirate real-time CELP coder", *IEEE Jour. on Selec. Areas in Commun.*, vol. 10, pp. 850–857, June 1992.

[13] W. B. Kleijn, "On the periodicity of speech coded with linear-prediction based analysis by synthesis coders", *IEEE Trans. on Speech and Audio Proc.*, vol. 2, pp. 539–542, Oct. 1994.

[14] V. Ramamoorthy, N. S. Jayant, R. V. Cox and M. M. Sondhi, "Enhancement of ADPCM speech coding with backward-adaptive algorithms for postfiltering and noise feedback", *IEEE Jour. on Selec. Areas in Commun.*, vol. 6, pp. 364–382, Feb. 1988.

[15] Y. Shoham, "Constrained-stochastic excitation coding of speech at 4.8 kb/s", in *Advances in Speech Coding*, edited by B. S. Atal, V. Cuperman and A. Gersho, Kluwer Academic Publishers, pp. 339–348, 1991.

[16] T. Taniguchi, M. Johnson and Y. Ohta, "Pitch sharpening for perceptually improved CELP, and the sparse-delta codebook for reduced computation", *Proc. IEEE Int. Conf. on Acoust., Speech and Signal Proc.*, Toronto, Canada, pp. 241–244, May 1991.

[17] I. A. Gerson and M. A. Jasiuk, "Techniques for improving the performance of CELP type speech coders", *Proc. IEEE Int. Conf. on Acoust., Speech and Signal Proc.*, Toronto, Canada, pp. 205–208, May 1991.

[18] W. B. Kleijn, "Improved pitch prediction", *IEEE Workshop on Speech Coding for Telecommunications*, Sainte-Adele, Canada, pp. 19–20, 1993.

[19] M. Yong and A. Gersho, "Efficient encoding of the long-term predictor in vector excitation coders", in *Advances in Speech Coding*, edited by B. S. Atal, V. Cuperman and A. Gersho, Kluwer Academic Publishers, pp. 329–338, 1991.

[20] I. A. Gerson and M. A. Jasiuk, "Efficient techniques for determining and encoding the long term predictor lags for analysis-by-synthesis speech

coders", in *Speech and Audio Coding for Wireless and Network Applications*, edited by B. S. Atal, V. Cuperman and A. Gersho, Kluwer Academic Publishers, pp. 211–216, 1993.

[21] W. B. Kleijn, R. P. Ramachandran and P. Kroon, "Generalized analysis by synthesis coding and its application to pitch prediction", *Proc. IEEE Int. Conf. on Acoust., Speech and Signal Proc.*, San Francisco, California, pp. I337–I340, March 1992.

[22] W. B. Kleijn, R. P. Ramachandran and P. Kroon, "Interpolation of the pitch predictor parameters in analysis by synthesis speech coders", *IEEE Trans. on Speech and Audio Proc.*, vol. 2, pp. 42–54, Jan. 1994.

[23] P. Kroon and W. B. Kleijn, "Linear predictive analysis by synthesis coding", in *Modern Methods of Speech Processing*, edited by R. P. Ramachandran and R. J. Mammone, Kluwer Academic Publishers, 1995.

[24] W. B. Kleijn, P. Kroon, L. Cellario and D. Sereno, "A 5.8 kb/s CELP algorithm for cellular applications", *Proc. IEEE Int. Conf. on Acoust., Speech and Signal Proc.*, Minneapolis, Minnesota, pp. II596–II599, April 1993.

[25] W. B. Kleijn, P. Kroon and D. Nahumi, "The RCELP speech coding algorithm", *European Trans. on Telecommunications*, vol. 4, pp. 573–582, 1994.

2

VECTOR QUANTIZATION OF LINEAR PREDICTOR COEFFICIENTS

John S. Collura

Department of Defense,
9800 Savage Road, Ft. Meade MD, 20755-6000, USA

1 INTRODUCTION

Quantization of data is performed to reduce the bit rate required either for storage of data, or transmission between two or more communicators. Linear prediction (LP) is an efficient way to represent the short term spectrum of the speech signal. The line spectral frequency (LSF) transformation provides benefits over than linear prediction representations such as reflection coefficients, arc sine reflection coefficients, or log area ratios. Vector quantization of the filter parameters allows for a larger reduction in the bit rate to represent a set of parameters over scalar quantization. This reduction comes at the expense of using more computational complexity and greater amounts of storage. To reduce the burden of both storage and computation, various techniques and procedures have been developed. These techniques include full, multistage and split vector quantization, as well as the adaptive, variable dimension and finite state vector quantizers [1][2][3][4][5][6][7][8].

This chapter is divided into several sections. The first section provides the reader with a short introduction to linear prediction and the line spectrum frequency transformation. The second section takes these LSF vectors and discusses the log spectral distortion and distance measures used to match vectors. Pros and cons of scalar quantization follow and lead into a discussion of vector quantization. This naturally leads to the structures used to implement vector quantizers, followed by a selection of methods used for training vector quantizers. The intricacies of constructing proper training databases and the importance of incorporating sufficient quantities of training material into the training procedures are discussed in the following section. Next, complexity trade-offs are discussed, followed by a section on the effects of channel errors on vector quantizers.

The single most comprehensive reference on vector quantization can be found in the book authored by Gersho and Gray [1]. Other excellent generic references on vector quantization include Makhoul et al. [2], and Gray [3].

2 LINEAR PREDICTION PARAMETERS

Linear Prediction estimates the value of the current speech sample based on a linear combination of past speech samples, where s is the sequence of speech samples, a_k is the kth predictor coefficient, p is the predictor order, and $\hat{s}$ is the estimated speech sequence.

$$\hat{s}(n) = \sum_{k=1}^{p} \alpha_k s(n-k) \tag{2.1}$$

Rearranging this equation, we define the long term prediction error as $e(n)$.

$$e(n) = s(n) - \sum_{k=1}^{p} \alpha_k s(n-k) = s(n) - \hat{s}(n) \tag{2.2}$$

Minimizing the mean squared prediction error with respect to the filter coefficients α_k of order p yields the following equation.

$$\frac{d}{d\alpha_j} E(e^2(n)) = \frac{d}{d\alpha_j} E[(s(n) - \sum_{k=1}^{p} \alpha_k s(n-k))^2] = 0 \tag{2.3}$$

This gives rise to the Linear Predictive (LP) analysis filter of the following form where z^{-1} is a one sample delay operator.

$$A(z) = 1 - \sum_{k=1}^{p} \alpha_k z^{-k} \tag{2.4}$$

This filter is sometimes known as a "whitening" filter due to the flat spectrum of the prediction error. This process removes the short term correlations from the signal. The information removed from the signal in the transmitter is restored in the receiver. One of the most interesting tutorials on the subject is found in [9], while a more extensive reference on linear prediction can be found in [10].

To obtain the line spectrum frequency "pair" representation for the LP analysis filter one must take the filter $A(z)$ and its time reversed counterpart $A(z^{-1})$ to create a sum filter, $P(z)$, and a difference filter, $Q(z)$, as shown below. If p is even, $P(z)$ has an extraneous real root at the half sampling frequency, while $Q(z)$ has an extraneous real root at 0 Hz. These extraneous roots can be ignored. The resulting line spectral frequencies are simply the alternating values of the roots of the sum and difference filters. These roots are spaced around the unit circle and have a mirror image symmetry about the real axis. Only the positive frequencies are considered here.

$$P(z) = A(z) + z^{-(p+1)} A(z^{-1}) \tag{2.5}$$

$$Q(z) = A(z) - z^{-(p+1)} A(z^{-1}) \tag{2.6}$$

There are a number of properties which make the LSF parameters desirable. Some of these properties include:

1. All of the roots of polynomials $P(z)$ and $Q(z)$ are simple, lie on the unit circle and are interlaced.

2. The minimum phase property of $A(z)$ can be preserved if the properties of 1. are intact at the receiver. This minimizes the effect of transmission errors and insures a stable filter for speech reconstruction.

3. The LSF's exhibit frequency selective spectral sensitivity. An error in a single LSF will be confined to the region of the spectrum around that frequency.

The monotonicity property allows wide latitude in how the parameters are quantized. For instance, dividing the vector into sub-vectors is made simple as long as the monotonicity property is preserved. This will insure that a stable linear filter will result when the quantized version of the filter is reconstructed. The same type of simple stability check for the α_k parameters does not exist. Frequency selective spectral sensitivity allows concentration of more bits on the most sensitive parameters. More intelligent channel error protection can be employed when coding with LSF parameters. An error in a single LSF can be easily corrected by a simple reordering or interpolation of the parameters followed by a bandwidth expansion of the resultant vector [27]. If an error occurs in the α_k parameters the effect is felt throughout the spectrum. The bandwidth expansion of the LSF's is necessary to prevent whistling artifacts. If the LSF's have less than 10 to 15 Hz separation they will have an associated strong resonance which approximates a pure tone sine wave at the average frequency of the pair.

Line spectral frequencies are a convenient and efficient representation for the linear predictive coefficients used in vector quantization. Just as important as the representation of the data is the mix of data collected into the training database. A later section will address the issue of how to put together the training database with as rich a mixture of appropriate conditions as possible. Once a parametric representation is established, a method for evaluating performance must be defined. The most widely accepted measure of performance used is the average log spectral distortion measurement. This and the Euclidean distance measure are discussed next.

3 DISTORTION MEASURES

One widely accepted measure of quantizer performance is the log spectral distortion between the quantized and unquantized vectors. Since the linear prediction vectors represent spectral information, it is straightforward to convert them to power spectra. The generic form of the distortion measurement is shown below where $S(\omega)$ is the power spectral density of the input signal, $\hat{S}(\omega)$ is the power spectral density of the quantized signal and B is the bandwidth of the measurement. This bandwidth varies, but Paliwal and Atal [4], used 3.0 KHz as the cutoff for the measurement when they introduced the standard of 1.0 dB average log spectral distortion.

$$\textbf{Log Spectral Distortion} = \sqrt{\frac{1}{B}\int_0^B \left[10\log S(\omega) - 10\log \hat{S}(\omega)\right]^2 d\omega} \quad (2.7)$$

The distortion was calculated discretely with a resolution of 100 Hz per sample, or 30 linearly spaced points from 0 Hz to 3.0 KHz. Their specific experiment was to quantize only the LSF parameters and then use the quantized parameters to generate synthetic speech with all other parameters held unquantized. The linear predictive residual error signal was used to excite the LP filter to produce quantized speech. Through informal listening tests, the average Log Spectral Distortion level at which spectrum quantization noise became audible was measured, and was found to be approximately 1.0 dB. Likewise a level of 2.0 dB was set for acceptable distortion, anything above this level is considered to be unacceptable and hence an outlying measurement. Atal and Paliwal caution that the number of outliers should be less than 2 percent for high quality spectrum quantization. The significance of this measurement is that under the conditions of the experiment, below 1.0 dB average Log Spectral Distortion, audible degradation due to spectrum quantization noise could not be detected.

The Average Log Spectral Distortion is considered to be one of the most effective measurements for determining the closeness of a pair of spectra. There are many theoretical motivations for this however, practical limitations on complexity prevent researchers from using this measure for quantizing spectra in real time devices. To overcome this limitation, the distance measure is introduced. Distance measures reduce the complexity relative to the log spectral distortion of searching vector quantizer code books by a ratio of at least 70 to 1, while incurring a small increase in spectral distortion. The use of distance measures yields a fairly good approximation to the code word selection as if done using a distortion measurement. The difference is that instead of measuring closeness of spectra, distance measures measure closeness of parameters. Perceptual weighting functions can be applied to the distance measurements to more closely approximate the selections as made by the distortion measurements. Whichever distance measurement is used, it is vital that there be a correlation back to a meaningful distortion measurement. Without this link, there is no guarantee that the best selection from the code book will be made.

Two of the more widely used distance measures are the squared Euclidean distance and the weighted squared Euclidean distance. The squared Euclidean distance between two vectors $\mathbf{x}$ and $\mathbf{y}$ is given by $D_E(\mathbf{x}, \mathbf{y})$, where p is the dimension of the vector.

$$D_E(\mathbf{x}, \mathbf{y}) = (\mathbf{x} - \mathbf{y})^T (\mathbf{x} - \mathbf{y}) \tag{2.8}$$

The weighted squared distance measure is given by $D_W(\mathbf{x}, \mathbf{y})$ where $\mathbf{W}$ is the p by p weighting matrix, and t denotes the vector transpose.

$$D_W(\mathbf{x}, \mathbf{y}) = (\mathbf{x} - \mathbf{y})^T \mathbf{W} (\mathbf{x} - \mathbf{y}) \tag{2.9}$$

If the weighting matrix $\mathbf{W}$ were the inverse covariance matrix of the input data $\mathbf{x}$, then this distance measure would be the Mahalanobis distance, pp 326 in [1]. The Mahalanobis distance is popular because it accounts for statistical properties of the input vector $\mathbf{x}$ when attempting to match them with code book values $\mathbf{y_i}$. Note that when the weighting matrix $\mathbf{W}$ is the identity matrix, $D_W(\mathbf{x}, \mathbf{y}) = D_E(\mathbf{x}, \mathbf{y})$. The distance measure which is used here is the squared Euclidean distance. This distance measure must be identical for both the clustering process as well as the quantization of the input vectors.

There are several cautionary statements that should be made here. Most importantly, the measurements should be consistent with respect to filtering when making comparisons to other work. Many claims are made where the system is said to include the entire bandwidth of 4.0 KHz, when in reality, the system filters the signal from 300 Hz to 3.3 kHz. Misleading results also arise from vector quantizers that are trained and tested on the same source material, or even from disjoint test sets from the same source. The utmost care must be exercised when selecting source material for training and testing vector quantizer code books.

Given a database or two (discussed later on), a distortion measurement, and a parametric distance measure, the next logical step is to combine these in a meaningful way to define or create a quantization scheme. This chapter is mainly interested in vector quantization, but for comparison sake, the scalar quantizer should be mentioned first.

4 SCALAR QUANTIZATION

Scalar Quantization is a process through which the elements of a vector are independently quantized. This process is based on the probability density function, (pdf) of the signal parameters as well as the linear dependencies between these parameters. The properties which make scalar quantizers desirable are minimum memory and computational requirements and a binary tree search capability. When scalar quantization is used with LSF parameters, the frequency selective spectral sensitivity allows for each parameter to be quantized based on the relative spectral sensitivity of that parameter. Hence, less important parameters are quantized with fewer bits, allowing for more important parameters to be quantized with additional bits. An interesting discussion on the scalar quantization of LSF parameters can be found in [12]. The performance of the scalar quantizer can be very good providing the correct training procedures are maintained. For instance, the 41 bit scalar quantizer used in the 2400 bps US Government Standard LPC10e speech coder is an excellent example of a scalar quantizer. On the disjoint database described in the vector quantization case, the 41 bit scalar quantizer measures a level of distortion of 0.504 dB average log spectral distortion over the entire 100,000 vector test database. The 34 bit scalar quantizer found in the US Government standard 4800 bps CELP algorithm measures a distortion level of 1.1924 dB average log spectral distortion on the same data set. The memory requirements for scalar quantizers depend upon the number of bits used to quantize each parameter. For the CELP algorithm which codes LSF parameters, the parameters require 3,4,4,4,4,3,3,3,3,3 bits for the respective elements of the parameter vector. This results in 112 memory locations, with a brute force search requiring only 112 comparisons, and with a tree search requiring only 34 comparisons. A breakdown for these numbers will be given later on in the section on complexity.

When channel bandwidth is limited, or other constraints require a restricted number of bits be used to describe the spectrum data, vector quantizers are required. Bit for bit, vector quantizers will outperform scalar quantizers. A detailed discussion of vector quantizers follows in the next section.

5 VECTOR QUANTIZATION

Vector Quantization is a process where the elements of a vector are jointly quantized. Vector Quantization is more efficient than scalar quantization by accounting for nonlinear dependencies and vector dimension as well as linear dependencies and the shape of the multidimensional probability density function [2].

Linear Predictive filter parameters in the form of Line Spectral Frequencies collectively form the vectors of interest in this chapter. For simplicity, we refer to these vectors as $\mathbf{x} = [x_1, x_2, \cdots x_p]^t$ for a pth order LP filter. The quantized version of the vectors is represented by the symbol $\mathbf{y}$, where $\mathbf{y} = q(\mathbf{x})$ or $\mathbf{y}$ is the quantized value of $\mathbf{x}$. Vector quantization is accomplished through a mapping of the continuous input parameter vector $\mathbf{x}$, into one of a set of discrete vectors $\mathbf{y}$ called code vectors. These code vectors are preselected through a clustering or training process to represent the training data, and stored in a table called a code book. Vector quantization is performed by comparing the input vector $\mathbf{x}$ to each of the code vectors $\mathbf{y_i}$ and selecting the code vector which achieves minimal difference in that

$$D(\mathbf{x}, \mathbf{y_i}) \leq D(\mathbf{x}, \mathbf{y_k}) \tag{2.10}$$

for all k, $k \neq i$.

The index which is assigned to the selected code vector is then transmitted to the receiver for reconstruction. The preferred measurement for the calculation of the difference is the log spectral error measurement. Due to computational considerations most vector quantizers use either the squared Euclidean distance or the weighted squared Euclidean distance measurement when searching the code books.

The nearest neighbor vector quantizer, also known as the Voronoi vector quantizer is probably the most common type of vector quantizer in use today. In nearest neighbor vector quantization, the partition of the training data is completely described by the combination of the code book and the distance measure. This means that no geometric representation of the partition needs to accompany the code book to allow searching that code book. The partition of the training data is implicit in the code book, while the distance measure defines the boundary regions between code vectors. The goal of the nearest neighbor vector quantizer is to minimize the average distortion for a given training set and code book size, $N = 2^L$. Here, L represents the number of bits used for transmission.

In speech coding, there are identical copies of the code book located in both the transmitter and receiver. The transmitter identifies the code vector with the minimum distance to the input vector and transmits the index or address of the code vector to the synthesizer. The synthesizer then simply performs a table lookup to obtain a quantized copy of the input vector. To illustrate this point graphically, refer to Example 1 below.

Example 1: Given a process such as linear prediction which produces vectors $\mathbf{x} = [x_1, x_2]^t$, and a code book training on a large number of independent occurrences of these vectors, match each input vector $\mathbf{x}$ with the "closest" code vector $\mathbf{y_i}$. To illustrate this process, refer to Figure 1, where the vector dimension p is 2, and the code book size L is 4 bits. This example is independent of the training algorithm as well as the distance measure. Since the code book is 4 bits large, there are 2^4 or 16 code vectors in the code book. The training process partitions the input data into $N = 2^L$ cells. The centroid of each cell is the code vector $\mathbf{y}$ which represents any input vector $\mathbf{x}$ that may fall within the bounds of that cell. If for instance, the input vector $\mathbf{x}$ falls anywhere within the region defined by cell i, then the centroid representation of cell i, $\mathbf{y_i}$ would be the quantized value of $\mathbf{x}$.

6 VECTOR QUANTIZER STRUCTURES

Vector quantizers take on many different structures ranging from the simple to the very complex. The majority of vector quantizers in use today fall within the realm of the full, the split, and the multistage vector quantizers. The most basic is the full vector quantizer. where the entire vector is considered as an entity for both code book training and quantization. Note that as the techniques for

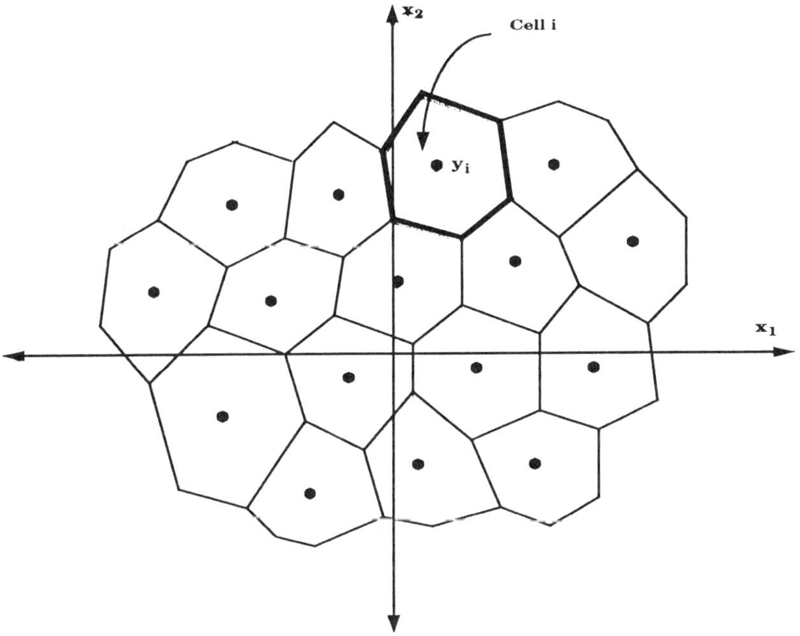

Figure 1 Voronoi partitioning of a 2 dimensional space

creating more elaborate code book structures are introduced, corresponding increases in the distortion level for the same bit rate are also introduced. The main reason for the different VQ structures is the necessity for lowering the complexity, bit rate and storage requirements for speech coders.

The Multi Stage VQ works by coarsely quantizing the input vector with the first stage code book and in doing so creating an error vector **e1**. This error vector is then finely quantized by the second stage, and if there are more than two stages, a second error vector **e2** is created by quantizing **e1**. This process continues until each stage of the code book has been used. The Split VQ structure divides the input vector into two or more sub-vectors which are independently quantized subject to constraints on monotonicity. The Split VQ is a special case of the Multi Stage VQ where different portions of the input vector at each stage are substituted with zeroes. These substitutions mask out the portions of the vectors where the data is not represented by the current code book.

It should be mentioned that the same distance and distortion measures that were employed for training the code book need to be used for searching those code books. As was stated above, the code book geometry need not be explicitly stated, as the distance and distortion measure combined with the code vectors implicitly define the geometry. It is absolutely crucial that these be consistent!

This next work will use Paliwal and Atal's measurement of 1 dB average log spectral distortion, as previously described as a goal. How this goal is achieved for a device which operates in adverse environments, or is small enough to be implemented in a hand held telephone is another matter.

6.1 Full VQ

Full vector VQ is the simplest form of vector quantization, implementation of which requires that the code book be trained and searched on whole or "full" vectors. The searching of this type of code book is performed by identifying the code vector $\mathbf{y_i}$ which is "closest" to the input vector $\mathbf{x}$, or $\mathbf{y_i} = q(\mathbf{x})$. This vector is selected as the representative or quantized vector. The address of that representative is transmitted and used for reconstruction of the quantized code vector. The two largest drawbacks for the full vector VQ are storage and computational complexities. To achieve an acceptable level of distortion would require a code book on the order of 20 bits [21]. This would require more than 1 million code vectors, each of dimension 10 for a total of more than 10 million words of memory. Searching a code book this large in real time is intractable, not to mention collecting a database of over 128 million vectors to adequately train the code book. To work around these problems split and multistage vector quantization are introduced.

6.2 Split VQ

The Split VQ structure divides the vector into two or more sub-vectors each of which are independently quantized subject to the monotonicity property. Code books must be created for each of the sub-vectors. Searching the code book which corresponds to the lower parameters proceeds as for that of the full vector VQ, i.e., with no constraints. Subsequent searching of each of the other code books proceeds in turn under the constraint that the search must be indexed to preserve the monotonicity property of the input vector. For a split-2 or two way split structure, the quantization process becomes $\mathbf{y1_i} = q1(\mathbf{x1})$, $\mathbf{y2_j} = q2(\mathbf{x2})$ where the $q2$ search is constrained to be monotonic when vector

$\mathbf{y2_j}$ is combined with $\mathbf{y1_i}$. In Paliwal and Atal [4], for a 10 dimensional LP coefficient vector a two way split would partition the vector between the 4th and 5th parameters. While for a three way split, the partition would be between parameters 3 and 4, and 6 and 7 respectively.

Splitting reduces the search complexity by dividing the vector into a series of sub-vectors, depending on how many bits are used for transmission. Since the lower frequency portion of the spectrum is more important from an intelligibility perspective, more bits should be concentrated for coding there. Given a two way split as an example, equal code book sizes devote an equal number of bits to describe each of the sub-vectors. Since the sub-vectors are of unequal dimension, the effect is to emphasize the code book of lower dimension. When the code book sizes are unequal, an extra bit is devoted to the code book for the lower frequencies. Similar logic is used to determine the code book sizes for a three way split.

Efficiency is one reason for selecting the split vector quantizer over the multistage vector quantizer. For the same data rate, the two way split vector quantizer requires 1/2 the computational complexity and 1/2 the storage capacity as the double stage vector quantizer. In comparing the three way split and the triple stage quantizer, a factor of 1/3 emerges.

6.3 Multistage Vector VQ

In contrast to the Split vector VQ techniques there is Multistage vector quantization. The main idea is to coarsely quantize the input vector $\mathbf{x}$ using a vector quantizer with a small number of code words, $\mathbf{y_1} = q1(\mathbf{x})$. Given this coarse quantization, an error vector $\mathbf{w} = (\mathbf{x} - \mathbf{y_1})$ is generated between the input vector and the quantized vector. This error vector is quantized using a code book which has been trained on similar data, $\mathbf{z} = q2(\mathbf{w})$. Additional stages are possible, but the discussion here will be limited to only the two stage case. This scheme is slightly inferior to the split vector quantizers. Intuition would indicate that not splitting the vector prior to quantization should result in a decoded vector which is lower in overall distortion, and hence should sound better in the reconstructed speech. This does not seem to be the case in practice. By splitting the vectors as in the split VQ above, a greater percentage of the quantization can be localized at the perceptually more important lower portion of the spectrum. Given an error corrupted channel, there is no easy way to insure the monotonicity property of the LSF parameters. For these reasons, the split vector quantizer is more desirable. An excellent speech coder which employs multistage vector quantization is found in [14].

6.4 Finite-State VQ

Finite state vector quantizers exploit the underlying content of the speech to provide the transmitter and receiver with mutual information which is not transmitted. There are two different equivalent types of finite state vector quantizers, the labeled state and the labeled transition. While it is outside the scope of this chapter to go into further detail of this type of vector quantizer, there are several excellent references on the subject. Of particular note is the paper by Farvardin, et al. [13] which combines both source and channel coding into their finite-state algorithm. Other good references include [1] and [8].

6.5 Variable Dimension VQ

As the name implies, the dimension of teh vector is considered to be variable. This may arise from coders which base how the spectrum is coded on the value of the pitch parameter. For higher pitch speakers there are fewer spectral harmonics than for lower pitched speakers. For these coders it is critical to encode the spectral harmonics correctly to insure high quality synthetic speech. The variable dimension vector quantizer attempts to match vector dimension to the number of spectral harmonics so as to provide the highest possible coding gain for each frame of spectrum data. Originally, the idea was to define a finite set of vector dimensions and create a code book for each. An elaborate coding scheme is required to switch between these code books, and the possibility of transmission errors makes large spectrum distortion a very real problem [7]. An alternative scheme was devised by Das and Gersho in [5] where they create a single universal code book with vectors of fixed dimension, yet representing the range of vector dimensions under consideration. This allows them to drastically reduce the amounts of training data that they would otherwise require for generating the independent code books for each vector dimension. Savings are also realized from reductions in both storage requirements and computational complexity.

6.6 Adaptive Vector Quantization

The adaptive vector quantizer allows for low probability code vectors which have not been used in a specified time period to be eliminated from the code book. These code vectors are replaced by higher probability code vectors selected from the current input data but not adequately represented in the existing code book. In effect, the current input data is added to the training set. These update code vectors are then transmitted to the receiver during periods of silence or low speech activity. In the limit, this technique can approximate a vector quantizer which was trained on the source material from the current user, allowing for a substantial improvement in distortion. The main drawback to this technique is that the transmission of the update code vectors across the channel without errors is a very unlikely occurrence. In [6] Paul tailors the transmission of the new code vectors to the current estimates of the channel error conditions.

7 VECTOR QUANTIZER TRAINING

Vector quantizers must first undergo a training procedure to generate a code book used to quantize the input vectors. The objective is to obtain a code book whose code vectors optimally describe the data. This may not happen as the various algorithms are subject to convergence on locally optimal solutions rather than globally optimal solutions. Training can produce a code book which describes the input data well, or poorly, depending on the design. A large number of vectors from diverse but representative sources are required to insure statistical significance of the resulting code book. This will be discussed later on in the databases section of this chapter. Before proceeding to algorithm descriptions, the term "cluster" must first be introduced. A cluster is a group of data points which have been classified into a set. This cluster is represented by a centroid of the members of that set, and is an average of these members. These representative centroids are referred to as code vectors of the code book.

Regardless of the type of algorithm used, most training algorithms incorporate the generalized Lloyd algorithm, known as the Lloyd iteration, in some form. The Lloyd iteration is described below.

1. Given a code book, find the optimal partition of the training data into the code vectors using the distance measures described above. When the "Nearest" code vector to the current training vector is identified, that

training vector is assigned to the cell represented by that code vector. The term nearest neighbor arises from the fact that the training vectors are assigned to whichever code vector is closest, defining its nearest neighbor.

2. For each cell and associated training set vectors described in 1., the new centroid is calculated. This centroid then becomes the new representative code vector for that cell.

The major drawback of using the Lloyd iteration is that it is highly susceptible to the location of the initial estimates for the cluster centroids. For this reason, many different algorithms exist for defining these initial estimates called seeds or seed points.

There are two major classes of algorithms used to train vector quantizing code books. These are the hierarchical algorithms and the divisive algorithms. Note that the hierarchial algorithms discussed in Gersho and Gray, pp 461,2 [1] are a particular type of vector quantization technique, and not a class of algorithms as is discussed here.

7.1 Divisive Algorithms

Divisive algorithms generally begin with all of the training vectors as members of the same cluster. As the name indicates, partitions are created which divide the data into successively smaller clusters, terminating when the correct code book size has been reached. Tree structured training procedures are good examples of divisive algorithms.

The K-means algorithm starts with a predetermined set of initial cluster centroids called seed points and iterates using the generalized Lloyd algorithm described above. The most frequently used stop rules are to allow the algorithm to proceed until either some maximum number of iterations is reached, or the error drops below some predetermined value, ϵ. This error is measured by calculating the average log spectral distortion between each vector in the training set and their respective cluster centroids. The most difficult part of this algorithm lies in the selection of the seed points as the algorithm is greatly influenced by these initial seeds. There are no guarantees that a global solution will result from a great deal of computation and much effort expended in the derivation of seed points. With a large enough training database, random indices can be generated where the vectors represented by the indices become the seed points for the algorithm. One flaw in this procedure is in the possibility that outlying data points may be selected as seed points which remain as code vectors in the resulting code book.

A major drawback of this algorithm is the tendency towards local minima rather than globally optimal solutions. To accommodate this drawback, the algorithm should be run many different times on different sets of training data. One possible method for accomplishing this is to divide the training set into disjoint groups of training and testing sets. The resulting code books are then tested, selecting the code book producing the minimum distortion over the broadest set of test conditions.

The term "tree structure" is drawn from the analogy between the growth of a tree in nature and the structure produced by the recursive process described for some of the divisive algorithms. In fact the clusters which are produced are referred to as leaf clusters, and the intermediate nodes of the tree structures are referred to as branches. Tree structured training algorithms begin with a single cluster composed of the entire training database. As the algorithms progress, a single cluster or a set of clusters is/are selected for splitting. Splitting is accomplished in two stages. First by generating seed points from within the selected cluster, and second by repartitioning the data with the seed clusters and all previous clusters. The algorithm recursively "grows" new clusters with each iteration, and terminates when the correct code book size has been reached. Upon completion of the algorithm, the centroids of the clusters become the code book entries. The most common of the tree structured algorithms is the Linde-Buzo-Gray algorithm described in [11]. If the underlying tree structure can be preserved, then a technique known as pruning can be applied by generating a cluster set that is much larger than required. Given this larger cluster set, certain clusters can be recombined into the original pre-split cluster such that the distortion level is minimally increased [15]. As before, when the correct number of clusters remain, the code book takes the form of the cluster centroids.

The Linde-Buzo-Gray (LBG) algorithm [11] begins with the training set as a single large cluster. New seed points are created by adding and subtracting a small fixed perturbation to each of the existing cluster centroids. Splitting is performed by applying the generalized Lloyd iteration on this new set of seed points, the number of clusters doubling with each iteration of the LBG algorithm. This type of tree is referred to as the uniform binary tree. One benefit is that a family of code books can be produced as a by-product of this algorithm, one code book for each level, $N = 1, 2, 4, \cdots 8$ and so on. Since the number of clusters double with each iteration, this algorithm is reasonably efficient.

Another type of tree structured algorithm "grows" the clusters one at a time, thus creating a nonuniform binary tree. The algorithm recursively splits the cluster with the largest distortion into two smaller clusters, the membership of which depends solely on that of the parent cluster. Periodically, a complete Lloyd iteration is required to prevent the reduction of the cluster set size below some preset minimum. This periodic Lloyd iteration will also prevent any bias towards outlying data. The mean, variance and standard deviation are measured and a small fraction of the standard deviation is both added and subtracted to the parent cluster centroid thus creating a new pair of seeds. This insures that the seeds lie within the original parent clusters regardless of the magnitude of the data. As the training set size is increased, this algorithm exhibits a quadratic increase in the amount of resources required for the Lloyd iteration.

Care must be exercised so that this class of algorithms has a sufficient amount of data for splitting. Outlying data points can cause low density clusters to have large distortion measurements, resulting in the incorrect selection of that cluster to split. In the limit, the algorithm will favor splitting off outlying data points into unity membership clusters. This will result in an undue emphasis on those outlying data point in the resulting code book [18]. To prevent the outliers from dominating the procedure, periodic Lloyd iterations should be computed, allowing the training vectors to freely partition into any of the clusters, rather than only the child clusters from a single parent. The best results are obtained when a generalized Lloyd iteration is performed after each splitting. Since each successive split is performed on the cluster with the largest measured distortion, the overall average distortion should be minimized.

7.2 Hierarchical Algorithms

In contrast to divisive algorithms, hierarchical algorithms begin with the training vectors classified as independent clusters which must be merged together to form larger clusters. With each iteration, groups of vectors are joined together into successively larger clusters. This joining may entail pairs of individual vectors, or in the limit, entire clusters may be joined together. The algorithms proceed until the correct number of clusters have been reached. Agglomerative algorithms are good examples of hierarchically structured algorithms.

Agglomeration simply means "to form into a rounded mass" [26]. Agglomerative clustering attempts to do just that. Initially, the algorithm has M clusters defined, one for each training vector, the algorithms proceed by merging these clusters until the correct number of code words remain. Agglomerative algorithms seek to combine the clusters so as to create seed points in the denser regions of the data space.

The pairwise nearest neighbor algorithm defined in [17] is a good example of an agglomerative algorithm. Given the initial cluster set and the training data, this algorithm calculates the pairwise distance between each cluster and all other clusters and combines this information with the number of vectors that each cluster represents to trade-off minimizing the distortion and affecting the fewest training vectors.

Another agglomerative algorithm begins with the initial cluster set defined as the training data and calculates the distance between each training vector and all other training vectors thus creating a similarity measurement. Vectors which are similar will have a smaller measurement than those which are not. The pair of vectors which have the smallest distance between them are identified as the nearest neighbors. An epsilon sphere is then placed around the centroid of these two points. This sphere represents a small neighborhood in the data space around the nearest neighbor pairs. Any vector from the training set which lies within this epsilon sphere is agglomerated into that cluster. If a minimum membership is not reached, the original pair of training vectors are removed from the data set, and the algorithm continues by identifying the next closest pair. This agglomeration phase is repeated until the correct number of valid cluster seeds points have been formed. These seeds comprise the initial code book required for the generalized Lloyd iteration. Note that the Lloyd iteration is performed on the entire original data set, including the rejected data points. There are several things which control the outcome of the seed point generation algorithm. The first is the selection of the minimum size for the seed clusters. If this is too large, and the ratio of the number of seeds to the overall training set size is low, the algorithm may terminate prematurely without generating the correct number of seed points required for the Lloyd iteration. Conversely, if the minimum distance is too small, then the algorithm may not achieve the placement of seeds in the denser regions of the training set. The second factor which is critical is the size of the epsilon sphere placed around the nearest neighbor pairs. If this is too large, then too many training vectors will be included in the seed clusters, in effect causing a coarser quantization of the training set into seed points. Again the algorithm will terminate without the correct number of seeds for the Lloyd iteration. If either case is true, then no code book will be formed and an enormous amount of computer time will have been wasted.

The Isodata algorithm of Ball and Hall [16] is worth mentioning here as it is an interesting hierarchical algorithm. The Isodata algorithm begins by assigning the first N training vectors as clusters. For each of the other training vectors in turn, the distance from that vector to all of the cluster centroids is calculated. The cluster which has the minimum distance to the new training vector incorporates that vector into its centroid. The algorithm continues until all of the training vectors are represented in the clusters. One improvement over the selection of the first N consecutive training vectors is to randomly select N training vectors as the initial cluster centroids. The only difference between this algorithm and the K-means algorithm is when each new training vector is classified into one of the existing clusters, that cluster centroid is recomputed prior to proceeding to the next vector.

7.3 Robust Training Approaches

Robust vector quantization techniques have become a topic of serious interest by speech coding researchers. One problem which continues to affect vector quantizers is the quantization of conditions which were not represented in the training data. These conditions might include microphone shaping, harsh acoustic background environments, filtering of the signal prior to digitization, etc. Vector quantizers do a very good job of partitioning the training data into code books. Where the training data has dense regions, coverage will be good, while for relatively sparse regions, the coverage is correspondingly sparse, or even nonexistent. Some approaches to robust vector quantization seek to intentionally place training data in these sparse regions. This can be done by either warping actual data, or by generating artificial data to cover these regions [22]. Another method used for robust vector quantization is to add information about the expected channel error environment to the training procedure [13]. A somewhat less drastic approach is to re-order the vector quantizing code books after they are completed. This re-ordering allows for single bit index assignment errors to be decoded into a vector which is spectrally "close" [19][20]. The use of vector quantizers in portable telephones mandates the use of error protection of the received code word indices. The combination of robust training, post processing of the index assignments and error corrective coding provides a reasonable amount of protection in typical communications grade channels.

8 DATABASES

The size and type of speech databases are of critical importance to the training process in vector quantization. The number of training vectors in the database has a direct effect on the type of structure the vector quantizer will have. This decision is based on the ratio of available training vectors to desired code vectors. If there are too few training vectors, and a large code book is required, then it may be advantageous to use a multistage or split structure where each step requires fewer code vectors. The rationale here is if this ratio becomes too small then the resulting code books will show signs of undertraining when tested on disjoint speech databases. Undertraining becomes apparent when the test data distortion exhibits an uncharacteristically large measurement. Testing on a disjoint data set is done to preserve the integrity of the test. If results were reported on within-set testing, then this is the absolute best that the particular algorithm can perform. If, however, the testing is performed on vectors which were completely unknown to the training algorithm then this provides an indication of how the vector quantizer might perform in the worst case. This is the main reason to make the conditions under which the test vectors are selected disjoint. If possible, these vectors should not be drawn from the same overall training set conditions. The only caution here is that the test data should not represent unrealistic conditions which have no probability of occurring in operational environments.

Vector quantization training procedures require a rich combination of source material to produce code books which are sufficiently robust for quantization of data not represented in the training set. Examples of some of the conditions which might enrich the training set include varying microphones, acoustic background environments, languages, gender, etc. The "Holy Grail" of collecting databases is to obtain as large and diverse a set of vectors as possible in order to represent a reasonable approximation to the expected input data being quantized. This goal is very difficult to reach as there are no guarantees that new or unforeseen applications may not arise. In general, a large diverse training set will provide a reasonably robust code book while at the same time providing a statistically significant basis for the mathematical models used to generate the code books used for quantization.

In [1], Gersho introduces the idea of the training set ratio $\beta = N/M$ where $M = 2L$ is the code book size (L is the number of bits of storage in the code book), and N is the size of the training set. A good rule of thumb is to use a β of at least 128. Note that β can be much larger if resources and time permit. As with any average, some code vectors will represent many more training vectors than others. For this reason, it is recommended that the designer incorporate more than the minimum of 128 training vectors per code vector. If a 10 bit code book were required, than a minimum data set size of approximately 2^{17} vectors would be required and 2^{19} would be recommended. For the data intensive algorithms, or when generating larger code books, this could be as high as 2^{20} or more vectors.

A method for demonstrating the relationship between the number of training vectors and the quality of a code book is as follows. For a given code book size of L bits, (in this case $L = 10$ bits has been selected) generate a set of code books varying only the amount of training data. Start at 2 training vectors per code vector and increase the training set size by doubling the number of training vectors for each successive code book. The limiting factor here is the total number of training vectors. In this case, the training set size is 2^{20} vectors so the upper limit is $\beta = 2^{20}/2^{10} = 1024$ training vectors per code vector. Evaluation of the performance of a given code book or set of code books compares the distortion resulting from quantizing both within-set and disjoint test vectors. The within-set test procedure measures the average log spectral distortion on the vectors used to train the code books. While the disjoint test measured the average log spectral distortion on a fixed set of 100,000 vectors. These disjoint vectors were not only disjoint from the training set, but drawn from vectors completely outside the training set conditions, i.e. different microphones, speakers, etc. The disjoint test vectors used in this chapter were drawn from the TIMIT test database for dialect regions 1 through 5. This database can be obtained from the Linguistics Data Consortium [24].

Figure 2 graphs the results of this test and an interesting point can be made. That is for the within set testing, the expected outcome of the tests are for the distortion measurement to be monotonically increasing for increasing training set size. There is a slight data anomaly between the cases $\beta = 2^6$ and $\beta = 2^7$ training vectors per code vector. The distortion for $\beta = 2^6$ is minutely larger than that for $\beta = 2^7$. This is counterintuitive as the expected result is for the curve to be monotonically increasing with increasing training set size. One possible explanation for this increase is that the code books are searched using the squared Euclidean distance measure, while the distortion is calculated based upon the average log spectral distortion measure. This anomaly serves to point out two interelated issues, first that there are differences between the two measures, and second that these differences are very small. This last point serves to reinforce confidence in using the squared Euclidean distance measure on LSF's instead of the average log spectral distortion measurement.

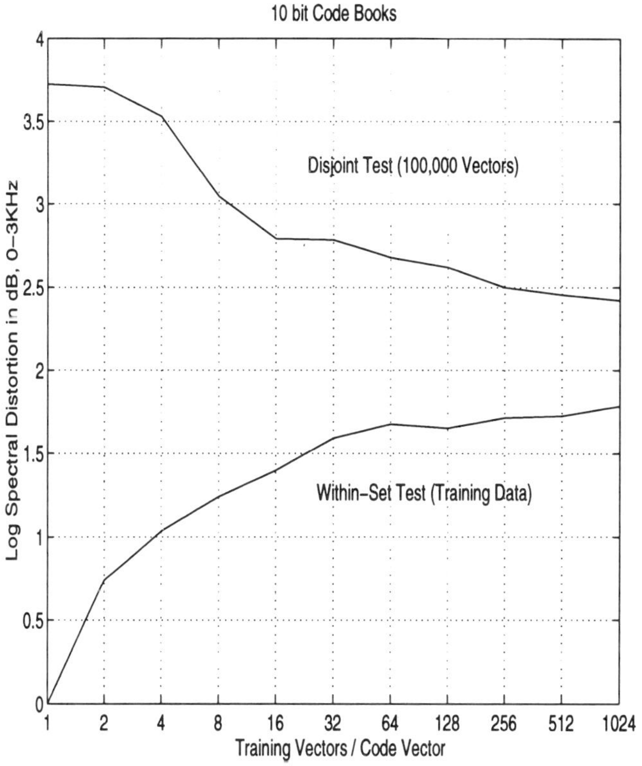

Figure 2 Within set testing vs. Disjoint testing, 10 bit code books.

Figure 2 indicates that a minimum of 128 training vectors per code vector are required to properly train a vector quantizing code book. This point is clarified by Figure 3 which graphs code book size versus distortion for full, split-2 and split-3 code books. Distortion was measured over the same disjoint set of 100,000 vectors as used for Figure 2. To make the comparison easier, the training set size was held constant at 2^{20} vectors. The most interesting information here is the fact that the traces for the three code book types intersect. Ideally, the distortion should be decreasing at the same rate, but when the number of training vectors per code vector decreases to less than 128, the code books become under trained. In Figure 3, observe the range of 13 to 16 bit full code books, and the range between 23 and 27 bit split-2 code books. For the full code book case, as the ratio of the number of training vectors to code vectors falls below $\beta = 128$, the distortion begins to level off to the point where a split-2 code book of the same size outperforms the full code book. For the split-2 code book, an equal split is used for equal code book sizes. For unequal code book sizes, the extra bit is applied to the lower code book. For the split-2 case, when the ratio drops to about $\beta = 256$ or so, the split-3 code books begin to outperform the split-2 code books. Observe also that this phenomenon occurs at higher thresholds when comparing the split-2 code books and the full code books. This is due to the splitting process destroying the inter vector dependencies.

Figure 2 also demonstrates the need for measuring the distortion of vector quantizing code books on disjoint data sets. No matter how much data is used to train the vector quantizing code books, when you test on the same source data, the results will always appear better than they actually are. Testing for Figure 2 was conducted using two distinct measures, the first is the squared Euclidean distance between parameter vectors used to search the code books, and the second is the average log spectral distortion used to measure the actual distortion as previously discussed.

Figure 3 shows a graph of code book size in bits versus average log spectral distortion for full code books and split code books with a split of two and three sub-vectors. All of the code books which were used to generate this graph were trained using 2^{20} training vectors. For comparison sake, the graph also displays the average log spectral distortion for one of the more commonly used scalar quantizers, the 34 bit CELP quantizer. The graph points out the deficiencies of training code books without adequate training data.

9 COMPLEXITY

The issue of complexity comes into play when system requirements call for a portable device with size weight and power constraints. Given such constraints, as well as those on data rate and channel error environments, compromises must be made. Scalar quantizers are very low complexity, for instance the US Federal Standard 1016 CELP algorithm which uses a 34 bit scalar quantizer on the 10 LSF parameters requires 3,4,4,4,4,3,3,3,3,3 bits for each vector parameter respectively. This requires only $(6 \times 8 + 4 \times 16)$ memory locations to store both the encoding and the decoding tables. Searching this scalar quantizer requires 112 comparisons for a brute force search, and 34 comparisons for a tree search, $(6 \times 3 + 4 \times 4)$. Unfortunately, most low data rate algorithms require a trade-off between low complexity high data rate scalar quantizers, and high complexity low data rate vector quantizers. For these vector quantizers,

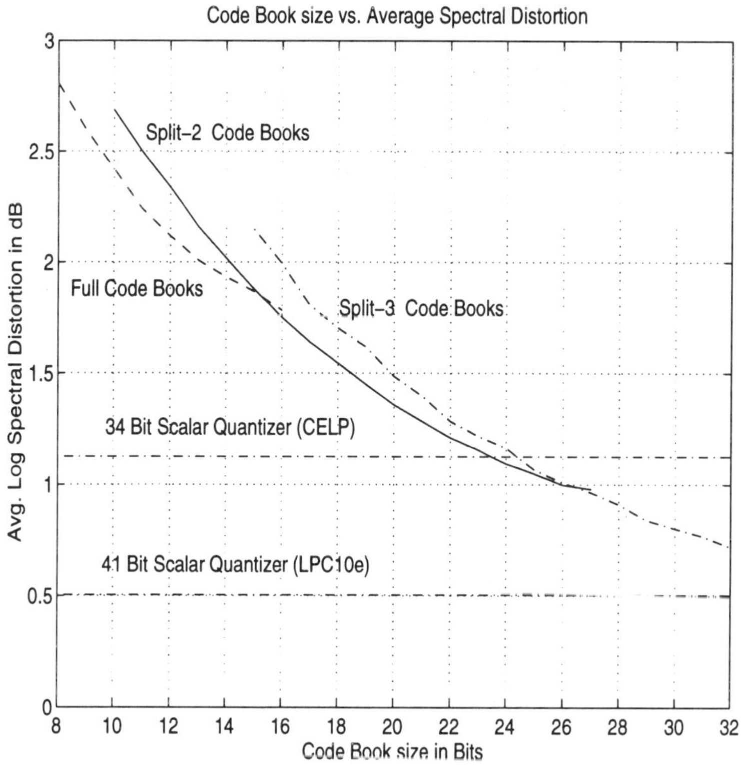

Figure 3 Code Book size vs. Av. Log Spectral Distortion

efficiency of transmission requires that the code books be an even power of 2 resulting in 2^L code vectors for an L bit code book. Given a pth order LPC vector, there are $p2^L$ elements for any particular code book. To get the number of bytes of storage required, multiply $p2^L$ by the number of bytes per element. To obtain a reasonably low level of spectral distortion with a full vector, the vector quantizer requires a code book of approximately 20 bits [2]. Complexity considerations preclude the use of such massively large code books in real time devices. Current state of the art hardware places practical limits on vector quantizer code book sizes to roughly 13 or 14 bits. A vector quantization scheme which somehow divides the work and storage complexity into smaller manageable partitions is needed. The split vector quantizer where the vector is partitioned into two sub-vectors can accomplish the task using 24 to 25 bits. This is done using a 13 bit code book for the lower 4 parameters, and a 12 bit code book for the upper 6 parameters. Note that 2^{13} plus 2^{12} is only

12,288 code vectors or 57,344 storage locations as opposed to 1,048,576 code vectors and 10,485,760 storage locations for the 20 bit full code book. These requirements are further reduced when using the split-3 vector quantizer with a (3,3,4) split. Assuming a 27 bit aggregate vector quantizing coding scheme equally split among the three code books requires $(3 + 3 + 4) \times 512 = 5120$ memory locations, with the same number of comparisons for a brute force search of the code books. Non-uniform binary trees can be constructed and used to dramatically decrease the amount of time required to search these code books with a minimal increase in distortion [23]. This split-3 structure provides a good trade-off between code book complexity and overall coder performance.

10 EFFECTS OF CHANNEL ERRORS ON VECTOR QUANTIZATION

In today's mobile communications environment, random Gaussian bit error rates on the order of 1 to 2 percent are common, as are overlaying Rayleigh or Ricean fading burst error conditions yielding long term average error rates of 5 to 10 percent. Traditional vector quantizers simply cannot withstand this onslaught of errors without help. This section takes a closer look at how the code vectors can be protected from random and burst channel errors. Ideally, one would like to deign vector quantizing code books such that any single bit error will result in a vector which is "close" to the intended code vector. Likewise, one would like to recover from a burst error situation where 90 percent of the frames have relatively few errors, and 5 percent of the frames might be completely lost.

Figure 4 shows the performance of the 26 bit split-2 vector quantizer trained on 2^{20} vectors and tested on the disjoint 100,000 vector TIMIT test sequence in Gaussian and Rayleigh fading error conditions ranging from 0.0 to 10.0 percent. As can be seen in the graph, there is a dramatic increase in the average log spectral distortion as a function of long term average bit error rate. The fading performance was done by generating bit error files via simulation of $\pi/4$ shifted QDPSK (quadrature differential phase shift keyed) modulation transmitted over Rayleigh fading channels with a 30 m.p.h. vehicular speed and a 900 MHz carrier frequency, typical mobile communications parameters. The long term average error rates in Figure 4 were arrived at by adjusting the long term Gaussian signal to noise ratio in dB (which adjusted the level of the constant Gaussian noise floor under which fading occurred) [25]. These bit files were then modulo-2 added to the binary channel indices for the 26 bit vector quantizer,

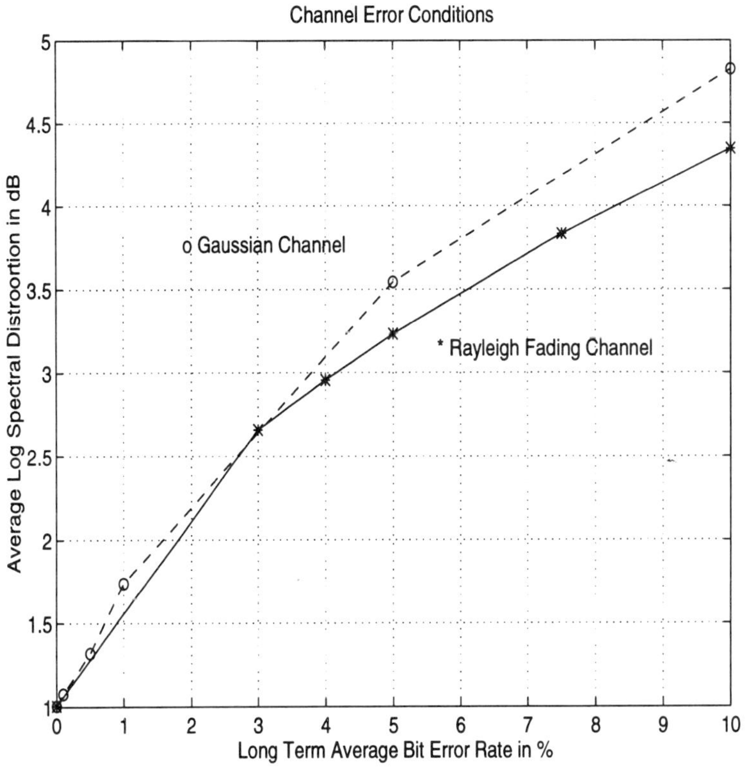

Figure 4 Bit Error Rates vs. Avg. Log Spectral Distortion

without interleaving of the indices. It should be noted that no attempt at index assignment has been made here to help mitigate the effects of channel errors. Indeed, if these are the random assignments for the code books it is a reasonable assumption that some improvement can be derived by ordering the code book indices through simulated annealing, or Hamming weighting as in [20]. An even greater increase in performance in channel error conditions can be derived if assumptions about the channel are incorporated into the training procedure as in [13] and [19]. An interesting observation to be made about Figure 4 is that if the Gaussian channel errors can be kept below about 0.5 percent, then even the random index assignments are reasonably robust.

11 CONCLUDING REMARKS

This chapter has attempted to introduce the reader to vector quantization. It is useful to reiterate several of the more important points made earlier.

LSF parameters have proven to be a desirable parameterization of the linear predictive filter. This parameterization lends itself to the use of the squared Euclidean distance measure as a much less complex matching criteria over the log spectral distortion measurement. Given the parameterization, and distance metric, a training database must next be collected. Proper selection and appropriate quantities of training data are necessary for the code book generation algorithms to produce high quality robust vector quantizers. Related to the amount of training data is the notion of from which sources the data is drawn. The goal here is to provide a rich variety of training material while at the same time restricting that material to be from *relevant* sources. A ratio, β was introduced which provides a good rule of thumb for deciding how much training data is required based upon the size of the code book to be generated. It was shown that for full code books, β should be at least 2^7, while for split-2 code books, β needs to be at least 2^8 training vectors to code vectors. Indeed, the designer might place constraints on the vector quantizer structure based upon the quantity of training data available. Due to practical limits on hardware capabilities the split-3 vector quantizer structure is the most widely used. The channel error environment might steer the designer into robust vector quantizer techniques which incorporate information about the error environment into the training procedure, the trade-off here is a greatly increased level of complexity. One such algorithm is the finite-state vector quantizer structure defined by Farvardin.

REFERENCES

[1] A. Gersho and R. M. Gray, *Vector Quantization*, Kluwer Academic, 1992.

[2] J. Makhoul, S. Roucos and H. Gish, "Vector quantization in speech coding", *Proc. of the IEEE,* Nov. 1985.

[3] R. M. Gray, "Vector quantization", *IEEE ASSP Magazine,* Apr. 1984.

[4] K. K. Paliwal and B. S. Atal, "Efficient vector quantization of LPC parameters at 24 bits/frame", *IEEE Trans. on Speech and Audio Proc.,* Jan. 1993.

[5] A. Das, A. V. Rao and A. Gersho, "Variable-dimension vector quantization of speech spectra for low-rate vocoders", *IEEE Data Compression Conf.*, 1994.

[6] D. B. Paul, "An 800 bps adaptive vector quantization vocoder using a perceptual distance measure", *Proc. IEEE Int. Conf. on Acoust., Speech and Signal Proc.*, Apr. 1983.

[7] J. P. Adoul and M. Delprat, "Design algorithm for variable length vector quantizers", *Proc. Allerton Conf. on Circuits, Systems and Computers,* Oct. 1986.

[8] M. O. Dunham and R. M. Gray, "An algorithm for the design of labeled transition finite state vector quantizers", *IEEE Trans. on Commun.*, Jan. 1985.

[9] J. Makhoul, "Linear prediction: A tutorial review", *Proc. of the IEEE,* 1975.

[10] J. D. Markel and A. H. Gray, *Linear Prediction of Speech*, Springer-Verlag, 1982.

[11] Y. Linde, A. Buzo and R. M. Gray, "An algorithm for vector quantizer design", *IEEE Trans. on Commun.*, Jan. 1980.

[12] F. K. Soong and B.-H. Juang, "Optimal quantization of LSP parameters", *IEEE Trans. on Speech and Audio Proc.*, Jan. 1993.

[13] Y. Hussain and N. Farvardin, "Finite-state vector quantization for noisy channels", submitted to *IEEE Trans. on Speech and Audio Proc.*

[14] B. Fette and C. Jaskie, "A 600 bps LPC voice coder", *Proc. of the IEEE Military Commun. Conf.*, McLean Virginia, Nov. 1991.

[15] E. A. Riskin and R. M. Gray, "A greedy tree growing algorithm for the design of variable rate vector quantizers", *IEEE Trans. on Signal Proc.*, Nov. 1991.

[16] G. H. Ball and D. J. Hall, "Isodata - an iterative method of multivariate analysis and pattern classification", *Proc. IFISP Congress,* 1965.

[17] W. H. Equitz, "A new vector quantization clustering algorithm", *IEEE Trans. on Acoust., Speech and Signal Proc.*, vol. 37, no. 10, pp. 1568–1575, Oct. 1989.

[18] J. S. Collura and T. Tremain, "Anomalous behavior associated with vector quantizer design", *IEEE Speech Coding Workshop,* Canada, Oct. 1993.

[19] P. Knagenhjelm, "How good is your index assignment?", *Proc. IEEE Int. Conf. on Acoust., Speech and Signal Proc.*, Minneapolis, Minnesota, Apr. 1993.

[20] P. Knagenhjelm, "Robust vector quantization", *Technical Report No. 12*, Department of Information Theory, Chalmers University of Technology, May 1993.

[21] P. Hedlin, "20 bit VQ codebooks for low data rate coding", *Proc. IEEE Int. Conf. on Acoust., Speech and Signal Proc.*, Adelaide, Australia, Apr. 1994.

[22] R. P. Ramachandran, M. M. Sondhi, N. Seshadri and B. S. Atal, "A two codebook format for robust quantization of line spectral frequencies", *IEEE Trans. on Speech and Audio Proc.*, May 1995.

[23] J. S. Collura and T. Tremain, "Vector quantizer design for the coding of LSF parameters", *Proc. IEEE Int. Conf. on Acoust., Speech and Signal Proc.*, Minneapolis, Minnesota, Apr. 1993.

[24] Linguistics Data Consortium, 441 Williams Hall, University of Pennsylvania, Philadelphia, PA, 19104-6305.

[25] D. J. Rahikka and D. Ray, "Reed-Solomon coding for CELP EDAC in land mobile radio", *Proc. IEEE Int. Conf. on Acoust., Speech and Signal Proc.*, Adelaide, Australia, Apr. 1994.

[26] Webster's II New Riverside University Dictionary, 1988.

[27] J. H. Chen, "A robust low delay CELP speech coder at 16 kbps", in *Advances in Speech Coding*, edited by B. S. Atal, V. Cuperman and A. Gersho, 1991.

3

LINEAR PREDICTIVE ANALYSIS BY SYNTHESIS CODING

Peter Kroon and W. Bastiaan Kleijn

Speech Coding Research Department,
AT&T Bell Laboratories,
Murray Hill, NJ 07984, USA

1 INTRODUCTION

The availability of inexpensive signal processing chips and a demand for efficient digital representations of speech signals have led to an increase of applications for speech coding. Some examples are: wired and wireless networks, voice encryption, videophones, simultaneous voice and data transmission, multimedia, announcements, and solid state answering machines. In the past 10 years, many digital speech coding standards have been defined for network and wireless applications. Most of these standards are based on the *linear prediction based analysis by synthesis* (LPAS) paradigm. LPAS coders provide state-of-the-art performance for bit rates in the range between 4 and 16 kb/s. This chapter will discuss this paradigm and related topics, and will focus on issues not discussed in some of the accompanying chapters in this book [1, 2, 3].

We start with a description of the basic principles of LPAS. Then we proceed with a generalization of this principle (GLPAS) and describe how this can be used to obtain a more efficient representation of periodic speech signals. In section 4 we discuss techniques for further improving the quality of the speech produced by coders based on the LPAS principle. Sections 5 and 6 discuss the various issues related to performance of a speech coder and their tradeoffs. Section 7 provides descriptions of standardized speech coders based on the LPAS paradigm.

2 PRINCIPLES OF LPAS

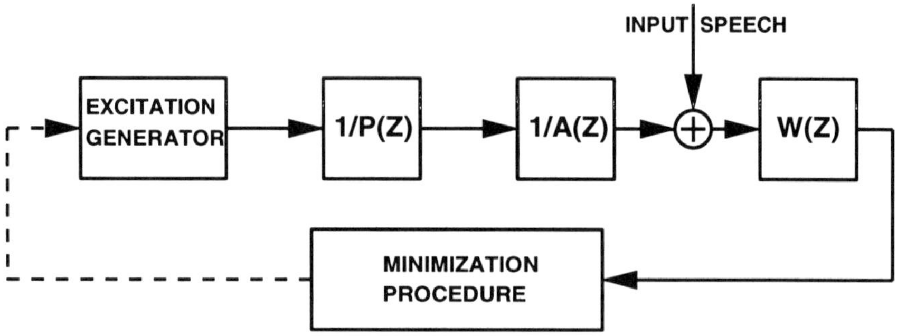

Figure 1 Block diagram showing the principle of linear predictive based analysis-by-synthesis (LPAS) coding.

Figure 1 shows the principle of linear predictive based analysis by synthesis (LPAS). The decoded speech is produced by filtering the signal produced by the excitation generator through both a long-term (LT) predictor synthesis filter $1/P(z)$ and a short-term (ST) predictor synthesis filter $1/A(z)$. The excitation signal is found by minimizing the mean-squared error over a block of samples. The error signal is the difference between the original and decoded signal. It is weighted by filtering it through a filter $W(z)$. Both short- and long-term predictors are adapted over time, but the short-term predictor is usually adapted at a slower rate than the long-term predictor and excitation parameters. Since the analysis procedure (encoder) includes synthesis (decoder), description of the encoder completely defines the decoder. Although most discussions in this paper are related to the configuration shown in Figure 1, many variations are possible. For example, the long-term predictor can be omitted, or the location of the long-term and short-term predictor can be interchanged (see for example [4][5]).

The filter $1/A(z)$ models the short-term correlations (spectral envelope) in the speech signal. A commonly used form is:

$$\frac{1}{A(z)} = \frac{1}{1 - \sum_{k=1}^{p} a_k z^{-k}} \tag{3.1}$$

where a_k are the short-term predictor coefficients and p is the order of the filter. Typical values for p are between 8 and 16. The predictor coefficients are determined from the speech signal using linear prediction techniques as described in [6][7]. The coefficients of the short-term predictor are adapted in time, with rates varying from 30 to as high as 400 times per second. To avoid transients during coefficient updates, it is common to use *spectral interpolation*. This technique interpolates the filter coefficients to a rate higher than their adaptation rate. Some short-term filter parameter representations that have good interpolation properties are line-spectral frequencies (LSF) and auto-correlation coefficients [8].

The filter $1/P(z)$ models the long-term correlations (spectral fine structure) in the speech signal [1]. Its most basic form is:

$$\frac{1}{P(z)} = \frac{1}{1 - bz^{-M}} \tag{3.2}$$

where M is the delay in samples and b is the long-term prediction coefficient. The value of M corresponds to the duration of the delay in samples. Typical durations are in the range from 2 to 20 ms. For periodic signals, the delay corresponds to the pitch period (or possibly an integral number of pitch periods). The delay is random for nonperiodic signals. The delay M and the coefficient b are determined either from the speech signal, or from the residual signal obtained after removing the short-term correlations. Typically, the long-term predictor coefficients are adapted in time at rates varying from 100 to 200 times per second. Increasing the order of the predictor results in better prediction, but more bits are needed to encode the additional coefficients [1]. For periodic segments, the main role of multiple coefficients is to provide interpolation between the samples if the period does not correspond to an integer number of samples. Instead of a higher-order predictor, one can use a first-order predictor with a noninteger delay, which allows a more efficient quantization of the predictor parameters [9]. A commonly used alternative structure for the pitch filter is the *adaptive codebook* [10]. In this approach, the long-term synthesis filter is replaced by a codebook that contains the previous excitation at different delays. The resulting vectors are searched and the one that provides the best match is selected. In addition, an optimal scaling factor can be determined for the selected vector. This representation simplifies the determination of the excitation for delays smaller than the length of the excitation frames.

To achieve a low overall bit rate, the average number of bits/sample for each frame of excitation samples has to be small. The *multipulse excitation coder* [11] represents the excitation as a sequence of pulses located at nonuniformly spaced intervals. The excitation analysis procedure has to determine both amplitudes and positions of the pulses. Finding these parameters all at once is a difficult problem and simpler procedures, such as determining locations and amplitudes one pulse at a time [11][12], are used. The number of pulses required for an acceptable speech quality varies from 4 to 6 pulses per 5 ms. For each pulse, both amplitude and location have to be transmitted, requiring about 7 to 8 bits per pulse [4]. In the *regular-pulse excitation coder* [13] the excitation signal is represented by a set of uniformly spaced pulses (typically 10 pulses per 5 ms). The position of the first pulse within a frame and the amplitudes of the pulses are determined during the encoding procedure. For a given position of the first pulse, all other pulse positions are known, and the amplitudes are found by solving a set a linear equations. The information to be transmitted are the position of the first pulse (2 bits) and the pulse amplitudes (3 to 4 bits per pulse).

Code-excited linear predictive (CELP) coders [14], which are the most common realization of the LPAS paradigm, use another approach to reduce the number of bits per sample. Here, both encoder and decoder store the same collection of C possible sequences of length L in a codebook. The excitation for each frame is described completely by the index to an appropriate vector in the codebook. In the configuration of Figure 1, this index is found by an exhaustive search over all possible codebook vectors, and selecting the one that produces the smallest error between the original and decoded signals. To simplify the search, it is common to use a gain-shape codebook in which the gain is searched and quantized separately. The index requires $(\log_2 C)/L$ bits/sample (e.g. 0.2 to 2 bits/sample), and the gain requires 2 to 5 bits for each codebook vector.

3 GENERALIZED LPAS

Linear predictive based analysis by synthesis (LPAS) based coders integrate the decoder into the encoder loop. This provides a direct feedback on the choice of the coder parameters, as is shown in the schematic representation of LPAS in Figure 2a. The (weighted) mean-squared error matching tends the make LPAS coders robust against a variety of nonspeech input signals. However, this property can also be a weakness. For example, slight mismatches in time-alignment of the original and decoded signal, which are not perceptually relevant will introduce a significant increase in error signal energy. As a result, the error minimization procedure effectively allocates all the excitation bits towards minimizing these (perceptually) irrelevant errors.

To counteract this shortcoming, an extension of the LPAS paradigm was proposed in [15]. This *generalized linear predictive based analysis by synthesis* (GLPAS) paradigm recognizes the inflexibility of the error criterion for certain

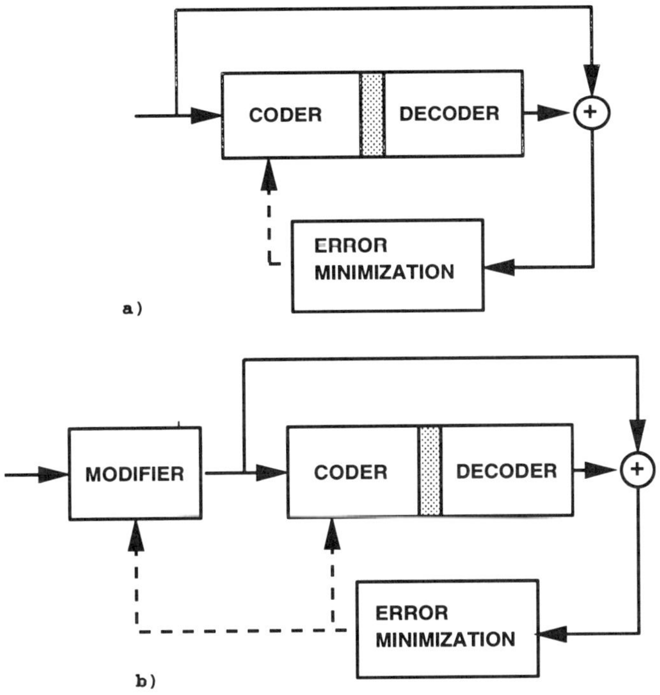

Figure 2 Simplified representation of LPAS (a) and generalized LPAS (b).

perceptually irrelevant differences and resolves this by allowing minor modifications of the original input signal. This is conceptually shown in Figure 2b. In this figure, the error minization procedure modifies the original, such that it optimally matches the speech produced by the decoder.

A good example of its use is a more efficient representation of the long-term correlation. For periodic speech segments the periodicity tends to vary slowly over time, and in conventional LPAS, the main reason for updating the delay frequently is to avoid time-misalignment between the original and decoded waveforms. By estimating only one delay value every 20 ms frame, and modifying the original signal, such that it matches the delay contour of the reconstructed signal, coding efficiency is significantly increased. Efficient implementations of this technique are described in [16][17].

4 IMPROVING SPEECH QUALITY

The perceived quality of the decoded signals can be further improved with a variety of methods. This section describes several of the most effective procedures. By using an *error-weighting filter* $W(z)$ at the encoder (see Figure 1), the error signal can be changed such that the coder is minimizing the error in the perceptually more relevant frequency regions. A *postfilter* at the decoder can be used to minimize the perceived distortion in the decoded speech. The parameters of both filters need not be transmitted and can be tuned for the application at hand.

The last subsection describes another way of improving the speech quality by increasing the bandwidth of the speech signal. It is shown that LPAS techniques can also be used for the coding of wideband speech signals.

4.1 Error Weighting Filter

In the paradigm shown in Figure 1, the coder parameters are selected such that the error energy between a reference and reconstructed signal is minimized. Minimizing a mean-squared error, results in a quantization noise that tends towards having equal energy at all the frequencies of the input signal. This is a direct consequence of Parseval's theorem which states that minimizing the mean-squared error in the time-domain is equivalent to minimizing the mean-squared error in the frequency-domain.

By using the properties of the human auditory system, one can try to reduce the perceived amount of noise. Frequency-masking experiments have shown that greater amounts of quantization noise are undetectable by the auditory system in the frequency bands where the speech signal has high energy [18]. To make use of this masking effect, the quantization noise has to be properly distributed among the different frequency bands. Spectral shaping of the noise can be achieved by minimizing a weighted error in the block diagram of Figure 1. Since we try to shape the noise as a function of the spectral peaks in the speech signal, it makes sense to derive the error weighting filter from the short-term predictor filter. A flexible form has been found to be [19]:

$$W(z) = \frac{F(z/\gamma_1)}{F(z/\gamma_2)}, \quad 0 \leq \gamma_2 \leq \gamma_1 \leq 1 \tag{3.3}$$

with

$$F(z/\gamma) = 1 - \sum_{k=1}^{p} f_k \gamma^k z^{-k} \tag{3.4}$$

In most applications $F(z)$ is chosen to be identical to $A(z)$. The parameters γ_1 and γ_2 are fractions between 0 and 1 that control the energy of the error in the spectral peak regions. Decreasing the value of γ increases the bandwidth of the zeros of $F(z/\gamma)$. Noise shaping increases the mean-squared error between the original and reconstructed speech, resulting in a reduction in segmental signal-to-noise ratio (SNR). The values of γ_1 and γ_2 must be determined by suitable listening tests. In Section 7 some typical values are listed.

4.2 Postfilter

Despite the error-weighting filter it is not always possible to mask the noise in the speech caused by the quantization of the excitation signal. Especially in the low-energy frequency regions, this quantization noise can dominate the speech signal. By using a separate post-processing technique after reconstruction by the decoder, the perceived noise can be further reduced. This operation, referred to as *postfiltering* trades off spectral distortion in the speech versus suppression of the quantization noise, by amplifying the spectral peaks and attenuating the spectral valleys [20]. Since the postfilter operation tends to change the energy of the signal, a gain normalization procedure which matches the energy of the output with the energy of the input is needed. Usually, this is done by computing the energy for a short frame (e.g. 5 ms) for both input and output, and scaling the output samples with the ratio of the two energy values. Smoothing of the scaling constant (e.g. low-pass filtering) further reduces any discontinuities in the output signal.

To have more flexibility in the shape of the postfilter, it is generally implemented in stages, as is shown in Figure 3. In this figure, the short-term and long-term spectral features are modified separately. In addition, a spectral shaping block is used to control the overall frequency response.

The short-term postfilter modifies the spectral envelope. This spectral envelope is usually based on the transmitted short-term predictor coefficients, but can also be derived from the reconstructed signal [21]. The latter method has advantages in that a different filter order can be chosen, and that the effect of spectral quantization errors is reduced. A typical short-term postfilter will be similar to the weighting filter of Equation 3.3 with different values for γ_1 and γ_2.

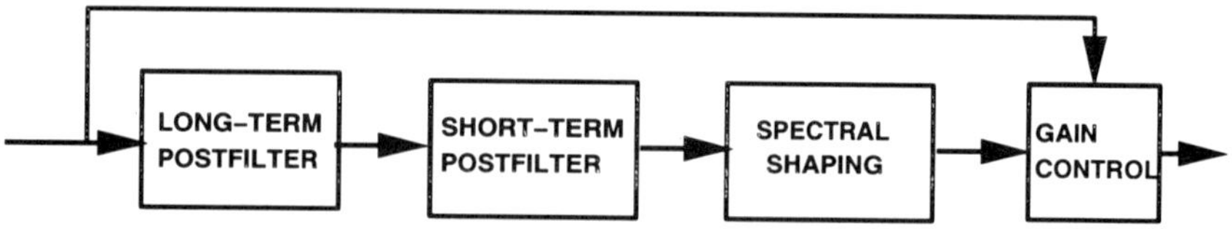

Figure 3 Generic block diagram of a postfilter.

Long-term periodicity enhancement can also be done as part of the postfilter, and it has been successfully used in different coders. This is represented by the long-term postfilter module in Figure 3. The parameters are either derived from the transmitted long-term predictor coefficients or are computed from the reconstructed speech. To avoid discontinuities in the postfiltered signal at time instances where the delay (M of Equation 3.2) changes, the long-term postfilter usually operates on the residual signal rather than the speech signal.

A postfilter can also be used to obtain an overall frequency shaping, such as removal of undesired low frequencies or some compensation of high frequency amplification as could be the case for speech coming from certain types of telephone handsets. This is indicated by the *spectral shaping* block in Figure 3. Most often, the spectral shaping is used to correct the spectral tilt introduced by the the short-term postfilter. In that case, this block is sometimes referred to as the *spectral tilt* control block.

4.3 LPAS for Wideband coding

The coding applications discussed in this chapter are generally aimed at telephone bandwidth speech (typically 200 - 3400 Hz). Slightly larger bandwidths are possible, but the 8 kHz sampling frequency used in digital telephony systems will limit the signal bandwidth to 4 kHz. However, a powerful technique to improve the quality of communication systems is to increase the bandwidth of the transmitted signal (both for lower and higher frequencies). Even if the wideband signal has more distortions than the telephone bandwidth signal, it is often rated as being of better quality. Many of the LPAS techniques discussed in this chapter can be easily applied to the encoding of wideband speech (typically 50 - 7000 Hz). Applications for wideband speech can be found in teleconferencing, ISDN, and multi-media. Since the update rates of the various parameters tend to remain the same, the 16 kHz sampling rate results in frames containing more samples and consequently a higher complexity. In addition, due to the increased bandwidth a higher order predictor (16 to 20) is needed. The error weighting filter needs some modification to accommodate the larger dynamic range between the low and high end of the spectrum [22]. Other approaches use LPAS techniques on band-pass filtered versions of the signal [23]. Such an approach tends to reduce the complexity and gives better control over the distribution of the quantization noise.

5 PERFORMANCE AND ROBUSTNESS

Two issues that are of importance for overall acceptance of a speech coder are its speech quality and its robustness against nonspeech inputs and channel errors. The latter is mainly an issue in wireless environments, where interference during wireless transmission causes errors in the information sent to the speech decoder. As a result, the decoded speech can contain artifacts such as pops and clicks, which have a dramatic impact on the perceived speech quality. In the next two subsections we will look at these issues in more detail.

5.1 Speech Quality

Since the primary goal of speech coders is to transmit speech sounds, their performance is optimized for such signals. Speech quality is typically determined using listening tests in which a panel of subjects rate the quality of processed speech samples. Commonly used tests are Mean Opinion Score (MOS) tests and Diagnostic Acceptability Measure (DAM) tests [24][25]. By testing with a large variety of speech material using many talkers and listeners, these tests can give some indication about the expected speech quality. However, in real-world applications such as cellular communications, other types of input signals such as car noise, babble noise, or music are also common. Especially when speech is present together with these signals, most speech coders introduce undesirable artifacts in the reconstructed output signal. Hence it is important to conduct listening tests that include these forementioned signals. In addition, conversational tests should be conducted to evaluate the long-term effects of certain distortions on the speech quality.

It is important to take into account this robustness requirement for nonspeech signals during coder design. However, it should be clear that high coding efficiency and a high level of robustness are conflicting requirements. Increasing the bit rate may be the best solution when an acceptable performance for music signals is desired. The reason for this is that no production model for music exists. Thus, the only options left are more accurate coding of the waveform and use of more sophisticated noise masking. Background noises can be handled most successfully through techniques which prevent them from entering the speech coder. Such an approach requires special microphones and possibly additional pre-processing.

Another type of degradation occurs when coders are operated in tandem. Examples are found in digital cellular systems during a mobile-to-mobile call, or when a mobile call is leaving a message on a voice mail system that uses speech coding. A similar problem arises when signals are summed together in a teleconferencing application. Since the summation is done in the time-domain, it can result in multiple encoding/decoding stages. By making this summation a function of active talkers, some of the problems can be avoided. Nevertheless, using coders which do not suffer severe degradation when operated in tandem reduce these problems.

In many applications, the signals processed by a coder are used by other systems. Examples are speech recognition systems for speech signals, tones generated by telephone buttons, and modem signals. The degradation in performance due to coding can be measured, and certain minimum requirements can be set.

5.2 Error Robustness

In most communication systems, two types of errors are typically seen: 1) *random* bit errors, and 2) *burst* errors. Burst errors are often the result of fading transmission channels and contain a high number of bit errors over a short interval (e.g. 20 to 40 ms). These burst errors tend to be more serious than random errors because they destroy many information bits at once, thereby introducing audible discontinuities in the signal. The use of interleaving techniques, in which the encoder information of one frame (e.g. 20 ms) is mixed with the information of the next frame, makes the burst errors look like random bit errors. This is at the expense of a delay and also assumes that the burst duration is shorter than the interleaving length. In packetized systems, a special type of burst error occurs, affecting one or more frames of coder information at once. These errors are referred to as *frame erasures*.

Error correction and detection techniques are aimed at protecting the most critical parameters, such as spectral information, pitch and overall gain. A typical error protection scheme would group the bits in a frame according to error sensitivity, and apply different levels of protection to each of these groups. The cellular standards GSM and IS54 both add an additional 60% of bits to the coder information to provide error correction. Adding this amount of error correction information typically reduces the effective net error rate of the protected parameters by an order of magnitude.

To withstand higher error rates it has been found effective to add checksum bits to the transmitted information, such that the decoder can determine if the transmitted information has been corrupted. Since the checksum has been protected by error-correction bits, it is likely that if the checksum does not match then the corresponding information bits are corrupted, and the frame is declared to be a *bad frame* (cf. [26]). An efficient way of recovering from a bad frame is frame substitution. In this method, the most correlated parameters (e.g. spectral information, pitch delay and energy), are repeated from the previous frame. All other parameters are either used as received or are also repeated. The latter approach is used for frame erasures. If more than one successive frame is in error, it is beneficial to apply some form of attenuation of the signal energy combined with spectral flattening. Properly implemented recovery strategies can minimize the effect of bit errors, and both IS54 and GSM have been shown to be operational under adverse channel conditions (e.g. 3% BER at different fading patterns corresponding to different vehicle speeds).

6 IMPLEMENTATION ISSUES

The speech quality of a speech coder is a function of bitrate, delay and complexity. In general, for a given bit rate, the quality can be improved by either increasing delay or complexity. For many applications delay and/or complexity are of great practical importance, and the following subsections address these issues in more detail.

6.1 Delay

The delay of a speech coder is defined as the time it takes for a given input speech segment to be reproduced at the decoder and is sometimes referred to as *end-to-end delay.* Delays less than 200 ms usually do not impair conversations.

However, because of improperly balanced communication lines, part of the transmitted signal can be returned to the sender, thereby introducing echos. These echos become noticeable if the delay increases beyond 10 ms. Echo cancellation techniques can suppress echos, but minimizing speech coder delay is often a more economical solution.

The end-to end delay of a speech coder is composed of several components, such as the algorithmic delay, the processing delay and the transmission delay. Since most coder implementations use all the available real-time of a signal processing chip, one can assume that in the worst case none of the processing can be done until all data for a given processing frame have been collected. Taking into account the time needed to transmit the bits, and the additional delay introduced by the decoder one finds that the total delay of a speech coding system is about 2 to 3 times the frame size. Error control techniques such as interleaving increase this delay.

In LPAS coders, the dominant factor in the algorithmic delay is the short-term predictor analysis. Typically, in systems that transmit the short-term predictor information, the update of the coefficients is done once every 20 to 30 ms, and the corresponding analysis windows will be of about the same duration. However, reducing the delay by more frequent updates of the short-term predictor results in a higher bit rate. Differential encoding techniques can avoid this problem (e.g. [27][28]). Another efficient procedure is not to transmit the short-term predictor coefficients, but to recompute them at the decoder side using previously reconstructed speech signals. This technique, which is more applicable at higher bit rates, is called backward adaptive prediction and is used in, for example, G.728 [21].

6.2 Complexity

There are two reasons for minimizing the complexity of a DSP implementation of a speech coder. First, a low complexity translates into a low power consumption, which is an important issue in mobile communications systems. Second, it allows implementation on a single device, which reduces cost. One way of measuring complexity is by counting the number of operations required by a certain algorithm. When using such a measure it should be taken into account that some operations are more complex than others (e.g. division versus multiplication). For fixed-point implementations, additional issues are the dynamic range of variables and the precision required to represent them.

Besides the number of arithmetic operations, the amount of memory needed by the algorithm also affects the cost. Both static (ROM) and dynamic (RAM) memory usage are an issue, but RAM tends to be more expensive. The amount of RAM needed is directly related to size of the buffers used and the number of intermediate states that need to be maintained. Sometimes RAM usage can be traded off against additional complexity by recomputing certain signals. Although specific numbers depend on the coder and target hardware, typical numbers for a medium complexity coder implemented on a fixed-point DSP are a complexity of 15-20 MIPS, about 1-2 kbytes of RAM and 10-20 kbytes of ROM (includes program space).

7 STANDARD SPEECH CODERS

Table 1 summarizes the most important standards that are based on LPAS. Most of these standards are related to cellular communication systems, except

Table 1 Speech coding standards based on LPAS (1994).

Name	System/Organization	Coder	Bitrate
GSM	European Cellular (TDMA)	RPE-LTP	13 kb/s
FS1016	Secure Voice Communications	CELP	4.8 kb/s
IS54	North American Cellular (TDMA)	VSELP	7.95 kb/s
IS96	North American Cellular (CDMA)	QCELP	1 to 8 kb/s
JDC-FR	Japanese Cellular FR (TDMA)	VSELP	6.7 kb/s
JDC-HR	Japanese Cellular HR (TDMA)	PSI-CELP	3.67 kb/s
G.728	ITU-T	LD-CELP	16 kb/s

for FS1016 which is used for secure voice communications, and G.728 which is intended for telephone network use. The creation of these standards has been responsible for the rapid proliferation of LPAS throughout the speech research community. Figure 4 shows the performance of the various standards for typical telephone bandwidth (200 - 3400 Hz) input speech without background noise. As a reference some other standards such as 64 kb/s μ-law G.711, 32 kb/s G.726 ADPCM and 2.4 kb/s FS1015 have been included.

Most of the digital cellular systems are based on a Time Division Multiple Access (TDMA) structure, using a larger frame (e.g. 40 ms) containing several speech-coder information channels. The so-called *full-rate* coders use two or more time slots within a TDMA frame. By reducing the number of slots needed

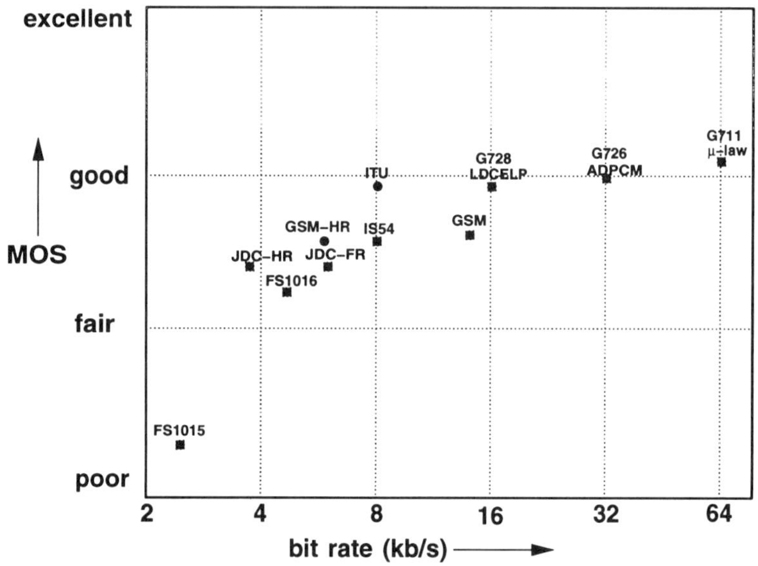

Figure 4 Subjective speech quality of various standard coders (squares). The circles represent targets for future coding standards. The quality adjectives are for the performance with speech without background noise, and are based on several (unpublished) subjective tests conducted by AT&T.

by a single speech channel the capacity of the overall system can be increased. As a result, all cellular standards based on TDMA have either defined or are in the process of defining a *half-rate* (HR) coder. Since the overall rate also includes error correction it is not necessary that the speech coder should operate at exactly half the rate of its full-rate counterpart. However, in practice most proposed candidates seem to follow this rule. The Japanese half-rate standard (PSI-CELP) is described in this section. The European half-rate standard will be finalized by 1995. The candidate coder is based on a VSELP coder as described in [29]. For the US cellular system it is unlikely that a half-rate candidate coder will be defined soon due to the strict performance requirements set by the governing standards body Telecommunication Industry Association (TIA).

The first standard coder based on the LPAS principle was defined by the European GSM committee. This coder is a 13 kb/s regular-pulse excitation coder with long term prediction (RPE-LTP) [13][30](Table 2). The regular-pulse excitation uses a pulse spacing of 3, resulting in 13 pulses every 5 ms. By using a weighting filter, whose denominator is based on the long-term averaged spectral characteristics a significant simplification in complexity is obtained. This simplicity makes it attractive for various other applications such as solid-state answering machines and multi-media. The 13 kb/s RPE-LTP is robust for many types of input signals, which is mainly due to its high bit rate. In the GSM system, the coder is operated in a discontinuous transmission mode, which turns off the coder when no speech is present. To avoid annoying artifacts at the decoder side, the background noise characteristics are transmitted at a low rate to allow for proper noise filling [31]. The standard is specified in a bit-exact way using 16 bit arithmetic. A bit-exact specification defines in detail all arithmetic operations and their required accuracy. As a result, each implementation that follows this specification, will produce the same output for a given input sequence. By providing a collection of test vectors as part of the standard, compliance can be easily verified.

Table 2 Characteristics of the GSM RPE-LTP 13 kb/s coder.

Parameter	Description	Frame (ms)	Bits (20 ms)
ST predictor	8 LARs (6,6,5,5,4,4,3,3)	20	36
ST interpolation	Log Area Ratios (LARs)	5	-
LT predictor delay	integer (range 40:120)	5	4×7
LT predictor gain	range 0.1:1.0	5	4×2
Excitation shape	regular pulse	5	4×2
Excitation gain	maximum absolute value	5	4×6
Pulse amplitudes	13 pulses, 3 bits/pulse	5	4×39
Perceptual weighting	simplified pole/zero	20	-
Postfilter	none	-	-

The FS1016 was the first standard based on CELP (see Section 2), and was completed in 1991 [32](Table 3). The standard is an algorithmic description which means that a given implementation could use different approaches to certain coder modules, as long as compatibility with other implementations of the standard is guaranteed. The adaptive codebook uses fractional delay values represented by an 8 bit table for subframes 1 and 3. For subframes 2 and 4 the delay is represented by a 6 bit offset relative to the 8 bit delay from the previous subframe. The fixed codebook vectors are obtained by shifting through a sequence of sparsely populated unity amplitude pulses. This structure of overlapping codebook vectors allows for more efficient search algorithms. The LPC parameters are quantized and interpolated as line spectral frequencies (LSF). The speech parameters require 4.6 kb/s, the remaining bits are used for

Table 3 Characteristics of the 4.8 kb/s FS1016 CELP coder.

Parameter	Description	Frame (ms)	Bits (20 ms)
ST predictor	10 LSFs (3,4,4,4,4,3,3,3,3,3)	30	34
ST interpolation	Line Spectral Frequencies	7.5	-
LT predictor delay	fractional delay (range 20:147)	7.5	8+6+8+6
LT predictor gain	range -0.993 : 1.020	7.5	4×5
Excitation shape	ternary shifted codebook	7.5	4×9
Excitation gain	scale factor	7.5	4×5
Perceptual weighting	$A(z)/A(z/0.8)$	30	-
Postfilter	short-term/tilt	30	-
Additional bits	1 sync, 1 unused, 4 FEC	30	6

synchronization and error protection. The error correction information allows the receiver to measure the quality of the channel, and apply smoothing of the coder parameters if bit errors are expected. The 4.8 kb/s FS1016 coder is used for secure communications within the government and the military. It has been proven to be robust for a wide variety of operating conditions.

The vector-sum excited linear predictive (VSELP) coder [26] is a CELP coder that uses a structured codebook based a limited number of basis vectors. Each codebook vector is generated from a linear combination of basis vectors using weights with either the value ± 1. As a result, the search complexity is low. Another advantage of this codebook is that single bit errors affect only one of the basis vectors, thereby maintaining a strong correlation between the originally transmitted codebook vector and the corrupted codebook vector. The coder uses two vector-sum codebooks and an adaptive codebook. The gains of each of the codebooks are encoded as energy ratios related to the overall excitation energy. The short-term spectral interpolation is done with the predictor coefficients, and requires a stability check for each interpolation. At the decoder a postfilter is used to further improve the quality. Within IS54 an additional 5 kb/s is used to protect about half of the speech coder information with a rate 1/2 convolutional code. The remainder of the information bits (mainly codebook indices) remain unprotected. A (protected) redundancy check allows the identification of bad frames, which initiate frame recovery procedures similar to the onces described in Section 5.2.

Table 4 Characteristics of the 8 kb/s IS54 VSELP coder.

Parameter	Description	Frame (ms)	Bits (20 ms)
ST predictor	10 reflection coefficients	20	38
ST interpolation	predictor coefficients	5	-
LT predictor delay	integer delay (range 20:146)	5	7
Excitation shape	2 vector sum codebooks	5	$4\times(7+7)$
Codebook gains	3-dimensional VQ	5	4×8
Energy	speech energy	20	5
Perceptual weighting	$A(z)/A(z/0.8)$	20	-
Postfilter	short-term/tilt	20	-

IS96 is the North American Cellular standard for a cellular system using Code Division Multiple Access (CDMA). It uses a variable rate CELP coder, referred to as QCELP [33], supporting four rates 8, 4, 2 and 0.8 kb/s. The coders for the different rates, referred to as rates 1, 1/2, 1/4 and 1/8, are all based on a CELP structure, with the lower rates being obtained by increasing the frame sizes. The larger frame size lowers the update rates of the coder parameters. At the lowest rate (rate 1/8), the adaptive codebook parameters are not transmitted and the fixed codebook is replaced by a noise generator. The rates are either determined by the system, or by a simple rate selector, which uses energy envelopes to determine the proper rate. In practice, the rate is 8 kb/s when speech is present and 0.8 kb/s when no speech is present, with the 4 and 2 kb/s rates used for transitions. The average bit rate is about 5 kb/s.

In Japan a digital TDMA cellular system was standardized in 1991 based on a 6.7 kb/s VSELP speech coder. This coder is similar to the IS54 VSELP, except that only a single fixed codebook is used. Its characteristics are listed in Table 6. The full-rate coder uses two time slots within a 40 ms TDMA frame. By using only one time slot combined with a half-rate coder, the capacity is doubled. Of all cellular standards, the JDC-FR operates at the lowest bit rate, and consequently provides the lowest speech quality (see Figure 4). In 1994 a 3.45 kb/s half-rate coder was defined based on pitch synchronous innovation (PSI-CELP) [34][35](See Table 7). This coder is taking advantage of the 40 ms duration of the time slot, to encode the spectral parameters using vector quantization techniques. To improve the performance for periodic signals, the

Table 5 Characteristics of the IS96 QCELP coder at 8 kb/s.

Parameter	Description	Frame (ms)	Bits (20 ms)
ST predictor	10 LSFs	20	40
ST interpolation	Line Spectral Frequencies	5	-
LT predictor delay	integer delay (range 17:143)	5	4×7
LT predictor gain	scale factor (range 0:2)	5	4×3
Excitation shape	center-clipped shifted cb	2.5	8×7
Fixed codebook gain	log energy difference + sign	2.5	8×3
Perceptual weighting	$A(z)/A(z/0.8)$	20	-
Postfilter	short-term/tilt	20	-

Table 6 Characteristics of the 6.7 kb/s JDC-FR VSELP coder.

Parameter	Description	Frame (ms)	Bits (20 ms)
ST predictor	10 reflection coefficients	20	36
ST interpolation	predictor coefficients	5	1
LT predictor delay	integer delay (range 20: 146)	5	4×7
Excitation shape	trained vector sum	5	4×9
Codebook gains	2-dimensional VQ	5	4×7
Frame energy	log speech power	20	5
Perceptual weighting	A(z)/A(z/0.8)	20	-
Postfilter	long-term/short-term/tilt	20	-

fixed codebook vectors are synchronized with the pitch information represented by the adaptive codebook. The excitation parameters are determined for 10 ms excitation frames using a *delayed decision* technique, which evaluates the future effects of a particular choice of excitation parameters.

The 3.45 kb/s coder is, according to subjective tests, equivalent in performance to the 6.7 kb/s VSELP coder. To enhance the performance for high levels of background noise (less than 10 dB SNR), a speech enhancement module is used.

In 1994 the International Telecommunications Union (ITU-T) standardized the G.728 Low-Delay CELP (LD-CELP) 16 kb/s coder [21], which has network quality and for certain conditions (e.g. channel errors) is better than G.726 (32 kb/s ADPCM). It is a CELP LPAS coder which features a backward adaptive

Table 7 Characteristics of the 3.45 kb/s PSI-CELP coder.

Parameter	Description	Frame (ms)	Bits (40 ms)
ST predictor	10 LSF	40	31
ST interpolation	Line Spectral Frequencies	20	-
LT predictor delay	fractional delay (range: 16:96)	10	4×8
Excitation shape	conjugate structure	10	4×10
Codebook gains	vector quantized	10	4×7
Power	speech power	40	7
Perceptual weighting	pole/zero + pitch	20	-
Postfilter	short-term/tilt/long-term	20	-

short-term predictor to allow for short algorithmic delay. The excitation codebook size is 5 samples, resulting in a coder delay less than 2 ms. The short-term predictor is of order 50 and no adaptive codebook is used. The coder has a trained gain-shape codebook with 128 shapes, a sign and 4 gain values. The absolute value of the codebook gains are estimated by a 10-th order log-gain predictor, which uses the energy of previous excitation vectors to estimate the gain of the current excitation vector. Both a floating-point and fixed-point description are defined in the standard, and test vectors are available for both versions to verify correctness of implementation. Its main characteristics are shown in Table 8.

Table 8 Characteristics of the 16 kb/s G.728 LDCELP coder.

Parameter	Description	Frame (ms)	Bits 0.625 ms
ST predictor	50-th order backward adaptive	4	-
Excitation shape	trained codebook	0.625	7
Codebook gains	(gain predictor) gain + sign	0.625	2 +1
Perceptual weighting	A(z/0.6)/A(z/0.8)	4	-
Postfilter	long-term/short-term/tilt	20	-

Another interesting activity sponsored by the ITU-T is the standardization of a network quality 8 kb/s coder. The coder is based on a combination of the two original candidate coders [27, 28], and it is expected that a standard will be defined by the end of 1995 and ratified in 1996. This coder is a CELP coder using a 10 ms LPC update frame. The excitation is determined for 5 ms subframes and is based on a linear combination of an adaptive codebook and fixed codebook contributions. The gains are jointly quantized with a 2-dimensional VQ. The decoder uses a postfilter based on both long-term and short-term prediction filters.

8 FUTURE OUTLOOK

In this chapter we discussed principles and applications of linear-prediction based analysis-by-synthesis (LPAS) coders. These coders provide state-of-the-art performance for bit rates in the range between 4 and 16 kb/s. The LPAS paradigm is very flexible, allowing trade-offs between bit rate, speech quality, complexity, delay, and robustness against channel errors and nonspeech sounds. As a result, LPAS coders have found use in such various applications as network telephony, cellular and personal communications, voice mail, answering machines, simultaneous voice and data transmission, and multi-media. It is to be expected that a number of speech-coding standards which are scheduled to be defined in the near future will also be based on LPAS. These future standards include new cellular communications standards, as well as the ITU 8 kb/s standard.

With the increased performance of speech coders, the range of bit rates which can be used in practical applications is moving downward. Soon coders with bit rates below 4 kb/s will be viable options for many applications. At these low rates, the basic principle of waveform-matching becomes less attractive, because it forces the coder to reconstruct aspects of the speech signal which are not perceptually relevant. On the other hand, this waveform matching property makes LPAS coders robust against a variety of input signals. Hence, a gradual change from the LPAS paradigm is a prerequisite to efficient speech coding at very low rates. The generalized LPAS paradigm can be seen as a first step in this direction. New coding paradigms such as waveform interpolation [36], can be viewed as a more radical change from LPAS. However, these coders still benefit from all the knowledge obtained with LPAS coding, and share the predictor-coefficient quantization techniques and spectral masking techniques developed for these coders.

REFERENCES

[1] R.P. Ramachandran. The use of pitch prediction in speech coding. In R.P. Ramachandran and R.J. Mammone, editors, *Modern Methods of of Speech Processing.* Kluwer Academic Publishers, 1995.

[2] J.S. Collura. Vector quantization of linear predictor coefficients. In R.P. Ramachandran and R.J. Mammone, editors, *Modern Methods of of Speech Processing.* Kluwer Academic Publishers, 1995.

[3] V. Cuperman and P. Lupini. Variable rate speech coding. In R.P. Ramachandran and R.J. Mammone, editors, *Modern Methods of Speech Processing.* Kluwer Academic Publishers, 1995.

[4] P. Kroon and E.F. Deprettere. A class of analysis-by-synthesis predictive coders for high quality speech coding at rates between 4.8 and 16 kb/s. *IEEE J. on Selected Areas in Communications*, SAC-6(2):353–363, February 1988.

[5] P. Kroon and B.S. Atal. Predictive coding of speech using analysis-by-synthesis techniques. In S. Furui and M.M. Sondhi, editors, *Advances in speech signal processing*, pages 141–164. Marcel Dekker Inc., New York, 1990.

[6] J.D. Markel and A.H. Gray. *Linear Prediction of speech.* Springer Verlag, Berlin, 1976.

[7] L.R. Rabiner and R.W. Schafer. *Digital processing of speech signals.* Prentice Hall, Englewood Cliffs, NJ, 1978.

[8] B.S. Atal, R.V. Cox, and P. Kroon. Spectral quantization and interpolation for CELP coders. In *Proc. ICASSP*, pages 69–72, 1989.

[9] P. Kroon and B.S. Atal. On improving the performance of pitch predictors in speech coding systems. In B.S. Atal, V. Cuperman, and A. Gersho, editors, *Advances in Speech Coding.* Kluwer Academic Publishers, Boston, MA, 1990.

[10] W.B. Kleijn, D.J. Krasinski, and R.H. Ketchum. Improved speech quality and efficient vector quantization in SELP. In *Proc. IEEE Int. Conf. Acoust., Speech, Signal Processing*, pages 155–158, April 1988.

[11] B.S. Atal and J.R. Remde. A new model of LPC excitation for producing natural-sounding speech at low bit rates. In *Proc. IEEE Int. Conf. Acoust., Speech, Signal Processing*, pages 614–617, Paris, France, April 1982.

[12] S. Singhal and B.S. Atal. Amplitude optimization and pitch prediction in multipulse coders. *IEEE Trans. Acoust., Speech, Signal Processing*, ASSP-37:317–327, March 1989.

[13] P. Kroon, E.F. Deprettere, and R.J. Sluyter. Regular-pulse excitation: a novel approach to effective and efficient multipulse coding of speech. *IEEE Trans. Acoust., Speech, Signal Processing*, ASSP-34(5):1054–1063, October 1986.

[14] M.R. Schroeder and B.S. Atal. Code-excited linear prediction (CELP): high quality speech at very low bit rates. In *Proc. IEEE Int. Conf. Acoust., Speech, Signal Processing*, pages 937–940, 1985.

[15] W.B. Kleijn, R.P. Ramachandran, and P. Kroon. Interpolation of the pitch-predictor parameters in analysis-by-synthesis speech coders. *IEEE Trans. on Speech and Audio Processing*, pages 42–54, January 1994.

[16] W.B. Kleijn, P. Kroon, L. Cellario, and D. Sereno. A 5.85 kb/s CELP algorithm for cellular applications. In *Proc. ICASSP*, pages II596–II599, 1993.

[17] W.B. Kleijn, P. Kroon, and D. Nahumi. The RCELP speech coding algorithm. *European Transactions on Telecommunications*, (5):573–582, 1994.

[18] B.C.J. Moore. *An introduction to the psychology of hearing*. Academic Press, London, second edition, 1982.

[19] M.R. Schroeder, B.S Atal, and J.L. Hall. Optimizing digital speech coders by exploiting masking properties of the human ear. *J. Acoust. Soc. Am.*, 66:1647–1652, December 1979.

[20] V. Ramamoorthy, N.S. Jayant, R.V. Cox, and M.M. Sondhi. Enhancement of ADPCM speech coding with backward-adaptive algorithms for postfiltering and noise feedback. *IEEE J. on Selected Areas in Communications*, SAC-6:364–382, February 1988.

[21] J.-H. Chen, R.V. Cox, Y.-C. Lin, N. Jayant, and M.J. Melchner. A low-delay CELP coder for the CCITT 16 kb/s speech coding standard. *IEEE Journal on Selected Areas in Communications*, pages 830–849, June 1992.

[22] C. Laflamme, J-P. Adoul, R. Salami, S. Morissette, and P. Mabilleau. 16 kbps wideband speech coding technique based on algebraic CELP. In *Proc. ICASSP*, pages 13–17, 1991.

[23] R.D. De Iacovo, R. Montagna, D. Sereno, and P. Usai. A two-band CELP audio coder at 16 kb/s and its evaluation. In B.S. Atal, V. Cuperman, and A. Gersho, editors, *Speech and Audio Coding for Wireless and Network Applications*, pages 141–146. Kluwer Academic Publishers, Boston, MA, 1993.

[24] S.R. Quackenbush, T.P. Barnwell, III, and M.A. Clements. *Objective measures of speech quality*. Prentice Hall, Englewood Cliffs, NJ, 1988.

[25] S. Dimolitsas. Subjective assessment methods for the measurement of digital speech quality. In B.S. Atal, V. Cuperman, and A. Gersho, editors, *Speech and Audio Coding for Wireless and Network Applications*, pages 43–54. Kluwer Academic Publishers, Boston, MA, 1993.

[26] I.A. Gerson and M.A. Jasiuk. Vector sum excited linear prediction (VSELP). In B.S. Atal, V. Cuperman, and A. Gersho, editors, *Advances in Speech Coding*. Kluwer Academic Publishers, Boston, MA, 1990.

[27] A. Kataoka, T. Moriya, and S. Hayashi. Implementation and performance of an 8 kb/s conjugate structure CELP speech coder. In *Proc. ICASSP*, pages II–93–II–96, 1994.

[28] R. Salami, C. Laflamme, and J-P. Adoul. 8 kb/s ACELP coding of speech with 10 ms speech frame: a candidate for ITU standardization. In *Proc. ICASSP*, pages II–97–II–100, 1994.

[29] I.A. Gerson and M.A. Jasiuk. A 5600 bps VSELP speech coder candidate for half-rate GSM. In *Proc. Eurospeech*, pages 253–256, September 1993.

[30] K. Hellwig, P. Vary, D. Massaloux, J.P. Petit, C. Galand, and M. Rosso. Speech codec for the european mobile radio system. In *Proc. GLOBCOM*, pages 1065–1069, 1989.

[31] CB. Southcott, D. Freeman, G. Cosier, and et al. Voice control of the pan-european digital mobile radio system. In *Proc. GLOBECOM*, pages 1070–1074, 1989.

[32] J.P. Campbell, T. Tremain, and V. Welch. The federal standard 1016 4800 bps CELP voice coder. *Digital Signal Processing*, pages 145–155, 1991.

[33] W. Gardner, P. Jacobs, and C. Lee. QCELP: A variable rate speech coding for cellular networks. In B.S. Atal, V. Cuperman, and A. Gersho, editors, *Speech and Audio Coding for Wireless and Network Applications*, pages 85–92. Kluwer Academic Publishers, Boston, MA, 1993.

[34] S. Miki, K. Mano, H. Ohmuro, and T. Moriya. Pitch synchronous innovation CELP (PSI-CELP). In *Proc. Eurospeech*, pages 261–264, September 1993.

[35] S. Miki, K. Mano, T. Moryua, and K. Oguchi. A pitch synchronous innovation CELP (PSI-CELP) coder for 2-4 kb/s. In *Proc. ICASSP*, pages II–113–II–116, 1994.

[36] J. Haagen and W. B. Kleijn. Waveform interpolation. In R.P. Ramachandran and R.J. Mammone, editors, *Modern Methods of Speech Processing.* Kluwer Academic Publishers, New York, 1995.

4

WAVEFORM INTERPOLATION

Jesper Haagen and W. Bastiaan Kleijn

Information Principles Research Laboratory,
AT&T Bell Laboratories, Murray Hill, NJ 07974, USA.

ABSTRACT

Waveform interpolation (WI) has proved to be an efficient procedure for high quality coding of speech at low bit rates. In this method, the speech signal is described by a sequence of characteristic waveforms, which are interpolated during reconstruction. Originally, the characteristic waveform was identified with a pitch cycle, and WI was applied to voiced speech segments only. A number of implementations, which use CELP for unvoiced signal segments, showed that the procedure can provide high performance. Recently, the method was extended to include unvoiced speech and background noise. To this purpose the characteristic waveform is decomposed into a slowly evolving waveform (representing the periodic component of the signal), and a rapidly evolving waveform (representing the other components of the signal). The rapidly evolving waveform requires high time resolution and only low quantization accuracy, while the slowly evolving waveform requires less time resolution and a more precise description. With this decomposition, switching between different coding models is avoided, and a robust coding method results.

1 INTRODUCTION

With the increased performance of signal-processing hardware, digital coding of speech signals has become attractive for a large number of applications, such as wireless communication, secure communication, and voice storage. As a result, much research has been aimed at the development of high-quality digital speech communication systems at low bit rates. Research in digital speech coding attempts to balance the demands for low bit rates, high reconstructed-speech quality, small coding delays, and low algorithm complexity and storage requirements. Currently, a significant effort is focused on obtaining high speech quality in the range between 2.4 and 4 kb/s.

Since the early eighties, much work in speech coding has concentrated on coding methods based on the analysis-by-synthesis paradigm, of which *Codebook Excited Linear Prediction* (CELP) [1] is most well-known. CELP was found to be a powerful procedure for coding speech down to approximately 4 kb/s, but at lower bit rates the quality has been found to deteriorate rapidly. The coding of voiced (quasi-periodic) speech segments especially causes this method difficulties at lower rates. CELP coders typically reproduce voiced speech with too much audible noise, which is a result of inaccurate waveform matching when the codebook size becomes too small. It appears that, at these low bit rates, the waveform matching procedure does not lead to an efficient use of the limited coding resources for the perceptually important features of the speech signal.

Speech coding based on *Waveform Interpolation* (WI) techniques is an attractive method to obtain high quality reconstructed speech even at very low bit rates. The WI method for speech coding was introduced in 1991 [2]. The fundamental strength of this method is that it efficiently exploits the nature of human speech production and perception. In particular, WI is known to work remarkably well for voiced speech. In combination with CELP coding for unvoiced speech, a number of promising coder configurations were developed e.g. [2, 3, 4, 5, 6, 7, 8, 9, 10].

Recently, it was found that WI can be used to describe both voiced and unvoiced speech. As will be shown in Section 3.1, the quality of the reconstructed signal (including unvoiced) converges to that of the original signal with increasing update rate. Such a WI system with an increased update rate can be quantized at a low bit rate by using a decomposition which is motivated by human perception [11]. The evolving waveform is decomposed into a slowly evolving waveform (SEW) and a rapidly evolving waveform (REW). The SEW represents the quasi-periodic component of the speech signal, and the REW represents the remaining nonperiodic component. For good perceptual performance, a relatively inaccurate description suffices for the REW, while a more precise description is required for the SEW. For clarity, the WI method using the SEW-REW decomposition will be referred to as generalized WI (GWI) in this chapter.

This chapter first describes the original WI method aimed at voiced speech, including all the basic principles (extraction, interpolation, and quantization), and then describes the fore-mentioned decomposition which allows the method to be extended to unvoiced speech. Descriptions of various practical implementations will be provided.

2 WI FOR VOICED SPEECH

The goal of the conventional WI method is the efficient and perceptually accurate description of the voiced speech signal. This reflects the decrease in performance for voiced speech of the CELP algorithm with decreasing bit rate. Most coding systems using WI employ a classification system which separates quasi-periodic and nonperiodic signal segments. The quasi-periodic segments are coded with WI, while the nonperiodic segments are coded with CELP. The WI employed by such systems will be discussed in this section. In Section 3 the method will be extended to include unvoiced speech and background noise.

WI is normally used in combination with Linear Prediction (LP) [12]. LP analysis results in a set of LP coefficients which specify the envelope of the magnitude spectrum with generally reasonable accuracy. By filtering the original signal with the prediction filter, an LP residual signal is obtained which retains the fine-structure of the speech spectrum and a remainder of the envelope of the spectrum. The fine-structure specifies the level of periodicity of the speech signal. The shape of the spectral envelope is largely determined by the vocal-tract shape, while many of the properties of the LP residual signal are determined by the glottal excitation. There are two major reasons to use LP analysis for WI. First, a large number of established quantization methods exist for the LP coefficients, e.g. [13, 14, 15]. Second, computationally efficient methods exist to extract waveforms from the residual signal. However, both extraction [3] and reconstruction [9] can be performed directly in the speech domain.

The WI coders describe the signal in terms of an evolving waveform. During voiced speech the waveform corresponds essentially to the shape of the pitch cycle. The extraction of the waveform requires reasonably accurate knowledge of the pitch track. Many pitch extraction methods exist, e.g. [16, 17, 18, 19], and any accurate method can be used. The appendix provides one method which has been found to work well in combination with WI.

Once the LP residual and the pitch track have been computed, the *prototype waveforms* (or prototypes) are extracted from the residual signal at regular intervals. These intervals are typically 20-30 ms in length. Each prototype is quantized in a perceptually relevant manner and transmitted to the receiver. At the receiver, the prototypes are dequantized. The dequantized waveforms are then interpolated to render a waveform for each sample of the excitation signal. This latter waveform has been called the *instantaneous* waveform [3]. The interpolation between the prototypes can be either linear or nonlinear. Finally, the excitation signal is reconstructed from the sequence of instantaneous waveforms. Figure 1 provides an overview of the WI scheme. The outer layer defines the basic analysis-synthesis system, whereas the inner layer performs the quantization and dequantization of the prototypes. The outer layer converts the one-dimensional input signal into a two-dimensional signal at the analyzer (left half of Fig. 1) and the inverse process is performed at the receiver (right half of Fig. 1). The inner layer operates on two-dimensional signals, which represent the waveform shape along one axis (the ϕ axis), and the evolution of this shape along the time axis. Sections 2.1 and 2.2 will describe the outer layer and Section 2.3 provides more detail about the inner layer.

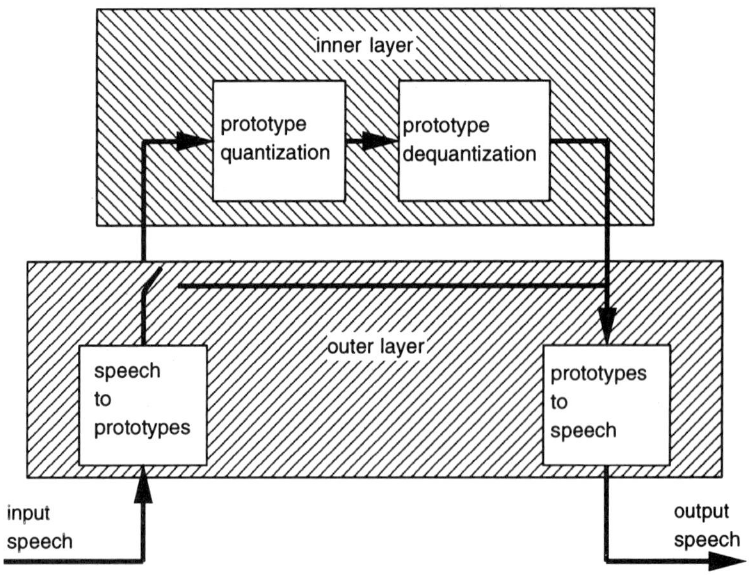

Figure 1 Overview of the WI method. The outer layer defines the basic analysis-synthesis system, whereas the inner layer performs the quantization and dequantization of the prototypes.

2.1 Extraction of waveforms

The waveform extraction process is performed in the analysis side of the outer layer of Fig. 1. For a quasi-periodic residual $r(t)$, a two-dimensional function $u(t_i, \phi)$, periodic in ϕ, is defined at each sampling instant t_i along an abstract phase axis ϕ. These functions define prototype waveforms which are representative pitch cycles of the evolving voiced-speech signal. First segments with a length of one pitch period are extracted from the quasi-periodic signal $r(t)$ at each t_i. These segments are normalized to have length 2π and then periodically extended. The choice of 2π is arbitrary, but it is convenient in a commonly used representation of the protototype waveforms which will be discussed later in this section. The normalization separates the pitch period and waveform shape as independent entities.

Let $p(t_i)$ be the pitch period at t_i, then one cycle of a two-dimensional signal $v(t, \phi)$ at t_i can be defined as

$$v(t_i, \phi) = r(t_i + \epsilon + p(t_i)\frac{\phi}{2\pi}), \qquad -\pi \leq \phi < \pi, \tag{4.1}$$

where ϵ defines an offset to minimize discontinuities caused by the periodic extension. In practice, the value of ϵ which minimizes the energy at the segment boundaries is a good choice. By aligning $v(t_i, \phi)$ with the previously extracted waveform, the new prototype is obtained

$$u(t_i, \phi) = v(t_i, \phi + \phi_i), \tag{4.2}$$

where

$$\phi_i = \arg\max_{\phi_i}\left(\int_0^{2\pi} v(t_i, \phi + \phi_i)u(t_{i-1}, \phi)d\phi\right). \tag{4.3}$$

As seen from Eqs. 4.2 and 4.3, the alignment is done by selecting the phase offset ϕ_i such that the cross-correlation between the waveforms is maximum. Thus, the main features in the waveforms appear for the same value of the phase ϕ. Figure 2 shows the evolving characteristic waveform for the case where the gain is normalized such that each waveform has unity average magnitude spectrum. The LP residual signal for the beginning of the word "hard" spoken by a female is shown in a) and the corresponding two-dimensional function $u(t, \phi)$ in b). The surface in b) illustrates the characteristic waveform features (one pitch cycle each) as a function of the phase along the ϕ-axis, and the evolution of the waveforms along the time axis.

In a practical implementation of WI, the original signal and the LP residual are band-limited signals, sampled in t. Thus, $u(t, \phi)$ is a band-limited, periodic

a)

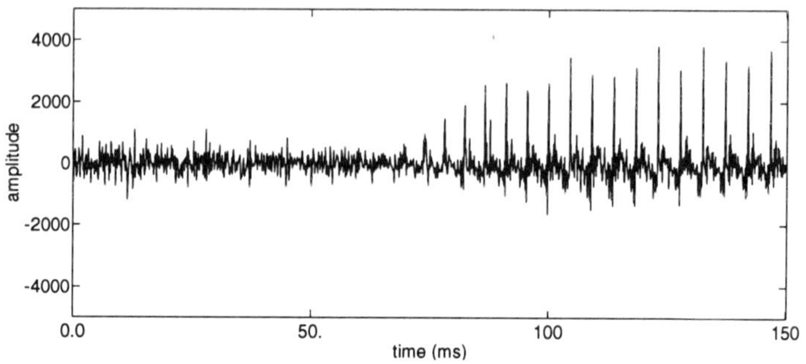

b)

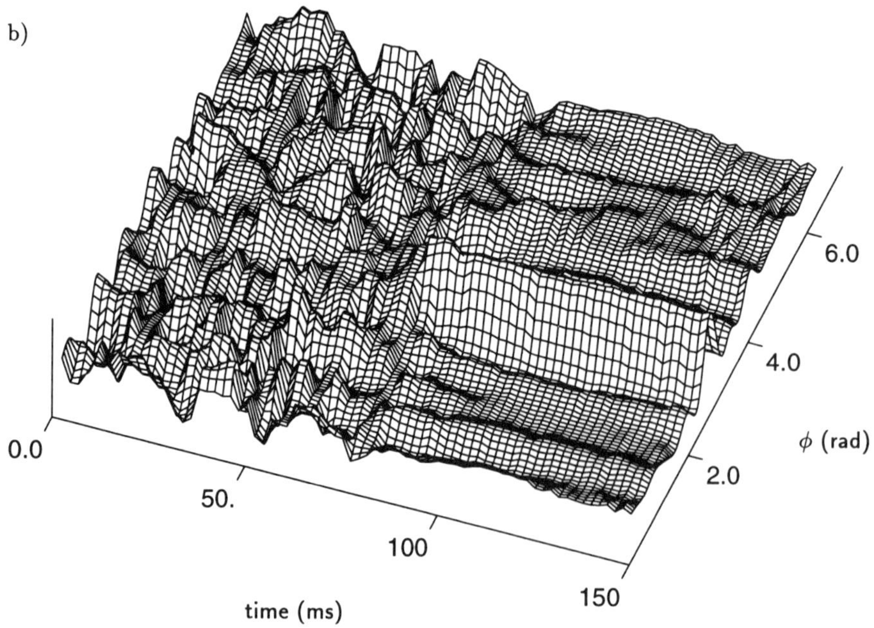

Figure 2 Evolution of the original LP waveform for the beginning of the word "hard": a) The LP residual and b) the surface describing the prototypes in the ϕ direction and the waveform evolution in the time direction.

function of ϕ, and it is convenient to describe it with a Fourier series (FS) which is a function of t

$$u(t,\phi) = \sum_{n=-N}^{N} c_n(t)e^{jn\phi}, \tag{4.4}$$

where $c_n(t)$ are the time-dependent FS coefficients. The earlier normalization of the pitch period to 2π was chosen to simplify the FS description. The number of complex FS coefficients corresponds to the number of harmonics in $u(t_i,\phi)$ (basically the ratio of the Nyquist frequency of the sampled speech signal and the fundamental frequency). At each extraction time t_i, a new FS is defined. Thus, in a practical system, the function $u(t,\phi)$ is described by a set of evolving FS coefficients, sampled at the extraction times t_i. The size of this set varies with the pitch period. WI procedures using a FS description are often referred to as frequency-domain techniques.

The WI method can also be implemented without time-scaling [20, 5]. The evolving waveform can now be denoted as a function $y(t,\tau)$, where t is time and τ defines the axis along which the waveform is displayed. In this case, zero-padding is used to make all prototypes of the same length. It is important that the zero-padding always extends from a location within the pitch cycle which has low signal power. A practical method is to keep the pitch pulse always at the origin ($\tau = 0$) for the current prototype, while simultaneously aligning the current prototype with the previous prototype. Zero-padding must then be performed for the region beyond $\tau > p(t)/2$ and $\tau < -p(t)/2$. Note that this method implies that the origin is redefined for the previous prototypes at each prototype update. In contrast to the function $u(t,\phi)$ the function $y(t,\tau)$ is usually represented as a two-dimensional signal sampled in both t and τ. These methods without time-scaling are sometimes referred to as time-domain techniques.

2.2 Reconstruction of the excitation signal

At the receiver end of the outer layer of Fig. 1, the transmitted and reconstructed prototypes $\tilde{u}(t_i,\phi)$ are used to generate an excitation signal $e(t)$ for the LP synthesis filter. Interpolation between the prototypes is used to define the instantaneous waveform at each sample of the output speech signal. Thus, the sampled, evolving waveform signal $\tilde{u}(t_i,\phi)$ is upsampled to the sampling rate of the speech signal (typically 8 kHz). The interpolation can be done in different ways as described in e.g. [20] where *continuous* and *block-wise* methods are discussed. The interpolation is often described by

$$\tilde{u}(t,\phi) = (1-\alpha(t))\tilde{u}(t_{i-1},\phi) + \alpha(t)\tilde{u}(t_i,\phi), \qquad t_{i-1} \le t \le t_i, \tag{4.5}$$

where $\tilde{u}(t_{i-1})$ and $\tilde{u}(t_i)$ are the previous and the current prototype waveforms respectively and $\alpha(t)$ defines the interpolation function. Continuous interpolation by a simple linear function from $\alpha(t_{i-1}) = 0$ to $\alpha(t_i) = 1$ is commonly used resulting in a smooth transition between the quantized prototypes. In block-wise interpolation, the instantaneous waveform shapes are kept constant over intervals which may correspond to one pitch cycle. This method leads to occasional discontinuities in the reconstructed excitation signal $e(t)$ when the instantaneous waveform is changed.

To get the one-dimensional LP-excitation signal $e(t)$ from the two-dimensional upsampled signal a relation between the phase and the time is necessary. This relation is governed by the time-dependent pitch period $p(t)$. For the prototypes as defined in Eq. 4.2, it is given by

$$\phi(t) = \phi(t_0) + \int_{t_0}^{t} \frac{2\pi}{p(t')} dt'. \tag{4.6}$$

The excitation signal is then obtained by

$$e(t) = \tilde{u}(t, \phi(t_0) + \int_{t_0}^{t} \frac{2\pi}{p(t')} dt'). \tag{4.7}$$

The operation of Eq. 4.7 is shown in Fig. 3. The curves crossing the surface parallel to the ϕ axis describe the prototypes. The diagonal curves ("phase track") correspond to the values of $\tilde{u}(t,\phi(t))$ for the phase function $\phi(t)$ given by Eq. 4.6. The excitation signal $e(t)$ is obtained by the projection of the signal values defined by the pitch track onto a plane orthogonal to the ϕ axis. This signal $e(t)$ is shown in front of the surface.

The procedure to get an LP excitation signal $e(t)$ is similar for the fixed time-scale case. The instantaneous waveforms can be generated by interpolation between the (zero-padded) prototypes. Again, continuous interpolation is commonly used for this purpose. If $y(t,\tau)$ is the surface representing the evolving waveform, then the excitation signal $e(t)$ corresponds to $y(t,\tau(t))$ for a particular function $\tau(t)$. Because there is no time scaling the relation $\tau(t)$ satisfies $d\tau = dt$ except for one discontinuity per pitch cycle. At the discontinuity the value of τ moves from $p(t)/2$ to $-p(t)/2$.

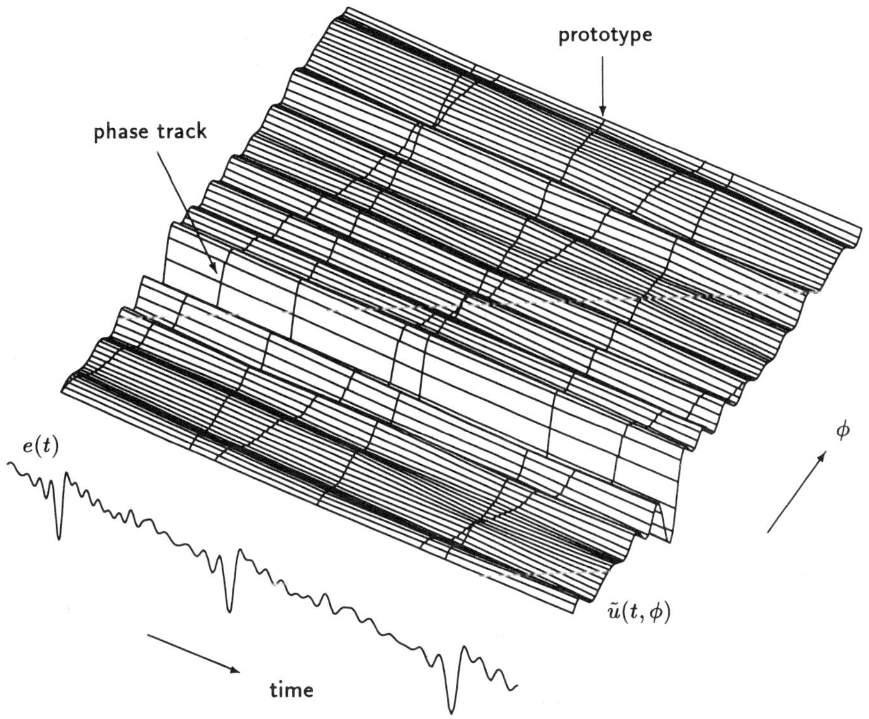

Figure 3 Construction of the excitation signal $e(t)$ from the reconstructed evolving waveform $\tilde{u}(t,\phi)$ using continuous interpolation with time-scaling. $e(t)$ is the projection of the signal values defined by the phase track onto a plane orthogonal to the ϕ axis.

2.3 Quantization of the prototype waveforms

The quantization of prototype waveforms can be done in both the time and the frequency domain. In this section attractive methods for low bit rate coding will be discussed.

Frequency-domain techniques

Several strategies can be used for quantization of the FS coefficients of Eq. 4.4. For higher bit rates (e.g. 4 kb/s) direct quantization seems advantageous [3].

Since adjacent prototypes are very similar in shape, a differential quantization scheme can be used. Using a multi-stage codebook approach (K stages) the reconstructed prototypes $\tilde{u}(t_i, \phi)$ for the current update can be expressed as

$$\tilde{u}(t_i, \phi) = \lambda_0 \tilde{u}(t_{i-1}, \phi) + \sum_{k=1}^{K} \sum_{n=-N}^{N} \lambda_k c_n^k(t_i), \tag{4.8}$$

where the gain λ_0 specifies the contribution from the previous quantized prototype $\tilde{u}(t_{i-1}, \phi)$ and λ_k for $k \geq 1$ is the gain of the contribution of codebook k, and c_n^k is the FS coefficient of index n of the selected codebook entry for codebook k. The codebooks can be designed and trained using the various methods known from CELP coding.

When lower bit rates (e.g. below 4 kb/s) are required it seems advantageous to split the FS coefficients into phase and magnitude, which can then be quantized separately. For the phase it is found that a fixed model, requiring no transmission of information, often works well. Assuming that the magnitude spectrum is flat, the phase spectrum determines the shape of the pitch cycles. A zero-phase model will result in symmetric pitch pulses which becomes more peaky with an increase of the number of harmonics. In general, such zero-phase pulses are perceived differently from a sequence of pulses with the same random phase spectrum. When studying the characteristics of actual waveforms extracted from speech, it becomes obvious that certain features are very common. A waveform with a low frequency pulse before the main pitch pulse (closure of the glottis) is more frequently seen than a random or symmetric behavior of the waveform. With the original magnitude spectrum, it is found that a fixed phase spectrum extracted in a typical voiced segment of a male speaker, results in natural sounding reconstructed speech. Male speech was selected for obtaining this fixed phase spectrum because it contains a large number of harmonics. Since the LP filtering approximately flattens the magnitude spectrum, a flat magnitude spectrum can be used for the excitation when very low bit rates are required. However vector quantization procedures for the magnitude spectrum can enhance performance significantly. Such vector quantization corrects for inadequacies of the spectral envelope description provided by the quantized LP coefficients. Good results are obtained by vector quantization of only the baseband (e.g. 0-1 kHz), in combination with a flat magnitude spectrum for the higher frequencies.

It is straightforward to apply noise masking as known from LP-based analysis-by-synthesis coders [21] to the FS quantization process. During quantization, the FS spectrum is weighted with a modified LP magnitude spectrum [3]. Other spectral manipulations such as using auditory masking models [22], noise injection, and LP postfiltering are also easily implemented within the WI structure. Note that a postfilter can be implemented by introducing a spectral envelope directly to the FS description of the reconstructed prototypes. In contrast to postfilters in CELP, this method requires no additional delay.

Time-domain techniques

In the time-domain techniques, the prototypes can be quantized by methods similar to those described in the previous section. Thus, differential quantization using multi-stage codebooks can be applied to the vectors describing the prototypes. At low bit rates, it is important to focus the quantization to perceptually relevant features of the waveforms only. Identification of the location of the pitch pulse within the prototype, leads to an efficient quantization method [5]. In this procedure, the prototype is windowed such that only the pitch pulse and a small segment prior to that remain. It was found that the signal properties prior to the pitch pulse are significantly more important for subjective quality than the signal properties after the pitch pulse. The total length of the windowed segment is typically 2.5 ms. The method is illustrated in Fig. 4, where the solid boxes contain the quantized prototypes and the dashed boxes contain the interpolated waveforms. It can be interpreted as a generalization of single-pulse excitation [6].

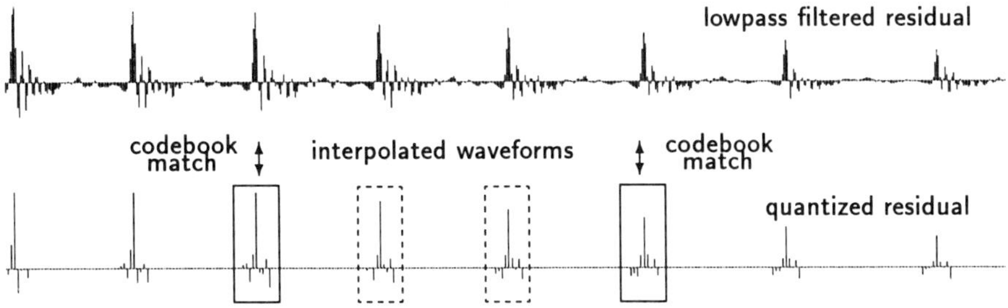

Figure 4 Time-domain coding of a lowpass filtered LP residual. The windows (solid boxes) used to extract segments of the prototypes are in this example centered around the main excitation pulses.

To improve the subjective performance further, matching of the windowed prototype segments can be limited to an e.g. 2 kHz baseband. In that case the unquantized prototype segments are extracted from a lowpass-filtered version of the residual signal. However, no downsampling is performed. Instead, the codebook consists of entries for which every other sample is zero. Due to spectral folding (the codebook entries are upsampled without lowpass filtering) this matching procedure provides a perceptually reasonable spectral structure for the frequency band between 2 kHz and the Nyquist frequency of 4 kHz, even though this band was not present in the signal which is quantized.

2.4 WI implementations for voiced speech

A number of different coder implementations using WI for voiced speech has been reported in the literature e.g. [2, 3, 4, 5, 6, 7, 8, 9, 10]. Both frequency-domain and time-domain coders have been developed and in the following, practical aspects of such implementations will be discussed.

Frequency-domain WI coding

The first practical implementation of WI was the PWI (Prototype Waveform Interpolation) coder. An example of this coder operating at about 4 kb/s [3] is now discussed. Table 1 shows the bit allocation for the PWI coder for quasi-periodic speech. The coder uses LP, updated at 50 Hz, using the quantization

Parameter	bits/20 ms
LP coefficients	30
Pitch period	7
Previous prototype gain	5
Codebook	2·8
Codebook gain	2·6
Total	70

Table 1 Voiced bit allocation for a 3.5 kb/s PWI coder.

described in [13]. Prototype waveforms are extracted from the LP residual and described by their FS coefficients at an update rate of 50 Hz. A differential quantization scheme with a 2-stage codebook (see Eq. 4.8) is used for quantizing the FS coefficients. For unvoiced speech the coder uses a CELP scheme but without a long-term predictor or adaptive codebook. The rate of the CELP coder is 4.1 kb/s. The complete coder was tested in a formal subjective MOS (Mean Opinion Score) test. In MOS tests inexperienced listeners judge sentence pairs processed by different coders to be either bad, poor, fair, good, or excellent, corresponding to scores of 1 to 5. For IRS weighted speech (this weighting describes a typical telephone transfer function) the WI coder obtained a score of 3.45, whereas the score for the GSM coder (13.2 kb/s) was 3.60 and the original speech obtained a score of 4.20.

Another coder implementation which uses a FS description of the prototype waveforms is discussed in [4] and called the TFI (Time-Frequency Interpolation) coder. This coder uses an efficient *matrix* quantization scheme for the LP parameters and separates the magnitude and the phase spectrum of the prototypes. The phase spectrum is fixed and thus requires no bit allocation. For a 2.4 kb/s implementation [4] three separate LP analyses, describing 60 ms of speech, are quantized together using 30 bits. This allows for a 200 Hz update rate for the magnitude spectrum of the prototypes. Such a rapid update rate is beneficial for voiced speech with aspiration noise, where more rapid changes in the waveforms are seen. This coder represents a first step towards the very high update rates used in the generalized WI discussed in Section 3.

Time-domain WI coding

The 2.4 kb/s coder reported in [23] will be used here as an implementation example of time-domain coding. While most WI-based coders have a two-way (voiced-unvoiced) classification, this coder has a three-way classification of speech segments into quasi-periodic, unvoiced, and nonperiodic peaky segments. WI is used for the quasi-periodic segments, a (baseband) CELP for the unvoiced segments, and multi-pulse for the nonperiodic peaky segments. Table 2 provides an overview of the bit allocation for voiced speech. The coder uses frames of 25 ms for LP analysis and quantizes the LP coefficients by a split vector quantization [13]. One prototype waveform is extracted for each LP analysis frame. In contrast to other WI implementations, this coder operates in a synchronous manner. Thus, the reconstructed waveform is aligned with the original signal, in a manner similar to the single-pulse coder [6]. The location of the main excitation pulse of the last pitch cycle within a frame is determined from the 2 kHz lowpass filtered residual. A varying number of bits is used to transmit this location since an initial estimate can be made knowing the last pitch pulse location. The approximate location of the other pitch pulses is computed by linear interpolation, and then refined using an additional 6-8 bits. The pitch pulse is windowed and quantized as explained in Section 2.4.

Parameter	bits/25 ms
LP coefficients	20
Voicing	3
Pitch period	7
Last pulse location	5-7
Other Pulse locations	6-8
Codebook	10
Codebook gain	5
Pitch postfilter	2
Total (25 ms)	60

Table 2 Voiced bit allocation for a 2.4 kb/s WI coder.

The interpolation is done in a continuous fashion. Together with a pitch postfiltering procedure, which introduces additional periodicity at higher frequencies, this procedure is found to yield high quality voiced speech.

3 A GENERALIZED WI MODEL

In the previous section it was emphasized that WI is very useful for modeling quasi-periodic signals such as voiced speech. Switching to another coding model for nonperiodic signals is inherently nonrobust and leads to problems for intermediate levels of periodicity and signals with background noise. In this section, it will be shown how WI can be generalized in a natural fashion by increasing the prototype update rate and exploiting the properties of the human auditory system for efficient coding. The effect of increasing the update rate is discussed in the next section. A generalized WI (GWI) method exploiting the increased update rate and human perception is discussed in Section 3.2.

3.1 Effect of changing the WI update rate

Interpolation of waveforms over long (20-30 ms) intervals constrains the evolution bandwidth of the waveform to be low. In other words, a low update rate forces the waveform to evolve slowly. If the waveform is under-sampled, WI will introduce a higher level of periodicity compared to the original signal, which manifests itself as buzziness in the output signal. This is the reason why conventional WI is applied only to voiced speech. To accommodate higher evolution bandwidths, which are characteristic for voiced speech with aspiration noise and unvoiced speech, a higher transmission rate of the prototypes is needed. Enhanced speech quality for voiced signals has been observed for a higher update rate [4].

The effect of increasing the update rate in WI systems is shown by the results of a formal MOS (see Section 2.4) test, which are provided in Table 3. In this test the residual signal is passed through the 'outer layer' in Fig. 1 only (i.e. unquantized mode). The WI method is applied here to both voiced *and* unvoiced speech. From the table it is seen that the MOS scores increase rapidly as a function of the update rate. The low scores at low updates are due to the buzzy speech quality of the unvoiced speech. To investigate further

Update Rate (Hz)	50	100	200	400
Mean Opinion Score	2.3	2.8	3.6	4.0

Table 3 MOS as a function of the update rate in WI systems.

the performance of the unquantized system at a high update rate, two formal AB-comparison tests were performed [24]. The WI system at an update rate of 400 Hz, was compared to the speech coded with the ITU G726 standard (32 kb/s ADPCM) and the original speech. Twelve randomly selected sentence pairs (7 sec each) of both "flat" and IRS weighted speech were used for the test. 10 persons not involved in the development of the WI system participated in the test. Compared to the ITU G726 standard the subjects preferred the WI analysis-synthesis system 68% of the time, and compared to the original the subjects preferred the WI analysis-synthesis system 40% of the time. The tests and informal listening confirm that the WI analysis-synthesis system converges towards the original quality for voiced, unvoiced, and background noise at high update rates. Interestingly, this is not the case for music; WI analysis-synthesis converts certain components of such complex signals into a smooth type of noise.

A higher update rate is contradictory to the requirements of lower bit rates. The entropy of the prototypes is too high, particularly during unvoiced speech, to be transmitted accurately at a low bit rate. The only way to lower this transmission rate while maintaining a high speech quality is by taking advantage of human perception. Section 3.2 will introduce an efficient method to do this. However, before that section a more straightforward (but less efficient) procedure will be discussed.

Experimental results show that the human auditory system has low time resolution (high frequency resolution) at lower frequencies and high time resolution (low frequency resolution) at higher frequencies. To exploit this, it seems natural to use a frequency dependent update rate of the prototypes. It is indeed found that a WI system where the higher frequencies are allowed to be updated more frequently than the lower frequencies facilitates a lower bit rate. As was shown in formal MOS testing, good quality speech can be maintained if the lower frequencies are updated at about 50 Hz and the higher frequencies are updated at about 200 Hz. Since the frequency resolution of the auditory system decreases with increasing frequency, the frequencies which require the highest update rate can be quantized with the lowest accuracy.

3.2 Decomposition of the waveform

To exploit the difference in human perception of a rapidly evolving waveform (as present in unvoiced speech) and a slowly evolving waveform (as present in voiced speech), a generalized WI (GWI) method was recently proposed [24]. In this method the two-dimensional signal $u(t, \phi)$ is filtered in the t direction. Such filtering of the evolving waveform was first introduced for WI by [10], where it was used to prevent aliasing when the prototypes are updated at a low rate (50 Hz). In GWI a high update rate is used (e.g. 480 Hz) and the filtering along the t-axis is used to decompose the evolving waveform into two components: a rapidly evolving waveform (REW) and a slowly evolving waveform (SEW). The boundary between the SEW and REW falls at an *evolution frequency* of about 20 Hz. In GWI, the SEW represents the quasi-periodic part of the one-dimensional LP residual signal whereas the REW represents the remaining nonperiodic component (the "noise").

Using the FS description, the decomposition of GWI is obtained by highpass and lowpass filtering the time sequence $c_n(t_i), c_n(t_{i+1}), c_n(t_{i+2}), \ldots$ (see Eq. 4.4) for each Fourier series coefficient (i.e. for each n). For the time-domain representation, the time sequence of corresponding samples in sequential prototypes is filtered. Figure 5 illustrates how the evolving waveform of Fig. 2 is split into a REW and a SEW. The figures correspond to a FS representation of the prototypes. The magnitude spectrum of the FS is normalized for the prototypes, so that it can be quantized separately. Because of this constraint on the sum of

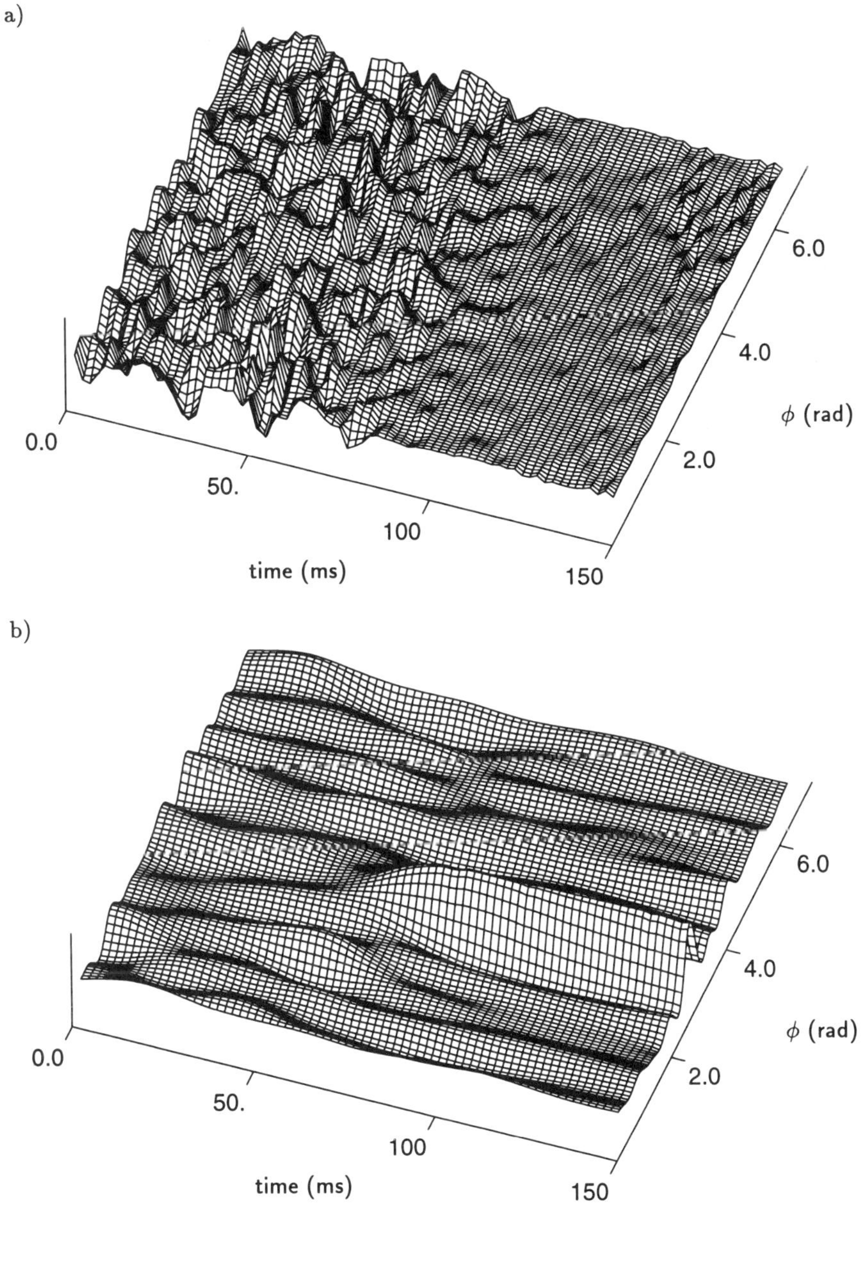

Figure 5 Decomposition of the residual into a) a rapidly evolving waveform (REW) and b) a slowly evolving waveform (SEW).

the REW and the SEW, the dominance of the REW during unvoiced speech, and the dominance of the SEW during voiced speech is clearly visible.

The REW describes the nonperiodic component of the signal. It is known that unvoiced speech, where this component dominates, can be quantized at a low bit rate in a *perceptually* accurate manner [25]. The quantization of the REW in GWI generalizes this result. Separating the REW into a magnitude and a phase spectrum it is found that the phase spectrum may be replaced by a random phase (which is changed at every update) without a significant decrease of the reconstructed speech quality. However, for this to be true, it is essential that the pitch period, $p(t)$, is no longer than about 5 ms for nonvoiced signals which have an insignificant SEW contribution. If the pitch period is longer in such nonvoiced signals, the energy fluctuations within the prototype can be weakly audible. For the REW magnitude spectrum it is found that a low-resolution quantization with a high update is sufficient.

Because of its low evolution bandwidth, the SEW can be downsampled to a low rate before quantization. Typically the update rate is chosen to be similar to that of the LP coefficients. Because of the similarity in the update rate to that of the prototypes of the original WI (aimed at voiced speech), the SEW can be quantized using the methods described in Section 2.3. Because of the LP filtering, the magnitude spectrum of the sum of the REW and the SEW is approximately flat. Furthermore, the spectrum has unity magnitude because of the normalization. Thus, an approximate magnitude spectrum for the SEW can be found by subtraction and smoothing of the REW spectrum from a unity magnitude spectrum. This is particularly useful at low bit rates. The bit allocation for an implementation at 2.4 kb/s is described in the following section.

Using the features described above, the GWI paradigm leads to highly independent ("orthogonal") parameters: LP coefficients, signal power, pitch period, REW and SEW. These parameters can be manipulated independently, making the method convenient for speech synthesis applications. In addition the independence means that the design of the quantizers is significantly simplified, and that, if needed, quantization tasks can be divided over various processors.

3.3 A GWI coder at 2.4 kb/s

Efficient GWI implementations have been developed which operate at low bit rates. These implementations operate on quasi-periodic signals, nonperiodic signals, and signals which are combinations of these two. Table 4 shows the bit allocation for a particular 2.4 kb/s GWI implementation. The 2.4 kb/s GWI

Parameter	bits/25 ms
LP coefficients	30
Pitch period	7
Signal power	2·4
REW	2·(3+1)
SEW	7
Total	60

Table 4 Bit allocation for a 2.4 kb/s GWI coder.

coder has 25 ms frames. It extracts prototype waveforms 12 times per frame (i.e. the update rate is 480 Hz). The ten LSF coefficients are quantized using a split vector quantization [13]. The procedure used for finding the pitch is described in the appendix. The signal power is first converted to the equivalent logarithm of the speech-domain power. It is then down-sampled to 80 Hz, and quantized differentially using a four-bit codebook. To quantize the REW, the magnitude spectrum of each sequential group of three prototypes is averaged first. For every second group, the magnitude spectrum is quantized by a three-bit (eight shapes) codebook. The codebook index of the intermediate groups is either set identical to that of the preceding or to that of the following group. Thus, the description of these intermediate groups require only one bit. At the receiver, the magnitude spectrum of each REW is determined from the transmitted bit stream, and a different random phase spectrum is added to each REW and the FS coefficients are computed.

Every 25 ms, an initial estimate of the SEW magnitude spectrum is found by subtracting a time-smoothed REW magnitude spectrum from a unity magnitude spectrum. A seven-bit codebook is then applied to increase the accuracy of the quantized SEW magnitude spectrum below 800 Hz. In other words the first 20% of the FS coefficients are refined with this codebook. At the receiver, linear interpolation is used to increase the sampling rate of the SEW to 480 Hz, making it identical to the sampling rate of the REW. For the SEW phase spectrum, a model which requires no transmission is used. A set of SEW phase spectra (extracted from a male speaker) are stored in a table and selected on the basis of the balance between the REW and the SEW signal power. When the waveform contains a high proportion of the SEW, then a phase spectrum resulting in a clearly defined pitch pulse is used in the construction of the SEW.

If the waveform contains a high proportion of REW, the energy of the SEW is evenly spread (along the ϕ axis). Note that the SEW evolves smoothly at transitions between these phase spectra because of the interpolation required for the upsampling.

Once both the SEW and the REW have been reconstructed, they are added. The resulting quantized prototypes are multiplied with the appropriate gain, and the waveforms are then upsampled to 8 kHz by interpolation. From this signal the LP-excitation signal is synthesized.

The 2.4 kb/s version of the GWI coder was entered in a survey of state-of-the-art 2.4 kb/s coders organized by the U. S. Department of Defense in the third quarter of 1994. Eight 2.4 kb/s coders were entered. Each coder was subjected to a large number of formal tests, involving a total of nine hours of processed speech. The tests include clean speech, speech with various types of background noise, as well as conditions with channel errors. Overall, the GWI coder was ranked as the best 2.4 kb/s coder in this survey. In all tests, the GWI coder was rated as either equivalent to, or better than the 4.8 kb/s FS1016 CELP coder [28], which was used as a reference. MOS, DAM (Diagnostic Acceptability Measure), and DRT (Diagnostic Rhyme Test) results for clean speech are shown in Table 5. The MOS scores are relatively high since the test contained mainly low-quality conditions.

	GWI (2.4 kb/s)	FS1016 (4.8 kb/s)
MOS	3.77	3.59
DAM	66.8	63.1
DRT	92.0	92.8

Table 5 MOS, DAM, and DRT results for clean speech.

The test results show that the 2.4 kb/s GWI coder is very robust against various types of background noise (e.g. car-, helicopter-, babble- and office-noise). The MOS scores for GWI and FS1016 in the case of car noise are 3.74 and 3.78, respectively. The reconstructed background noise tends to have a steady character. This contrasts with coders using classification, which may switch between the various coding modes, causing an annoying nonsteady character of the reconstructed background-noise signal.

4 CONCLUSION

To meet todays requirements for robustness and high quality reconstructed speech at low bit rates, it is necessary to focus the limited coding resources on perceptually relevant features of the speech signal. This chapter has described a coding method aimed at satisfying these requirements. In this waveform-interpolation (WI) coder the speech signal is described as an evolving waveform. WI was originally developed for voiced speech, where the waveform evolves slowly, allowing a low transmission rate. However, more recently it was noted [11] that any evolving waveform can be decomposed into a slowly evolving waveform (SEW), which dominates during voiced speech, and a rapidly evolving waveform (REW), which dominates during unvoiced speech. These two waveforms have very different requirements for perceptually accurate quantization. As a result, a generalized WI (GWI) which exploits the separate quantization of the SEW and REW requires less bits than the quantization of their sum, the prototype waveform. In contrast to the earlier WI coders, which use a voicing decision, the GWI allows the simultaneous encoding of quasi-periodic and non-periodic signal components. The avoidance of voicing decisions and switching between different modes results in high-quality low bit rate coding for signals with and without noisy background.

APPENDIX A

A PITCH CALCULATION ALGORITHM

In GWI and WI a pitch track is normally obtained by updating and interpolating the calculated pitch period for each LP analysis frame (e.g. each 20-30 ms). To make the WI method work successfully it is very important that the pitch track is sufficiently accurate. In principle, the reliability can be increased by increasing the window size over which the pitch period is estimated. However, this is only the case if the pitch period can be considered to be relatively constant. High resolution methods (fractional pitch-period estimation) can also be used to obtain higher reliability in the actual pitch period [27, 18] if the pitch period is constant. Both these procedures fail if the pitch period changes as a function of time.

One method which increases the estimation reliability with increasing window size even when the pitch period is changing was presented in [19]. A modified version of this algorithm, which is used in the 2.4 kb/s GWI coder described in Section 3.3, will now be provided.

For a segment of the signal $s(k)$ with K samples, let the correlation be defined by $R(d) = \sum_{k=0}^{K} s(k)s(k-d)$. The correlation is computed for three such signal segments: past, current, and future. The pitch period in the current frame is determined by a composite correlation function which depends on the correlations of the past, the current, and the future. The composite function is found by first weighting each correlation function by the respective maximum peak values (normalized correlations) which is a measure of the "reliability" of the peak. For each candidate pitch period value, d, a contribution from the past and the future is added to the current correlation curve. The contributions are found by windowing the correlation curves by a window positioned such that the peak appears at the measured pitch period. Within each window the maximum value is found and added to that of the composite correlation. The composite function can be expressed as

$$\begin{aligned} R_{composite}(d) &= R_{current}(d) + \\ &\quad max\{w(i) \cdot R_{past}(d+i)\} + \\ &\quad max\{w(i) \cdot R_{future}(d+i)\}, \qquad -l(d) \leq i \leq l(d) \end{aligned} \tag{A.1}$$

where the $R's$ are the correlation functions, $w(i)$ is the window, and $l(d)$ is the length of the windows. The length of the windows depends on the candidate pitch period d. Note that the individual correlation functions are added coherently, i.e. the peaks of $R_{current}$ are reinforced by nearby peaks of R_{past} and R_{future}. This allows the method to provide reliable pitch-period estimates even if the pitch period changes as a function of time.

The procedure can be generalized to have N signal segments rather than three. However, instead a hierarchical structure is used. First, the method is applied to three short segments (one third of a frame), resulting in a first-stage composite correlation function for a frame (or "supersegment"). The procedure is then repeated for three frames, rendering a second-stage composite correlation function. This second-stage correlation function is then used to obtain the final pitch-period estimate.

REFERENCES

[1] B. S. Atal and M. R. Schroeder, "Stochastic Coding of Speech at Very Low Bit Rates," Proc. Int. Conf. Comm., Amsterdam, pp. 1610-1613, 1984.

[2] W. B. Kleijn, "Continuous Representations in Linear Predictive Coding," Proc. Int. Conf. Acoust. Speech Sign. Process., pp. 201-204, IEEE, 1991.

[3] W. B. Kleijn, "Encoding Speech Using Prototype Waveforms," IEEE Trans. Speech Audio Process., Vol 1, No. 4, pp. 386-399, 1993.

[4] Y. Shoham, "High-Quality Speech Coding at 2.4 to 4.0 kbps based on Time-Frequency Interpolation," Proc. Int. Conf. Acoust. Speech Sign. Process., pp. II 167-170, IEEE, 1993.

[5] J. Haagen, H. Nielsen and S. Duus Hansen, "Improvements in 2.4 kbps High-Quality Speech Coding," Proc. Int. Conf. Acoust. Speech Sign. Process., pp. II145-II148, IEEE, 1992.

[6] W. Granzow, B. S. Atal, K. K. Paliwal and J. Schroeter, "Speech coding at 4 kb/s and Lower Using Single-pulse and Stochastic Models of LPC Excitation," Proc. Int. Conf. Acoust. Speech Sign. Process., pp. 217-220, IEEE, 1991.

[7] G. Yang, H. Leich, and T. Boite, "Voiced Speech Coding at Very Low Bit Rates based on Forward-Backward Waveform Prediction (FBWP)," Proc. Int. Conf. Acoust. Speech Sign. Process., pp. II179-II182, IEEE 1993.

[8] I. S. Burnett, and R. J. Holbeche, "A Mixed Prototype Waveform/CELP Coder for Sub 3 kb/s," Proc. Int. Conf. Acoust. Speech Sign. Process., pp. II175-I178, IEEE 1993.

[9] M. Leong, "Representing Voiced Speech Using Prototype Waveform Interpolation for Low-rate Speech Coding," Master's Thesis, McGill University, Montreal, 1993.

[10] Y. Tanaka and H. Kimura, "Low-bit-rate speech coding using a two-dimensional transform of residual signals and waveform interpolation," Proc. Int. Conf. Acoust. Speech Sign. Process., pp. I173-I176, IEEE, 1994.

[11] W. B. Kleijn and J. Haagen, "Transformation and Decomposition of the Speech Signal for Coding," IEEE Signal Processing Letters, Vol. 1, pp. 136-138, September 1994.

[12] J. Makhoul, "Linear Prediction: A tutorial review", Proceedings of the IEEE, vol. 63, Apr. 1975.

[13] K. K. Paliwal and B. S. Atal, "Efficient Vector Quantization of LPC Parameters at 24 Bits/Frame," IEEE Trans. Speech Audio Process., Vol. 1, No. 1, pp. 3-14, 1993.

[14] W. F. LeBlanc, B. Bhattacharya, S. A. Mahmoud, and V. Cuperman, "Efficient Search and Design Procedures for Robust Multi-Stage VQ of LPC Parameters for 4 kb/s Speech Coding," IEEE Trans. Speech Audio Process., Vol. 1, No. 4, pp. 373-385, 1993.

[15] R. Hagen and P. Hedelin, "Robust Vector Quantization in Speech Coding," Proc. Int. Conf. Acoust. Speech Sign. Process., pp. 13-16, IEEE, 1993.

[16] W. Hess, "Pitch Determination of Speech Signals," Springer Verlag, Berlin, 1983.

[17] D. P. Prezas, J. Picone, and D. L. Thomson "Fast and Accurate Pitch Detection Using Pattern Recognition and Adaptive Time-Domain Analysis," Proc. Int. Conf. Acoust. Speech Sign. Process., pp. 109-112, 1986.

[18] Y. Medan, E. Yair, and D. Chazan "Super Resolution Pitch Determination of Speech Signals," IEEE Trans. Signal Process., Vol. 39, No. 1, pp. 40-48, 1991.

[19] W. B. Kleijn, P. Kroon, L. Cellario, and D. Sereno, "A 5.85 kb/s CELP Algorithm for Cellular Applications," Proc. Int. Conf. Acoust. Speech Sign. Process., pp. II596-II598, IEEE 1993.

[20] W. B. Kleijn and W. Granzow, "Methods for Waveform Interpolation in Speech Coding," Digital Signal Processing, Vol 1, No. 4, 1991, pp. 215-230.

[21] M. R. Schroeder, B. S. Atal, and J. L. Hall," Optimizing Digital Speech Coders by exploiting Masking Properties of the Human Ear," The Journal of the Acoustical Society of America, vol. 66, no. 6, pp. 1647-1652, 1979.

[22] D. Sen, D. H. Irving, and W. H. Holmes, "Use of an Auditory Model to improve Speech Coders," Proc. Int. Conf. Acoust. Speech Sign. Process., pp. II411-II414, IEEE, 1993.

[23] J. Haagen, "Digital Speech Coding at 2.4 kbit/s: New Strategies for Quantization of the Residual," Ph.D. thesis, Technical University of Denmark, 1993.

[24] W. B. Kleijn and J. Haagen, "A General Waveform Interpolation Structure for Speech Coding," Signal Processing VII: Theories and Applications, pp. 1665-1668, European Association for Signal Processing, 1994.

[25] G. Kubin, B. S. Atal, and W. B. Kleijn, "Performance of Noise Excitation for Unvoiced Speech," Proc. IEEE Workshop on Speech Coding for Telecommunications, pp. 35-36, 1993.

[26] B. S. Atal and B. E. Caspers, "Beyond Multipulse and CELP towards High Quality Speech at 4 kb/s," Advances in Speech Coding, pp. 191-201, Kluwer Academic Publishers, 1991.

[27] P. Kroon and B. S. Atal, "Pitch predictors with high temporal resolution," Proc. Int. Conf. Acoust. Speech Sign. Process., pp. 661-664, IEEE, 1990.

[28] J. P. Campbell, V. C. Welch, and T. E. Tremain, The DOD 4.8 kbps Standard (Proposed Federal Standard 1016). In B. S. Atal, V. Cuperman, and A. Gersho, editors, *Advances in Speech Coding*, pp. 121-133, Kluwer Academic Publishers, Dordrecht, Holland, 1991.

5

VARIABLE RATE SPEECH CODING

Vladimir Cuperman and Peter Lupini

School of Engineering Science, Simon Fraser University, Burnaby, B.C. Canada V5A 1S6

vladimir@cs.sfu.ca, lupini@cs.sfu.ca

1 INTRODUCTION

An important goal in the design of voice communication networks and storage systems is to maximize capacity while maintaining an acceptable level of voice quality. Conventional speech coding systems use a fixed bit rate regardless of factors such as local speech statistics, transmission channel conditions, or network load. One method of maximizing capacity while maintaining an acceptable level of speech quality is to allow the bit rate to vary as a function of these factors. Variable rate speech coders exploit two important characteristics of speech communications: the large percentage of silence during conversations, and the large local changes in the minimal rate required to achieve a given speech reproduction quality.

Variable rate coders can be divided into three main categories

- **source-controlled** variable rate coders, where the coding algorithm determines the data rate based on analysis of the short-term speech signal statistics.
- **network-controlled** variable rate coders, where the data rate is determined by an external control signal generated by the network in response to traffic levels.
- **channel-controlled** variable rate coders, where the data rate is determined by the channel state information (such as estimated channel SNR)

The first two categories were defined in [1]. Channel controlled variable rate coders are used in systems where a fixed aggregate rate is divided between the speech coder and the channel coder under the control of a channel state estimate with the objective of optimizing the speech quality for the end user.

Historically, network-controlled variable rate coders were developed first and applied to digital speech interpolation (DSI) using embedded ADPCM; embedded coding is presented in Section 2. Section 3 examines source-controlled variable rate speech coding, including digital cellular communications applications. Section 4 presents research work in the area of channel-controlled variable rate coding. Finally, Section 5 examines the application of variable rate speech coding to ATM networks.

2 NETWORK-CONTROLLED VARIABLE RATE CODING

The simplest way to convert a fixed rate coder into a variable rate coder is to use two rates: a fixed rate for active speech and a zero rate for silent input. In a two-way voice communication, each direction is active about 36-44% of the time [2, 3], so there are considerable rate savings available using this approach. Active speech is detected using a voice activity detector (VAD) which detects the active speech patterns (talk spurts) on each voice channel.

This idea was used initially in the analog Time Assignment Speech Interpolation (TASI) systems where one way circuits in a multiplexed carrier system were time-shared by a number of users significantly larger than the fixed number of circuits [4]. The analog TASI systems were then replaced by digital transmission systems based on the same concept and called Digital Speech Interpolation systems. DSI systems were successfully applied first to submarine telephone cables and later to satellite communications.

The capacity gain achieved through statistical multiplexing, known as the DSI gain, can reach 2.5 for a voice activity factor (VAF) of 40%. DSI gains as large as 3 were reported by Yatsuzuka based on a system which combines DSI with ADPCM variable rate coding [3].

A common problem in DSI systems is the blocking or "freezing" which appears when the number of users with active talkspurts exceeds the system capacity. Network-controlled variable rate can be used to alleviate this problem by reducing the rate available to all users rather than blocking completely some of the users' talkspurts. The design particularities of DSI systems make it more convenient to reduce the rate by dropping bits from a fixed rate bit stream without switching the rate at the encoder. This is possible in embedded speech coding.

An embedded speech codec produces a fixed rate bit stream in which lower rate substreams are "embedded" in the bit stream of the higher rate substreams. The encoder state (filter memories, etc.) is determined by the lowest rate substream, hence transmitter and receiver will have the same state even if the bits used only for the higher rate substreams are dropped by the network. Figure 1 shows a block diagram of an embedded coder which produces e-bit codewords. In response to network traffic conditions, the controller removes some of the e bits produced by the coder, transmitting only c bits. If the embedded coder is properly designed, the speech quality at the decoder will be close to that obtained by using a fixed rate c bit coder.

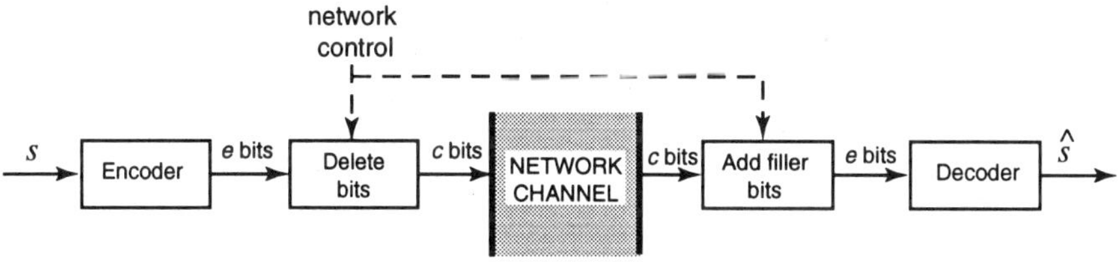

Figure 1 Embedded Coder Block Diagram

Pulse Code Modulation (PCM) quantization provides a straight forward example of embedded coding. If all but the c most significant bits are stripped from an e-bit PCM codeword and replaced with zeros at the decoder, an output signal can be obtained which is close to the output of a fixed rate c bit PCM encoder. The degree to which the quality of an embedded lower rate encoding can approach the quality of a fixed rate coder operating at the same rate depends on the codec structure and on the choice of quantizers; usually there is some degradation associated with the constraint imposed by the embedded codec structure.

Embedded coding was initially introduced for ADPCM by Ching and described by Goodman [5]. A description of embedded ADPCM and the CCITT standard G.727 is presented in section 2.1. More recently embedded coding was introduced for CELP coders by Iacovo and Sereno [6]. Their system embeds

6.4 kb/s and 8 kb/s bit streams into a 9.6 kb/s bit stream by representing the residual of the short term and long term synthesis filters as a sum of three different vector quantizer outputs. Only the contribution from the first VQ is used to update filter memories.

2.1 Embedded ADPCM

Although PCM can be used directly as an embedded coder, DPCM cannot. The reason can be seen clearly in figure 2 which shows a block diagram of an ADPCM encoder and decoder. If some of the e bits are dropped before transmission through the channel, the predictor at the decoder will have different input from the predictor at the transmitter, resulting in a large drop in performance. It was found that using unmodified DPCM as an embedded coder produces SNR degradations of 3.5 dB to 10 dB [5].

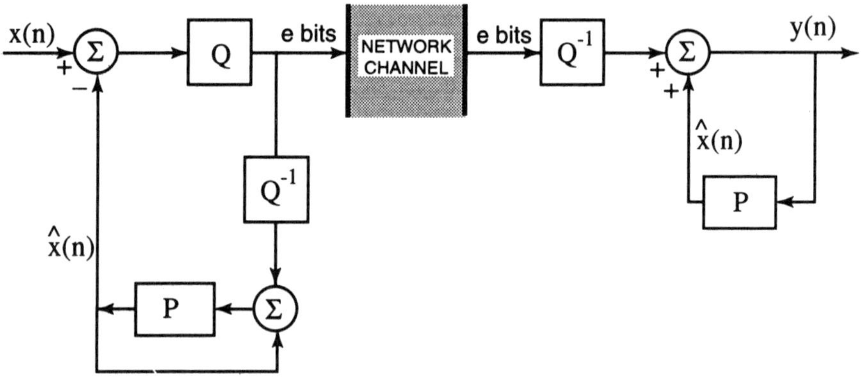

Figure 2 ADPCM Encoder/Decoder Block Diagram

To keep the encoder and decoder states identical, the encoder must make predictions based only on the signal with the lowest possible rate. The decoder also drops the maximum number of bits before performing the prediction, regardless of how many bits were actually dropped by the channel. The embedded ADPCM block diagram is shown in figure 3.

Because the lowest possible rate signal is used for prediction, the gain of the predictor is in general lower than with the conventional ADPCM system. This results in a lower SNR when using an embedded ADPCM system as compared

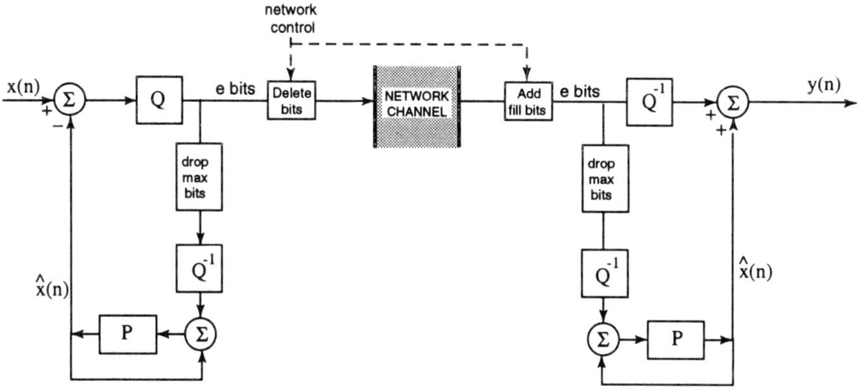

Figure 3 Embedded ADPCM Encoder/Decoder Block Diagram

to a conventional ADPCM system operating at the same rate. Goodman reported a 0.7 dB penalty when using embedded ADPCM coding [5].

The CCITT standard G.727 describes a set of embedded ADPCM encoders, each with different maximum and minimum rates. Note that because minimum bit rate signals are used for prediction, the minimum bit rate must be specified as part of the standard. For example, G.727 describes a codec with allowable rates of 40 kb/s, 32 kb/s, 24 kb/s, and 16 kb/s, as well as a codec which allows rates of only 40 kb/s and 32 kb/s. Other codecs with maximum rates of 32 kb/s and 24 kb/s are also specified in the standard.

3 SOURCE-CONTROLLED VARIABLE RATE SPEECH CODING

In some speech coding applications, the bit rate can be controlled by the coder itself rather than through external control signals. This type of control makes it possible to exploit the large local changes in the minimal rate required to obtain a given speech quality by adjusting the rate based on frame by frame speech analysis.

Source-controlled variable rate speech coders have been applied to digital cellular communications and to speech storage systems such as voice mail and voice response equipment. In both cases replacing the fixed-rate coders by variable-rate coders results in a significant increase in the system capacity at the expense of a slight degradation in the quality of service.

The relatively new IS-95 North American Telephone Industry Association (TIA) standard for digital cellular telephony adopted in 1993 is based on code division multiple access (CDMA) and variable rate speech coding. In CDMA all users share the same frequency band and the system capacity is limited by the interference generated by users. The amount of interference generated by a user depends on the average coding rate and any average rate decrease translates directly into capacity increase. Such a system inherently exploits the advantage of source controlled variable rate coding. The variable rate speech coder used in conjunction with the CDMA IS-95 standard is known as QCELP [7] and has been adopted as the TIA standard IS-96.

Conventional time division multiple access (TDMA) systems such as the digital cellular system of the IS-54 TIA standard allocate a fixed time slot to each unidirectional voice communication regardless of voice activity and use fixed rate speech coding. Vector sum excited linear prediction (VSELP) at 8 kb/s is used in IS-94. However, new TDMA versions such as enhanced TDMA (E-TDMA) [8] are designed to exploit variable rate in order to increase capacity. Finally, packetized networks based on such techniques as Packet Reservation Multiple Access (PRMA) [9] and Asynchronous Transfer Mode (ATM) can exploit variable rate in order to reduce the overall network load and to increase capacity.

One of the earliest examinations of source-controlled variable rate coding was presented by Dubnowsky and Crochiere [10] in which they analyze some of the theoretical concepts for practical variable rate transmission including the problem of buffer underflow and overflow when a variable rate bit stream is converted to fixed rate. A practical source-controlled coder is presented in [10] using multi-rate ADPCM with rate selection based on input signal variance estimation.

Kou et al [11] present an analysis of DSI using source-controlled variable rate subband coders. These coders process blocks of speech by iteratively assigning bits to each subband until a constant noise level is obtained.

In [12], Nakada and Sato present a variable rate codec which uses two quantizers. The first quantizer is a conventional quantizer in the prediction loop with the number of bits determined by the input signal level. The second quantizer analyzes the output of the first quantizer and discards excessive bits.

Yatsuzuka et al [13] present a source-controlled variable rate codec which includes a highly sensitive speech detector based on short-time energy, zero-crossing rates, and sign bit sequences of the input signal. Speech is further classified into narrow-band spectrum, wide-band spectrum, and low-power sounds for rate selection in an ADPCM codec.

Many of the source-controlled variable rate coders presented so far contain classifiers which use one or two parameters to make rate decisions. Often this leads to a general classification of voiced, unvoiced, and silence. In QCELP, for example, an energy based classification is used which tends to map voiced frames onto the highest rate, although there are significant exceptions from this mapping (some voiced phonemes, for example nasals, can have relatively low energy). One recent approach to variable rate coding by Paksoy et al [14] does attempt direct phonetic classification. The classification algorithm uses eight parameters to classify speech subframes as either voiced or unvoiced. The subframe classes are then analyzed to obtain four rate classes: all voiced, all unvoiced, unvoiced to onset transition, and onset to voiced transition. The codec rates range from 750 bps to 5.75 bps, with a reported average rate below 3 kb/s.

The rest of this section will describe QCELP [7] followed by a multimode variable rate CELP codec based on our own research [15].

3.1 Variable Rate Speech Coding Based on QCELP

The QCELP speech codec is based on Code Excited Linear Prediction (CELP) [16], an analysis-by-synthesis coding procedure. The QCELP encoder selects for each 20 ms speech frame one of four rates: 8 kb/s, 4 kb/s, 2 kb/s, or 1 kb/s. These rates are sometimes referred to as full rate, half rate, quarter rate, and eighth rate. The algorithm estimates the background noise energy and selects the rate based on the difference between the current frame energy and the estimated background noise energy. High energy signals such as strong voiced sounds are encoded at 8 kb/s, silence intervals are encoded at 1 kb/s, and intermediate energy sounds are encoded at 2 kb/s or 4 kb/s.

Figure 4 shows a block diagram of the QCELP decoder, with the transmitted parameters shown as dashed lines. For all rates except the eighth rate, the excitation is specified by a codebook index and gain; the eighth rate uses a transmitted seed value for pseudorandom sequence generation. The excitation is passed through a pitch (long-term) synthesis filter which is implemented as an adaptive codebook having as index the pitch lag. The output of the pitch filter is then transformed into speech using a short-term LPC filter. The LPC coefficients are derived from the ten transmitted Line Spectral Pair (LSP) frequencies. A post filter and gain control algorithm are used to enhance the reconstructed speech quality.

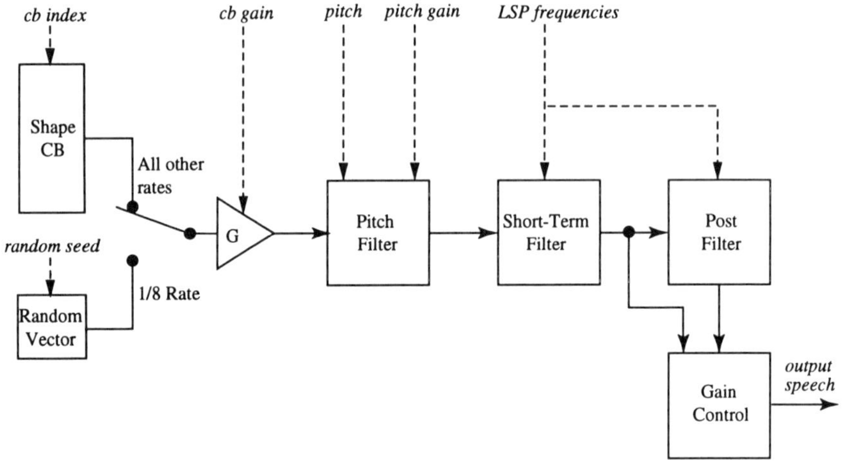

Figure 4 QCELP Decoder Block Diagram

The basic QCELP structure is scalable, which minimizes the complexity by allowing an integrated implementation of all rates. Instead of using different structures for each rate, the four rates are obtained by changing the number of bits assigned for the quantization of LPC parameters and changing the update interval for the adaptive codebook and stochastic codebook parameters. LPC quantization uses 40 bits at 8 kb/s, 20 bits at 4 kb/s and 10 bits at the two lower rates. Adaptive codebook parameters are quantized using 7 bits for the pitch lag and 3 bits for the pitch gain and are updated four times per frame at 8 kb/s, two times per frame at 4 kb/s, and once per frame at 2 kb/s; the adaptive codebook is not used at 1 kb/s. Excitation codebook parameters are quantized using 7 bits for the cb index (4 bits for the eighth rate random seed) and 3 bits for the gain. The excitation codebook parameters are updated eight times per frame at 8 kb/s, four times per frame at 4 kb/s, twice per frame at 2 kb/s, and once at 1 kb/s.

QCELP achieves good communications quality at a relatively low average coding rate. A new standard for CDMA is under development by TIA. The new standard will offer toll quality at the rate of 8 kb/s and has as an objective a low average rate (about 2.4 kb/s) with little quality degradation.

3.2 A CELP Multimode Coder

In this Section we present a multimode CELP coder which operates as a source-controlled variable rate coder at several bit rates [15]. It is optimized for rates of 16 kb/s, 8 kb/s, and 4 kb/s. A 925 b/s configuration is included for silent frames. The rate selection is performed on a frame by frame basis using a classifier. The codec can also operate in network-controlled mode if the classifier is replaced by an external network signal.

The codec uses a modular design in which the general structure and coding algorithm is the same for all rates. All configurations are based on the system with the highest bit-rate. The lower bit-rates are obtained by varying the frame/subframe sizes, using codebook subsets for quantization, and in some cases disabling codec components.

Figure 5 shows a block diagram of the encoder. Each input speech frame is analyzed by the frame classifier in order to determine the desired coding rate. The appropriate configuration is selected by specifying the allowed ranges for the shape and adaptive codebook indices (indicated by control signals). The short-term filter parameters are then computed and the excitation signal is formed as a summation of gain-scaled vectors from a two stage shape codebook and a single-tap adaptive codebook. Vectors are selected using an analysis-by-synthesis search based on a perceptually weighted MSE distortion criterion. The details of each major system block are discussed below.

The frame classifier is used only for the source-controlled mode of the variable rate coder. The purpose of the frame classifier is to analyze each input speech frame and determine the appropriate rate for coding. Ideally, the classifier will assign each frame to the lowest coding rate which still results in reconstructed speech quality meeting the requirements of the given application.

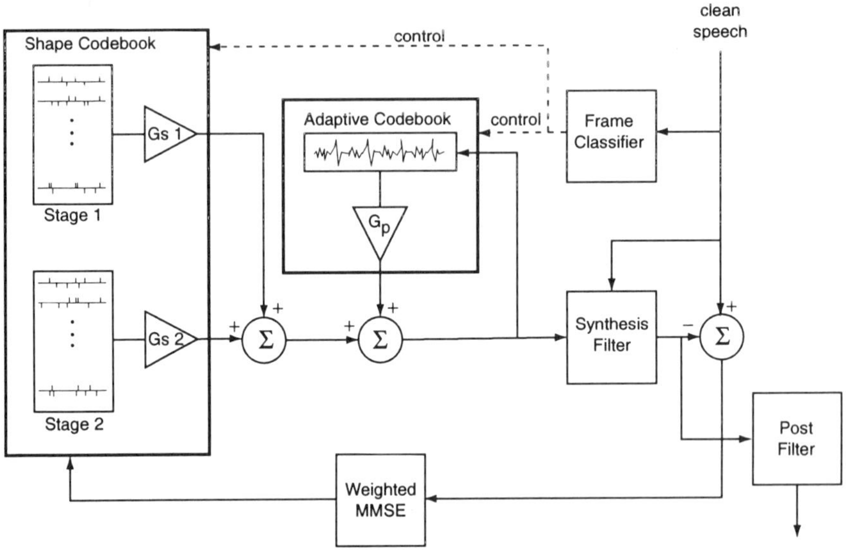

Figure 5 Block Diagram of Variable-Rate CELP Codec

Based on analysis of several different frame classification methods we found that classification based on the normalized autocorrelation coefficient at the pitch lag worked well. The classifier was made more robust to rapid voiced phoneme changes by computing the autocorrelation over several small subframes within a frame. For example, a frame may be encoded with the highest rate if more than 3/4 of the subframes have a normalized autocorrelation coefficient above a pre-defined threshold. Zero-crossings, low-band energy, and the short-term autocorrelation function are also used by the classifier to reduce the probability of assigning low rates to voiced frames.

The short-term predictor $1/A(z)$ is a tenth order LPC all-pole filter. A perceptual weighting filter of the form $H(z) = A(z)/A(z/\gamma)$ is derived from $A(z)$. Band-width expansion and high-frequency compensation are used during the LPC analysis.

The adaptive codebook consists of past excitation sequences. Lags from 20 to 147 samples are searched closed-loop to find the codebook entry which minimizes the weighted mean-squared error.

The shape excitation vector is selected using analysis-by-synthesis search of a multi-stage shape codebook. Each stage consists of stochastic codes, which are overlapped by two samples to reduce search complexity. Complexity is further reduced through the use of sparse ternary codes. The shape codebook data for each class is identical – different excitation bit-rates are obtained by searching subsets of the largest codebooks.

We use an adaptive post-filter presented in [17] which consists of a short-term pole-zero filter based on the quantized short-term predictor coefficients followed by an adaptive spectral tilt compensator. The pole-zero filter is of the form $H(z) = A(z/\beta)/A(z/\alpha)$. We use $\beta = 0.5$ and $\alpha = 0.8$. An automatic gain control is also used to avoid large gain excursions.

Table 1 gives the detailed bit allocations for the short term predictor (STP), adaptive codebook (ACB), and the shape codebook (SCB) for each rate. The 4k_S class is used for source-controlled mode, and the 4k_N class is used for network-controlled mode. The bits marked as "auxiliary" can be used for error detection and synchronization.

PARAMETER	16k	8k	4k_S	4k_N	0.925k
Frame Size (ms)	20	20	20	40	40
Subframe Size (ms)	2.5	5	5	10	40
STP bits	28	28	28	28	28
ACB bits	8x7	4x7	-	4x7	-
ACB Gain bits	8x5	4x6	-	4x5	-
SCB bits	8x12	4x10	4x8	4x10	-
SCB Gain bits	8x12	4x9	4x4	4x10	5
Classification bits	2	2	2	2	2
Auxiliary bits	2	2	2	2	2
Total Bits	320	160	80	160	37
Bits/s	16000	8000	4000	4000	925

Table 1 Bit Allocations for Each Class

Note that during source-controlled operation the 4 kb/s system does not use an adaptive codebook in order to take advantage of the fact that frames encoded at 4 kb/s will generally be unvoiced. In network-controlled mode the 4 kb/s rate can be requested at any time and therefore an adaptive codebook is included.

Initial tests of the system performance were based on analysis of frame SNRs for each fixed-rate system as compared with the variable rate system. Ideally, the variable rate system would perform as well as the fixed rate system for any given class. For example, during a transition from unvoiced to voiced, the frame SNRs should quickly increase from typical 4 kb/s SNRs to typical

16 kb/s SNRs. Figure 6 shows a typical unvoiced-voiced transition sequence, along with the SNRs obtained by the different systems for each frame. During each frame shown, the variable rate system performs close to the fixed rate system which is mapped to that class.

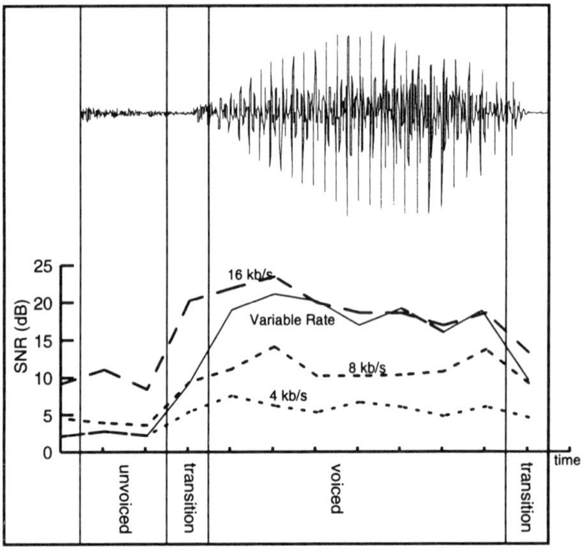

Figure 6 SNR Performance During Typical Unvoiced-Voiced Transition

Informal MOS tests were used to evaluate the overall subjective performance of the variable rate system. The following systems were tested:

1. Vector Sum Excited Linear Prediction (VSELP) at 8 kb/s [18]
2. Fixed-16 (from 16 kb/s class of variable rate system)
3. Fixed-8 (from 8 kb/s class of variable rate system)
4. SCV-CELP (Source-controlled variable rate CELP)

Table 2 shows the results of the MOS test for all speakers, male speakers and female speakers. Table 3 gives the classification mix generated by the variable rate system, and the average bit rates for the male, female, and all MOS files. The variable rate system operating at an average rate of about 8 kb/s performed significantly better than the fixed 8 kb/s system (0.3 on the MOS scale) and very close to the fixed 16 kb/s system (less than 0.1 difference on the MOS scale).

SYSTEM	FEMALE	MALE	BOTH
VSELP-8	3.66	3.62	3.64
Fixed-16	4.11	4.14	4.13
Fixed-8	3.79	3.78	3.78
SCV-CELP	4.09	4.01	4.05

Table 2 Mean Opinion Score Test Results

SPEAKER	% 16k	% 8k	% 4k	% 0.925k	BR (bps)
Female	41.2	5.6	32.7	20.6	8538
Male	30.3	9.3	40.0	20.5	7382
Both	36.3	7.0	36.3	20.4	8009

Table 3 Variable Rate Class Statistics for Mean Opinion Score Test Files

Figure 7 shows the average bit rate of the variable rate CELP codec as a function of the percentage of silence in the speech material. Our test data set had approximately 20% silence and therefore the average rate for all files is almost exactly 8 kb/s. The figure illustrates that even when there is no silence in the encoded material, the average rate of the system is under 10 kb/s – still well below the maximum rate of 16 kb/s. For typical voice activity rates of 35% to 45% (shaded region), the codec achieves average rates of approximately 4 kb/s to 5 kb/s.

4 CHANNEL-CONTROLLED VARIABLE RATE CODING

In traditional digital communication systems, the available channel bandwidth is divided between a fixed-rate source coder, a fixed-rate channel coder (error protection), and control functions (protocols). For example, the TIA digital cellular standard IS-54 allocates 8 kb/s for voice coding, 5 kb/s for error protection, and 3 kb/s for control functions. The allocations to the speech coder and the channel coder are designed according to some "nominal" channel conditions. However, in mobile wireless communications, channel conditions can vary widely due to fading and shadowing and fixed rate-allocation systems may become ineffective in maintaining a given quality of service.

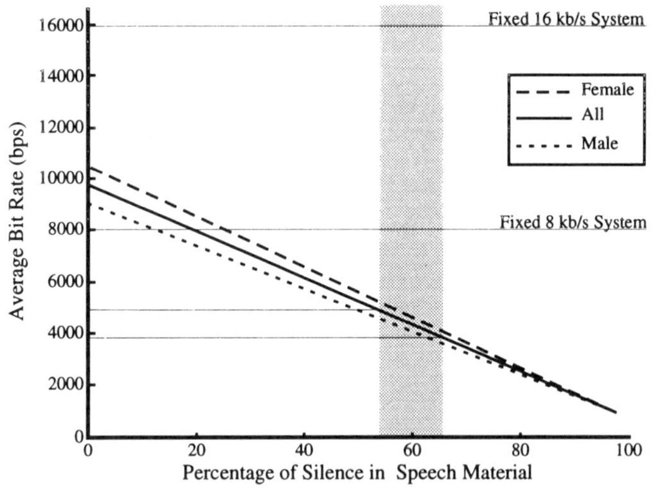

Figure 7 Average Bit-Rate as a Function of Percentage of Silence in the Speech Material

A significant improvement in the quality at the end user may be obtained by using a variable rate assignment whereby a fixed aggregate rate is divided between the speech and the channel coders according to some channel state information like the estimated channel SNR. Such a system uses channel-controlled variable-rate coders and may require a low-rate feedback channel for the channel state information. An early example of a channel-controlled variable rate system using an aggregate rate of 32 kb/s was presented by Goodman and Sundberg [19].

This section presents an example of the potential benefit of adaptive combined speech/channel coding for systems operating in the mobile radio environment [20]. The block diagram of the system shown in Fig. 8 consists of a variable rate speech coder, a variable channel coder, a channel estimator, and a feedback channel. The aggregate rate to be divided between the source and channel coders is 12.8 kb/s; another 3.2 kb/s are reserved for control information giving a total rate of 16 kb/s per user (compatible to the IS-54 rate assignment).

For an optimal adaptive channel-controlled system, the speech and channel coding rates have to be adjusted continuously, which implies that an infinite number of speech and channel coder rates are required. As a trade-off between system complexity and performance, a number of rate combinations that have good performance over the typical range of channel SNRs were selected. Ta-

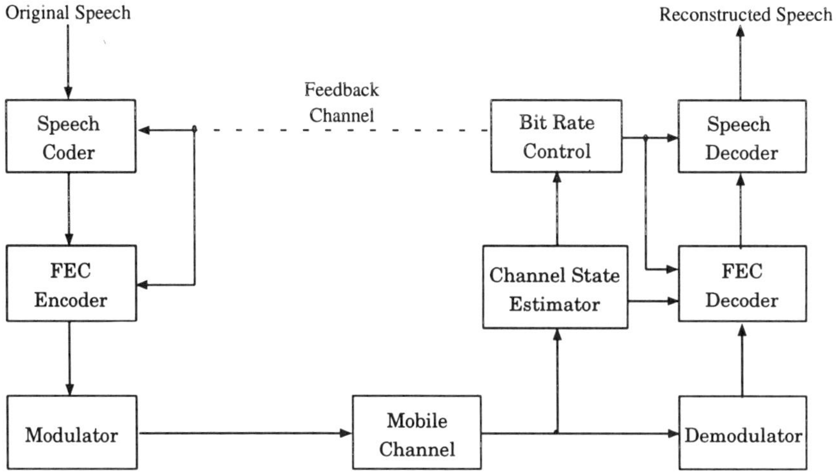

Figure 8 Block diagram of the variable rate system with feedback channel

ble 4 summarizes the rates of the different combined speech and channel coders considered in this example [20]. A multi-mode (switchable rate) CELP speech coder similar to that described in [15] is used. For the variable rate channel coder, the rate compatible punctured convolutional codes (RCPC) of rates 1/3 and 2/3 with constraint length 5 are used.

Coder	Aggregate Rate [bit/s]	Speech Coding Rate [bit/s]	Channel Coding Rate [bit/s]
1	12800	12800	0
2	12800	9600	3200
3	12800	8000	4800
4	12800	5000	7800

Table 4 Rate assignments for speech and channel coders

The system simulation assumes that the encoded binary digits modulate a carrier signal using the QPSK format. The modulated signal is transmitted over a mobile radio channel modeled as a Rayleigh flat fading channel. Rate adjustment commands are sent periodically by the receiver to the transmitter through the feedback channel based on an estimate of the channel SNR. Channel-controlled variable-rate is implemented by selecting the rat assignment that has the best performance at a given channel SNR. More details about this system can be found in [20].

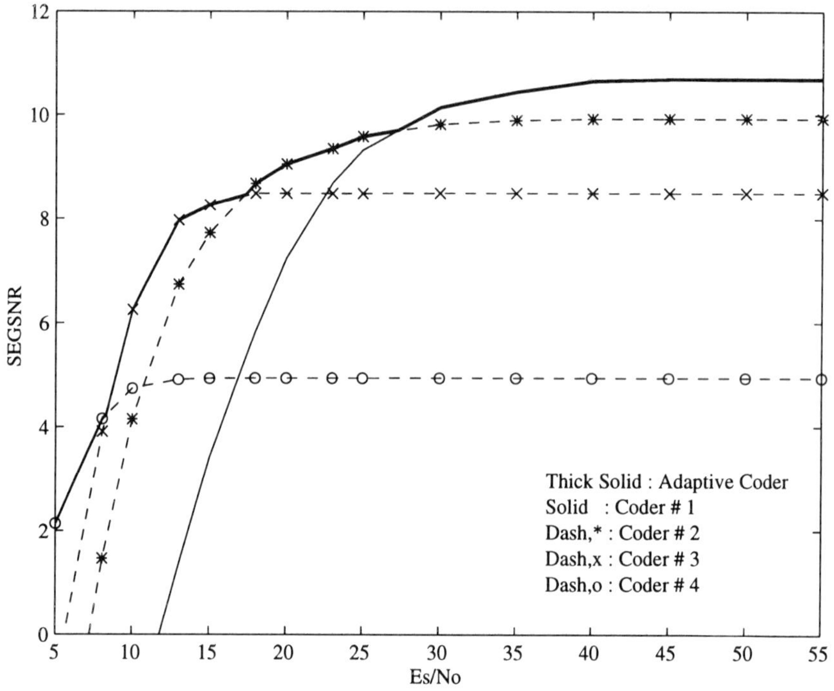

Figure 9 Performance of combined multimode speech and RCPC coders for combined rate = 12.8 kbit/s

Figure 9 shows the performance of the coders described in Table 4 plotted versus the channel symbol SNR, E_s/N_o. The fixed rate coders show significant degradation in performance if the channel SNR is mismatched with respect to the nominal (design) channel SNR. Particularly, the systems with little error protection perform badly at low channel SNRs, while the systems where a large rate was allocated for error protection show degraded performance at high channel SNR. The adaptive variable-rate system performance follows the "envelope" given by the best fixed-rate systems at each channel SNR showing significant improvement over the fixed-rate coders. Improvements of up to 1.2 dB in segmental SNR (SEGSNR) of the speech signal and up to 0.8 on the Mean Opinion Score (MOS) scale in informal subjective quality for a combined rate of 12.8 kbit/s were found. The adaptive channel-controlled variable rate system achieved a MOS of 3.7 on a fading channel while the fixed-rate system showed MOS scores lower than 3.0 [20].

5 VARIABLE RATE SPEECH CODING FOR ATM NETWORKS

Much of the current variable rate speech coding research effort is being directed towards applications for Asynchronous Transfer Mode (ATM) networks. ATM networks have been proposed in order to provide a common format for both bursty high-speed data and real-time traffic such as voice and video. To accomplish this, ATM uses short fixed-length packets called cells. ATM cells are 53 bytes long, including a 5 byte header and 48 byte information field. The use of short cells is a compromise between the demands of non real-time data and real-time voice and video. Short cells mean larger overhead, but ensure that time sensitive data will not have to wait long for completion of a cell transmission already in progress before gaining access to the channel.

The nature of the ATM cell format imposes some restrictions on the way in which variable rate speech coding can be used to increase channel capacity. Bits cannot simply be packaged into variable sized cells. Instead, variable rate coders must vary the number of cells transmitted for a given speech interval, or vary the size of the interval encoded in a single cell. For example, a coder operating at rates of 48 kb/s and 16 kb/s might use cells which encode 8 ms and 24 ms of speech respectively. An alternative would be to always encode 24 ms of speech, but use 3 cells for the high rate and 1 cell for the low rate.

There is an important advantage in using multiple cells. Through hierarchical packaging of the bits (eg. putting MSBs in one cell and LSBs in another cell) priorities can be attached to the cells which can be used by the ATM network to control the traffic flow. Low priority cells can be dropped when the network becomes congested without severely reducing the received speech quality. The technique by which the ATM network drops cells is known as cell discarding.

Cell discarding in ATM is an important congestion control technique allowing the network to operate effectively at higher loads. However, the largest gain in capacity will come from voice compression methods and DSI. For example, Sriram et al [21] report that at the 45 MB/s and 150 MB/s ATM link speeds, the use of DSI combined with 32 kb/s ADPCM achieves better than 4:1 compression over 64 kb/s PCM, and cell discarding further increases capacity between 10% and 15%.

One example of a variable rate coding scheme designed specifically for ATM has been presented by Kondo and Ohno [22]. The proposed codec classifies speech as either silence, unvoiced, or voiced. Voiced speech is classified using an embedded ADPCM encoder operating at 48 kbps (6 bits per sample) maximum rate with an embedded 24 kbps (3 bits per sample) rate. One ATM cell is used to encode the 3 MSBs for a block of samples and is tagged as a high priority cell. A second cell is used for the LSBs and is tagged as low priority. Unvoiced speech is similarly packaged except that a 32 kbps ADPCM encoder with an embedded 16 kbps rate is used. Because the MSBs and LSBs for this mode are 2 bits per sample each, longer speech blocks are used to fill the fixed length ATM cell. Silent frames are encoded at 16 kbps using a single cell consisting of the MSBs of the 32 kbps encoder; the LSB cell is not transmitted.

As the likelihood of ATM becoming a dominant technology in digital communications continues to grow, so does the demand for better ways to encode speech using this fixed length packet technology. The main advantage of ATM – its flexibility in handling diverse traffic such as voice, video, and data – was achieved by compromising on the ideal cell length for any single traffic type. Speech coding systems designed specifically for ATM will have to deal with this compromise. For example, achieving higher compression through rate reduction is not straight forward. As the rate decreases, longer encoding frames must be used to fill the fixed length ATM packets. This has implications for both delay and recovery from cell loss due to corrupted headers or buffer overflow. However the fixed length packets and virtual channels associated with ATM make variable rate compression attractive. In short, the popularity of ATM has opened many new research opportunities in the field of variable rate speech coding.

REFERENCES

[1] A. Gersho and E. Paksoy, "An overview of variable rate speech coding for cellular networks," in *Proc. of the Int. Conf. On Selected Topics in Wireless Communications*, (Vancouver, B.C., Canada), 1992.

[2] P. T. Brady, "A technique for investigating on-off patterns of speech," *Bell Systems Technical Journal*, vol. 44, pp. 1–22, 1965.

[3] Y. Yatsuzuka, "High-gain digital speech interpolation with adaptive differential PCM encoding," *IEEE Transactions on Communications*, pp. 750–761, April 1982.

[4] K. Bullington and J. M. Fraser, "Engineering aspect of TASI," *Bell Systems Technical Journal*, vol. 38, pp. 353–364, 1959.

[5] D. J. Goodman, "Embedded DPCM for variable rate transmission," in *Procedings of the ICC*, pp. 42.2–1–42.2–5, 1980.

[6] R. D. De Iacovo and D. Sereno, "Embedded CELP coding for variable bit-rate between 6.4 and 0.6 kbit/s," in *Proc. IEEE Int. Conf. Acoust., Speech, Signal Processing*, (Toronto, Canada), pp. 681–683, May 1991.

[7] P. Jacobs and W. Gardner, "QCELP: A variable rate speech coder for CDMA digital cellular systems," in *Speech and Audio Coding for Wireless and Network Applications* (B. S. Atal, V. Cuperman, and A. Gersho, eds.), Kluwer Academic Publishers, 1993.

[8] S. Kay, "Extended-TDMA: A high capacity evolution of us digital cellular," in *International Conference on Universal Personal Communications*, (Dallas), pp. 07.04/1–3, September 1992.

[9] D. J. Goodman, "Cellular packet communications," *IEEE Transactions on Communications*, vol. 38, pp. 1272–1280, August 1990.

[10] J. Dubnowski and R. Crochiere, "Variable rate coding of speech," *Bell Systems Technical Journal*, vol. 58, pp. 577–600, March 1979.

[11] K. Y. Kou, J. O'Neal, Jr., and A. Nilsson, "Digital speech interpolation for variable rate coders with application to subband coding," *IEEE Transactions on Communications*, vol. 33, pp. 1100–1108, November ???? 1985.

[12] H. Nakada and K. I. Sato, "Variable rate speech coding for asynchronous transfer mode," *IEEE Transactions on Communications*, vol. 38, pp. 277–284, March 1990.

[13] Y. Yatsuzuka, S. Lizuka, and T. Yamazaki, "A variable rate coding by APC with maximum likelihood quantization from 4.8 kbit/s to 16 kbit/s," in *Proc. IEEE Int. Conf. Acoust., Speech, Signal Processing*, pp. 3071–3074, April 1986.

[14] E. Paksoy, K. Srinivasan, and A. Gersho, "Variable rate CELP coding of speech with phonetic classification," *European Transactions on Telecommunications*, September 1984.

[15] P. Lupini, N. B. Cox, and V. Cuperman, "A multi-mode variable rate CELP coder based on frame classification," in *Proc. International Conference on Communications*, (Geneva), 1993.

[16] M. Schroeder and B.Atal, "Code-excited linear prediction (CELP): High-quality speech at very low bit rates," in *Proc. ICASSP*, pp. 937–940, 1985.

[17] J. Chen and A. Gersho, "Real-time vector APC speech coding at 4800 bps with adaptive postfiltering," in *Proc. ICASSP*, pp. 2185–2188, 1987.

[18] I. Gerson and M. Jasiuk, "Vector sum excited linear prediction (VSELP) speech coding at 8 kbps," in *Proc. ICASSP*, pp. 461–464, 1990.

[19] D. Goodman and C. E. Sundberg, "Combined source and channel coding for variable-bit-rate speech transmission," *Bell System Technical Journal*, pp. 2017–2036, Sept. 1983.

[20] E. Yuen, P. Ho, and V. Cuperman, "Variable rate speech and channel coding for mobile communications," in *Proc. 43rd IEEE/VTS Vehicular Technology Conference*, 1994.

[21] K. Sriram, R. S. McKinney, and M. H. Sherif, "Voice packetization and compression in broadband ATM networks," *IEEE Journal on Selected Areas in Communications*, vol. 9, pp. 294–304, April 1991.

[22] K. Kondo and M. Ohno, "Variable embedded ADPCM coding scheme for packet speech on ATM networks," in *Proc. IEEE Globecomm*, (San Diego, CA), pp. 523–527, Dec 1990.

PART 2

SPEECH RECOGNITION

6

WORD SPOTTING

Jan Robin Rohlicek

<rohlicek@bbn.com>

BBN HARK Systems Corporation
A subsidiary of Bolt Beranek and Newman Inc.
Cambridge, Massachusetts 02138

ABSTRACT

Word spotting has been an active area of speech recognition for over twenty years. Although it initially addressed applications requiring the scanning of audio data for occurrences of particular keywords, the technology has become an effective approach to speech recognition for a wide range of applications. The term "word spotting" is now used to refer to a variety of techniques that are useful in speech recognition applications where relevant information, such as a command, must be recognized even when it is embedded in irrelevant speech input or other audio interference, or when the desired information may not be present. The related areas of filler modeling and out-of-set rejection share many of the same underlying technical problems and approaches to word spotting. Depending on the particular application, different types and combinations of word spotting techniques are appropriate and effective. Most recently, a variety of statistical modeling techniques have provided higher accuracy than previous approaches. Many of these techniques share aspects, such as use of hidden Markov models (HMMs) and statistical language models, with other areas of speech recognition. This chapter presents a survey of various approaches to word spotting and related areas, suggests appropriate applications of these approaches, and identifies unresolved research problems.

1 INTRODUCTION

In this chapter, we discuss several areas of speech recognition that are related to *word spotting*. Typical speech recognition approaches have addressed the problem of selection from an enumerable set of hypotheses, possible word or phoneme sequences, that best accounts for an input utterance. Several speech recognition problems are not of this type however, including the following:

- *Keyword and Phrase Detection:*

  Detect occurrences of a particular keyword or phrase in audio data possibly containing the keyword or phrase. If possible, provide a score that characterizes the certainty that the keyword actually occurred. An application of this is in automatic sorting of audio data by its topic based on the keyword content [31, 39, 22], for example, selecting radio news stories that contain the words "health care" in them. Another application is in *open microphone* systems where the recognizer is always listening, waiting for the user to say a known utterance or a specific "trigger word" such as "Computer, turn on the lights." In such an open-microphone system, extraneous speech not preceded by the word "Computer" must be ignored.

- *Embedded Keyword or Phrase Recognition:*

  Select from a set of keywords or phrases the most likely to have been contained, along with other words or sounds, in a particular utterance. This problem is like typical speech recognition but the additional "garbage" around the words of interest must be ignored. An application is selection from a set of city names contained in an utterance in response to "What city would you like to travel to?" A typical response might be "I'd like to fly to Newark New Jersey please." Another fielded application is in automated telephone operator services where utterances such as the following are processed: "I'd like to make a collect call please" [42].

- *Out-of-Set Rejection:*

  Determine whether or not an utterance corresponds to one of a specified set of word sequences. For example, in the recognition system expecting to select from a set of city names, when the the user says "I don't know" the system should not make a forced choice of one of the cities but rather should determine that none was present in the utterance.

Although the work on these problems has progressed somewhat independently, the current solutions are very similar. Their key common characteristic is that they involve not only a selection of the best matching word sequence but also an element of detection of the presence of particular word sequences or the determination of their location. State-of-the-art approaches share many of the statistical modeling techniques used in automatic transcription, including Hidden Markov Models (HMM), and statistical language (word sequence) modeling. In addition, a variety of techniques have been developed to address specific aspects of word spotting.

In the next section, each of these problems and their interrelationships are described more fully. Section 3 contains a survey of published approaches to word spotting. Section 4 provides a discussion of the available approaches and Section 5 contains a discussion of several open issues that continue to be areas of research.

2 WORD SPOTTING PROBLEMS

In this section, we describe more fully some of the various problems outlined in the introduction. The focus is on the relationship of these problems and on how these problems differ from typical speech recognition problems. In Sections 3 and 4, specific approaches to address these problems are discussed.

2.1 Speech Recognition

Before describing word-spotting related problems, we consider the problem addressed by typical speech recognition systems. The characteristics of this problem, which can be termed "closed-set transcription", are as follows:

- A complete vocabulary of allowable words is known and models for the words are predetermined. The set of "words" may be augmented by models for silence and other speech sounds (such as a cough, or pause fillers "um" and "ah").
- An input utterance is assumed to come from a possibly constrained sequence of words and modeled sounds. For example, in isolated word recognition, the utterance is assumed to begin and end with silence and contain exactly one of the words in the vocabulary. In continuous speech recognition, the word sequence is assumed to be a string of words from the vocabulary.
- The goal is to select the best matching word sequence to the entire input utterance.

Because of the closed vocabulary, one can, in principle, enumerate the set of hypotheses, each of which is a finite sequence of words from the vocabulary. Recognition corresponds to a classification problem where we select the hypothesis $\hat{H}$ with score $S(\hat{H})$ such that $S(\hat{H}) \geq S(H)$ for all H. In statistical speech recognition, the score used is the probability $S(H) = \mathrm{P}(H, X)$, where X is a representation of the input utterance, and $\hat{H}$ is the maximum *a posteriori* (MAP) estimate of the utterance spoken.

The performance of this type of system can be characterized by the probability, P_c, of selecting the correct hypothesis. In recognition of multi-word sequences, an average error rate taking account of not only errors in individual words but also deletions of words and insertions of extra words can be computed in addition to overall utterance accuracy.

2.2 Keyword and Phrase Detection

In keyword spotting, the problem is one of detecting occurrences from a particular set of words or phrases. Detection of the phrase "health care" in news broadcasts would be a typical use of a type of word spotting algorithm. This is in contrast to the speech recognition problem described in Section 2.1 where the entire utterance is accounted for by a sequence of known words. In keyword detection, we attempt to detect the presence of the keyword without necessarily explicitly or completely accounting for the rest of the input. The problem is therefore one of detection rather than simply classification.

Receiver Operating Characteristic

In word spotting, as in any detection problem, there is a trade-off between the ability to detect true occurrences of a keyword and generation of false acceptances (false alarms) when the keyword is not present. The declarative by the detector of the presence of a keyword is often referred to as a "putative" keyword event. In a detection problem where there are distinct trials that must be classified as containing a particular event or not, the performance of the system can be characterized by the probability of correctly detecting the event of interest in a trial, P_d, versus the probability of incorrectly declaring the event when it was not truly there, P_f. Often, a system is designed to select explicitly the tradeoff between these two probabilities. The curve tracing out this tradeoff is known as the Receiver Operating Characteristic (ROC) curve as shown in Figure 1. This concept was introduced in signal detection theory [27] and is discussed in any current text in the area. A scalar measure of the quality

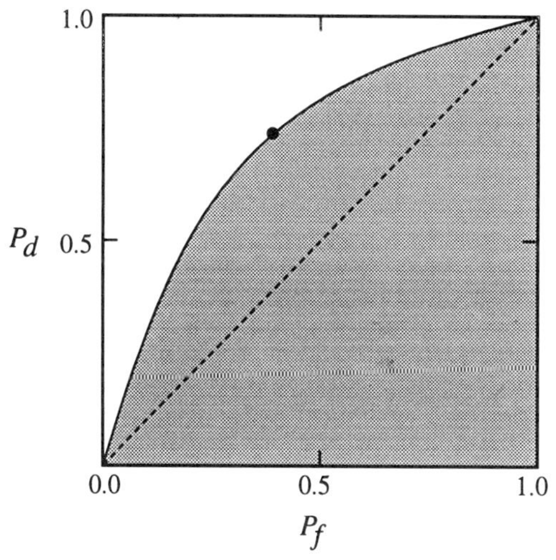

Figure 1 The Receiver Operating Characteristic (ROC) curve traces a set of possible operating points for the detector. The point (0.0, 0.0) corresponds to a detector that never declares that the event is present and therefore creates a zero probability of either detection or false acceptance. The point (1.0, 1.0) corresponds to always declaring that the event is present.

of a detector is the area under the curve. For reference, a detector that ignores its input and randomly selects a fraction θ of the trials has an operating point $P_d = P_f = \theta$ and therefore has an ROC curve that is a straight line $P_d = P_f$.

A traditional ROC curve would appropriately characterize a system where there are separate intervals in the data for which we are trying to determine whether a particular keyword was present. For example, in an air-traffic-control monitoring application where individual radio transmissions containing the word "cleared" are to be detected, the ROC curve characterizes the behavior of the system.

In the case where there are not distinct intervals to be considered, the concept of an ROC curve is modified to replace the probability of false acceptance with a *rate* of false acceptance, λ_f (false acceptances per unit time). In order to compare ROC curves, an average detection probability over a range of false acceptance rates can be computed. This metric has been called the "Figure of Merit" in recent publications and has been used as a means of comparing various word spotter configurations.

Keyword Scoring and Sorting

One type of keyword detector is based on first computing a score that the keyword is present in a particular interval. The detector declares that the keyword is present if the score is greater than a particular threshold. By varying this threshold, various operating points can be selected. If the score is the likelihood ratio

$$S = \frac{\mathrm{P}(X \mid \text{keyword present})}{\mathrm{P}(X \mid \text{keyword absent})}$$

then thresholding this score results in the binary decision of whether the keyword was present or not.

In some applications, explicit detection is not necessary; the problem is one of sorting data so that data containing the keyword is more likely to be at the head of the list. For these applications, selection of the threshold is not necessary. An experimental ROC curve can be computed by considering the achieved P_d and P_f or λ_f for all possible threshold values.

Detection of Multiple Keywords

When the goal is to detect any of a set of M keywords, one approach is to construct M independent detectors. With such an approach, for any trial, multiple keywords may be declared as being present. In many applications, we may have *a priori* knowledge regarding how many keywords can be present; a simple case is when we know that there is at most one keyword.

Suppose we have M detectors each set to the same P_d with false acceptance probabilities $P_f^{(n)}$. The overall P_d does not depend of M. However, the expected number of false acceptances grows with M as does the probability that there are one or more false acceptances

$$\mathrm{E}N_f = \sum_m P_f^{(m)}, \; P_f = 1 - \prod_m \left(1 - P_f^{(m)}\right).$$

For large M, this growth in false acceptance probability is not acceptable. This is particularly true when we are detecting phrases where M grows combinatorially, for example if we are detecting the presence of any dollar amount under \$100 such as "five dollars and ten cents."

The general approach to this problem is to limit the number of putative events in some way. To understand the effect of this approach consider the ROC curve shown in Figure 2. Use of M detectors independently does not change

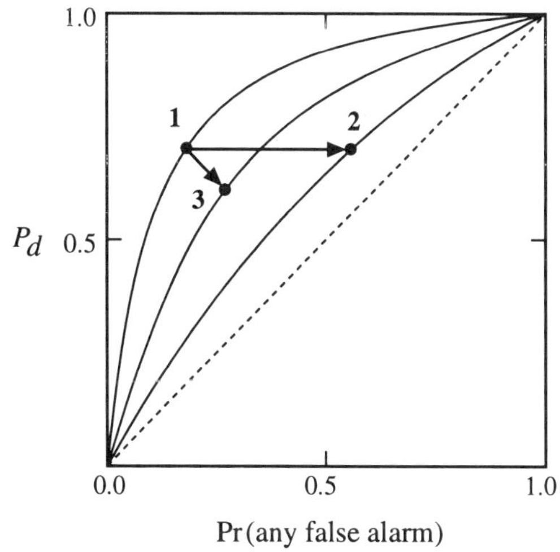

Figure 2 Effect of the detection of multiple keywords. Curve 1 is the ROC curve of each detector. Curve 2 illustrates the effect of using M independent detectors. Curve 3 illustrates the effect of selecting a subset of putative events. The arrow indicates the change in a particular operating point as multiple detectors are used. Note that when a subset of putative events are reported, there is less increase in probability of false acceptance at the expense of some reduction in detection probability.

P_d, but P_f grows. If on the other hand, the set of putative events is restricted in some way, the number of false acceptances should be reduced. However, this is generally only at the expense of lower detection probability. Various strategies for censoring the list are available, including keeping only a single putative event and keeping only the best of temporarily overlapping events. The appropriate choice of approach depends on the nature of the application including prior information about the number of true events expected, and the postprocessing to be performed on the putative event set.

Another choice for restricting the putative event list is to declare either no event was present or the single best event. This can be addressed by a combination of embedded recognition and out-of-set rejection, each of which is described in the next sections.

2.3 Embedded Keyword or Phrase Recognition

Many applications require an approach which is a combination of speech recognition and keyword detection. They correspond to utterance classification problems for which the criterion is the presence of a particular keyword or phrase. We call these embedded recognition problems. For example, we may want to classify utterances according to the most likely name present in the utterance: "I need information about President Clinton." In addition to detecting each possible name, the best matching name must be selected. The assumption that exactly one name is present essentially limits the false acceptances to those utterances in which the correct name is not the best scoring hypothesized event.

2.4 Out-of-Set Rejection

In both complete transcription and embedded keyword recognition, rather than classification as one of the possible word sequences or embedded keywords, we may also want to reject utterances that do not come from the expected set. Rejection of complete utterances is equivalent to not detecting a valid phrase with the added assumption that the phrase accounts for the whole utterance. Note that rejection encounters the same difficulty as multiple keyword or phrase detection — as the number of alternatives grows, P_f will by necessity increase for a fixed P_d.

Rejection for embedded keyword recognition is equivalent to assuming that either zero or one keyword is present. In addition to selecting the highest scoring keyword, a decision of whether to accept or reject the word must be made based on the predetermined operating point on the ROC curve.

2.5 Rejection and Accuracy Tradeoff

In any recognition system with rejection, there are three types of error that can be made. In general it is possible to trade of some types of errors for other types. The types of errors are:

- *Substitution:* The utterance is valid, but an incorrect hypothesis is declared. This is either a substitution error, or a combination of a miss and a false acceptance in the keyword detection problem.

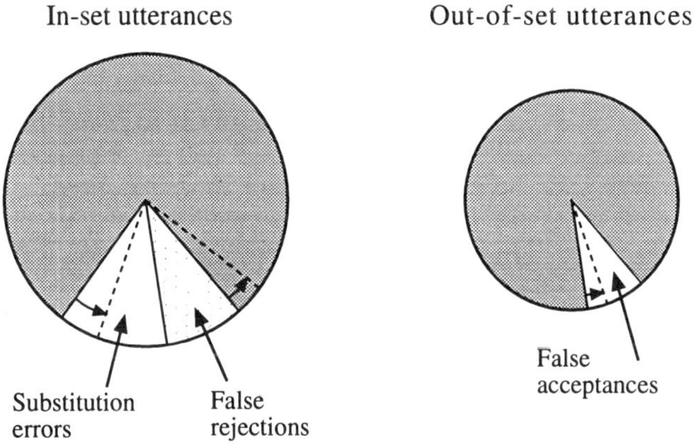

Figure 3 Tradeoff between false acceptance, substitution, and false rejection. The two charts represent the in-set and out-of set utterances respectively. Errors for in-set (valid) utterances are either substitions or false rejections. For out-of-set (invalid) utterances, errors are false acceptances. The dotted lines show the general tradeoff where substitution and false acceptance errors can be reduced at the expense of more false rejections.

- *False Rejection:* The utterance is valid but no hypothesis is declared.
- *False Acceptance:* The utterance is not valid but an incorrect hypothesis is declared.

In various applications, these types of errors may have very different costs and the appropriate balance can be selected in the same way that an operating point is selected in the simple detection problem.

Two aspects of this tradeoff should be considered. In general the probability of substitution errors can be reduced at the expense of an increase in false rejection. Second, in many systems, the same mechanism may be used to control rejection of valid utterances which are likely to result in substitution errors and rejection of utterances that are not valid. Therefore, in general, it is difficult to reduce the number of false acceptance errors without also increasing the number of false rejection errors. This tradeoff is illustrated in Figure 3.

3 SURVEY OF APPROACHES

Word spotting approaches can be divided into those which explicitly attempt to model, or otherwise account for, all the input data in which the keyword may be embedded and those that do not. The earliest approaches, as well as some recent ones are of the latter type. Approaches that only model the keywords typically amount to variations of "sliding" a model across the data to detect well-matching regions. Other techniques use explicit models of all the data to make a decision based in one way or another on the relative match of a keyword to the model of the "filler."

In the remainder of this section, these two types of approaches are discussed. In each case, alternative algorithms that have been published are presented. Section 4 contains further discussion of some of these techniques.

3.1 Sliding Model Methods

The earliest methods of modeling keywords for speech recognition, as well as for word spotting, were based on template matching. A reference example of a keyword is extracted from training data, and is matched against an unknown utterance. To allow for the natural time variability in speech, various time warping algorithms were devised. The most successful of these algorithms are based on dynamic programming, also known as "dynamic time warping" (DTW). Variants of these DTW techniques use multiple reference examples or "averages" of multiple examples.

In this discussion of sliding model methods, we first discuss DTW-based approaches. Then an HMM-based approach, very much in the spirit of the DTW algorithms, is described. Finally, recent approaches based on Neural Network (NN) techniques are described.

Dynamic Time Warping

Early speech recognition systems approached the problem in two steps: locating the endpoints of a word followed by classifying the word. Taking into account the possible non-uniform time warping between occurrences of words in a training set and the test data was addressed in a computationally efficient manner using dynamic programming [33]. The difficulty with this speech recognition approach is that it is not always possible to locate the endpoints of a word, for example when an isolated word is spoken in high-noise environments or when the word is embedded in a continuously spoken utterance. Although it is possible to locate manually the endpoints in training data, the same is not possible for the test data in an automated system.

In [6] a computationally efficient dynamic programming approach is used to address the situation when the endpoints of a keyword are not know. At any time t_e, the match score of the best warp and start time t_s of the reference template is computed. To deal with the length-dependence of the score resulting from diffent possible start times, a "decay factor", $0 < \gamma < 1$, is introduced. Specifically, a dynamic programming lattice $L(t, j)$ for input time t and template reference time j is maintained using the iteration

$$L(t,j) = \begin{cases} S(x[t], r[1]) & \text{if } j = 1 \\ \max\limits_{(\alpha,\beta)} (1-\gamma)L(t-\alpha, j-\beta) + \gamma K(\alpha,\beta)\, S(x[t], r[j]) & \text{otherwise} \end{cases}$$

where (α, β) are in the set $\{(1,1),(0,1),(1,0)\}$ and $S(x, r)$ is the match score between input x and reference template sample r. The warping penalty $K(\alpha, \beta)$ chosen was

$$K(\alpha,\beta) = \begin{cases} 1 & \text{if } (\alpha,\beta) = (1,1) \\ 0 \le k \le 1 & \text{otherwise} \end{cases}$$

A keyword ending at time t is associated with a high value of $L(t, J)$, where J is the length of the reference template. This algorithm was used in subsequent work [9, 10] in which multiple reference templates for a particular word were combined in the search.

Other DTW-based work focused on the problem of uncertainty in the endpoints of a test utterance. In [25] various approaches for introducing a fixed tolerance (time window) for possible start and end times are considered. This approach to automatic endpoint determination is used for word spotting by periodically restarting the dynamic programming algorithm with a sequence of overlapping start-time windows. Keyword detection is again based on thresholding a time-warp-normalized match score.

Hidden Markov Models

In [41], an HMM technique for word spotting addressing a problem of embedded keyword recognition with rejection is presented. The dynamic programming computations described above (e.g., [25]) are replaced with a likelihood computation on subintervals in the data. For all ending times j, the average observation likelihood along the most likely state sequence, the Viterbi path, through the keyword model for input from time i to j is computed in one of two ways: weighting each time sample equally and weighting the time interval in each model state equally. A secondary classification scheme is used to reject keyword hypotheses; if multiple keywords survive, then the one with the highest state-weighted average observation likelihood is selected.

Neural Networks

Neural networks have been used in two basic modes for word spotting. The simplest involves sliding a window across the input data and computing a fixed-length representation of that window which is fed into a decision or scoring network. An example of this is found in [12] where a neural network is trained to detect the presence of entire words. The second mode for word spotting involves sliding model techniques where the time warping of a model (network) to the data is explicitly performed using a dynamic programming approach. Two such approaches are described below. Finally, another mode of using NN for word spotting is in a secondary processing where a primary wordspotter not only hypothesizes the possible locations of keywords, but also provides a time alignment that is used to construct a fixed-length representation. These approaches are discussed is Section 3.3.

Although Time-Delay Neural Networks (TDNNs) have been used in various speech recognition tasks where the endpoints of the word of interest are known (e.g., [37]), use of a TDNN alone has not been successful in word spotting where the word is embedded in continuous speech. An approach presented in [44, 45] takes advantage of a sliding window to compute matches to particular pieces of a keyword model, and uses an explicit Dynamic Programming stage (also thought of as a series of TDNN states) to compute the best overall match of the input data to the keyword. A similarly motivated approach is presented in [21] where Radial Basis Functions (RBFs) are used to compute state scores that are then processed in a DP stage. In both these cases, although the model is trained using examples of the target keywords, the nature of the discriminative training procedure uses speech intervals not containing the keywords in order to provide the "negative examples" needed. Because of the discriminative nature of the NN, all the speech is implicitly accounted for. In the next section, we continue with systems where there is an explicit model for the non-keyword speech.

3.2 Modeling of Complete Input

A critical component of current word spotters is modeling of the complete input rather than simply modeling the input when the keyword is present. The utility of this type of explicit modeling of an entire utterance rather than just the portion of interest was shown in problems of automatic endpointing of isolated words in a "silence" background [7]. Explicit "silence" templates are used to determine optimal endpoints to address deficiencies of energy-based techniques in high-noise environments and when speech begins with a weak sound. The techniques for automatic endpointing in silence progressed to include techniques for automatic endpointing in the presence of speech rather than simply background noise [41]. Other early word spotting techniques approached the problem directly [2, 16] and also saw the need for modeling the complete signal.

In the remainder of this section, we first describe some early DTW-based approaches and then describe several HMM techniques that have been reported.

Dynamic Time Warping

A deficiency of the direct application of DTW methods for forced-choice recognition to the word spotting problem is that the scores obtained for an instance of a keyword is not directly comparable to other instances. A poorly matching speaker is not distinguishable from a poorly-matching word. In [16, 5] an innovative DTW-based word spotting system is described. Two relevant characteristics of the system were:

1. A continuous speech recognition (CSR) approach was used in which explicit "filler" templates are used to account for non-keyword speech
2. An approximation to a likelihood ratio test is used to compare the "probability" of a keyword generating a sequence of input samples versus the filler model generating that data.

An interesting aspect of the set of fillers used in that work is that they were chosen to be pieces of the keywords. Other techniques, such as using representative pieces of general speech, provided inferior results. For many years, this word spotting approach represented the state of the art in word spotting algorithms tailored for continuous monitoring applications.

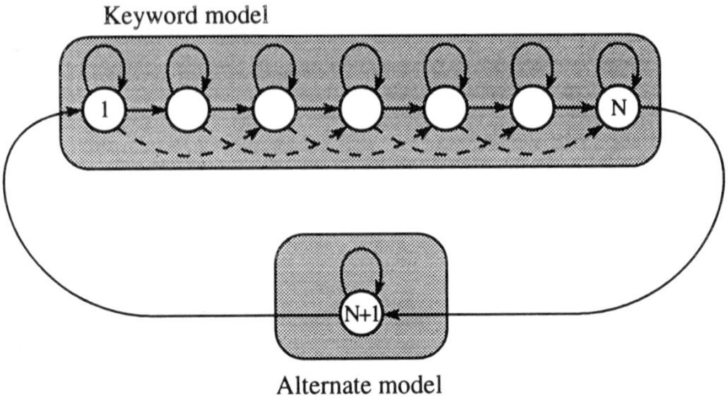

Figure 4 The HMM word spotting network used in [2]. The dotted transitions between keyword states are optional. The alternate model, referred to as the "Null State" is used to account for all non-keyword speech.

Hidden Markov Models

Hidden Markov models (HMM) have, in the past five years, gained acceptance as the preferred approach to word spotting. It is interesting that almost twenty years ago, a conceptually equivalent HMM approach had been developed [2]. In that work, a model of a "left-to-right" HMM is constructed for a keyword. In addition, a single state is inserted to account for all non-keyword speech. The entire HMM network is constructed in a form equivalent to that shown in Figure 4. Word spotting is then reduced to detection of passage though the sequence of keyword states.

Two decoding algorithms were presented in [2]. The first, corresponding to the CSR approach, uses the Viterbi algorithm (Dynamic Programming) [11] to find the most likely state sequence given an input sequence. The Viterbi algorithm amounts to computing the iteration

$$L(t, j) = \max_i \ L(t-1, j)\, a_{ij}\, b_j(x[i])$$

where $b_j(x)$ is the probability of observing input x at state j and a_{ij} is the transition probability from state i to state j taking on non-zero values for the arcs indicated in the figure. The most likely sequence of states is recovered by recording the values of i that maximize the terms at each time. Keywords are declared to have occurred whenever the most likely state trajectory for the entire utterance passes though the keyword.

The second approach uses the Forward-Backward algorithm [3]. Given the entire input sequence, the probability of occupying any state at each time can be computed. Following the presentation in [3] two iterations are computed

$$\begin{aligned}
\alpha(0,j) &= \begin{cases} 1 & j = N+1 \\ 0 & \text{otherwise} \end{cases} \\
\alpha(t,j) &= \sum_i \alpha(t-1,i)\, a_{ij}\, b_j(x[t])\,, \quad t = 1,2,\ldots,T \\
\beta(T,i) &= 1 \\
\beta(t,i) &= \sum_j a_{ij} b_j(x[t])\beta(t+1,j)\,, \quad t = T-1,\ldots,1
\end{aligned}$$

Based on the results of the iterations, one can compute

$$\gamma(t,i) \equiv \mathrm{P}\left(s[t]{=}i, x[1],\ldots,x[T]\right) = \alpha(t,i)\,\beta(t,i)$$

where $s[t]$ is the state at time t. The probability of occupying any state i at time t can then be computed as

$$\tilde{\gamma}(t,i) = \frac{\gamma(t,i)}{\sum_j \gamma(t,j)}\,.$$

The probability of occupying the keyword can be computed as

$$1 - \tilde{\gamma}(t, N+1) = \sum_{j=1}^{N} \tilde{\gamma}(t,j)\,.$$

This probability is compared to a threshold. Note that both these approaches require data past the end of the keyword to be known.

In [28], an independently developed HMM system, conceptually very similar to the earlier work [2], and incorporating a more explicit model of non-keyword speech was presented. As in [2], each keyword was independently trained based on excised examples of the keywords. In order to provide results in a "causal" manner, only a forward iteration was used to compute the probability of occupying each state at any time without considering future data. Specifically, the posterior probability of occupying the last state of keyword k, e_k, at time t given only data up to time t was computed as

$$\mathrm{P}\left(s[t]{=}e_k)|x[1],\ldots,x[t]\right) = \frac{\alpha(t,e_k)}{\sum_j \alpha(t,j)}.$$

Various configurations of filler (or "alternate" model) were considered. The simplest configuration consisted essentially of a single state with a mixture

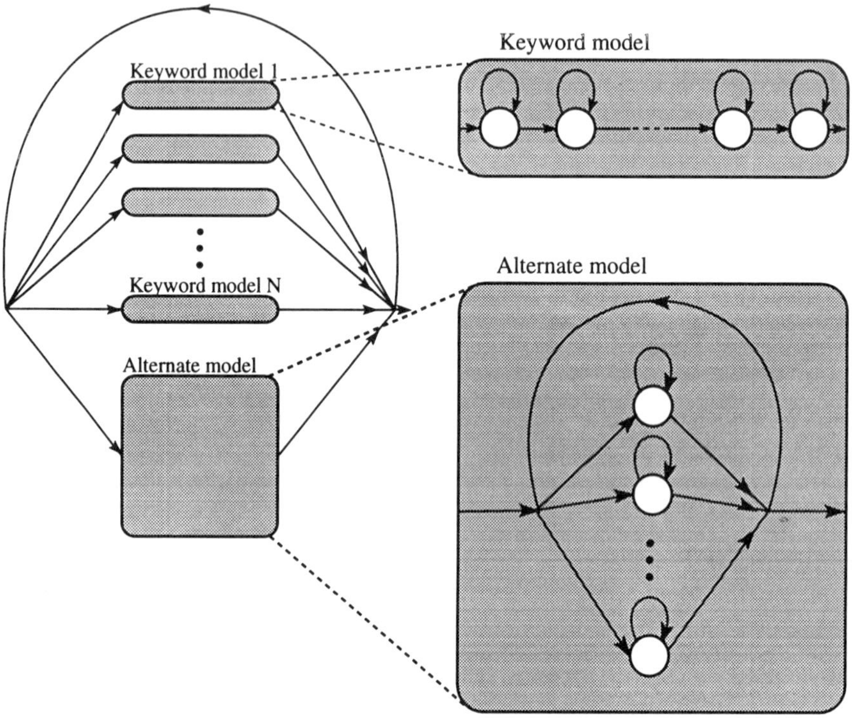

Figure 5 The word spotting network used in [28] consists of a parallel arrangement of N keywords and an alternate model. The keyword models are simple left-to-right networks. The alternate model consists of a parallel arrangement of states, one for each state in each keyword using that state's observation distributions and self-transition probabilities.

distribution composed of components corresponding to all other state distributions. Performance was greatly improved when the filler model consisted of a parallel loop of copies of all the states of the keywords, as shown in Figure 5.

Other HMM-based word spotting evolved from techniques for automatic endpointing [40]. In [42] the silence state used in endpoint detection is replaced by a "garbage" state whose output distribution is determined from the typical filler occurring before or after the keyword. An HMM network (shown in Figure 6) is used in a forced-choice recognition mode. Various filler models were considered including a parallel connection of the most common words, a single pooled "other" word, and unsupervised clustering of the other words.

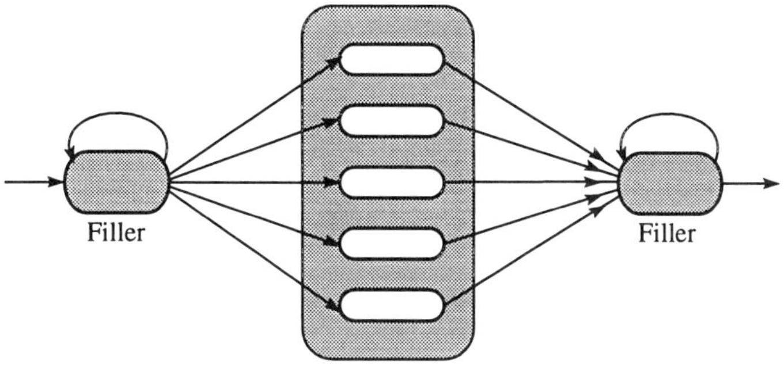

Figure 6 An HMM network equivalent to that reported in [40]. Silence and speech filler models are present before or after the keyword. Exactly one keyword is assumed to occur between the fillers. The highest scoring state sequence determines the keyword.

Other HMM approaches have adapted techniques from large-vocabulary continuous speech recognition (CSR) to word spotting, much in the way that DTW-based CSR approaches were used for word spotting [16]. Phonetically based training for the keyword models and a filler model composed of context-independent phoneme models were used in a Viterbi recognition mode [30]. To provide a score for the keywords detected along the Viterbi path, duration normalized observation likelihoods were used. In [32], discriminative training was used to improve performance. In [29, 18, 38], phonetic training, a phonetically based filler model, and use of a moderate and large number of words in the filler model, and use of a statistical language model were considered. A combination of Forward-Backward scoring, as in [2, 28], and phonetic training is presented in [29]. Performance with this scoring approach is somewhat higher than with the Viterbi approach. More important, this scoring approach allows explicit threshold setting for each keyword. It is unlike the Viterbi approach, where the ROC operating point is determined implicitly by the transition probabilities in the HMM. Explicit scoring based on a duration-normalized observation likelihood computed for each keyword occurrence is addressed in [38]. Statistical language modeling was introduced into word spotting approaches using explicit scoring [29] as well as other systems based on the Viterbi approach [38].

Including explicit models for non-keyword vocabulary improves performance relative to a more general filler loop. However, it appears that a moderate number of words that are either confusable with or that occur in the vicinity of keywords are most important; little gain is obtained by explicitly modeling

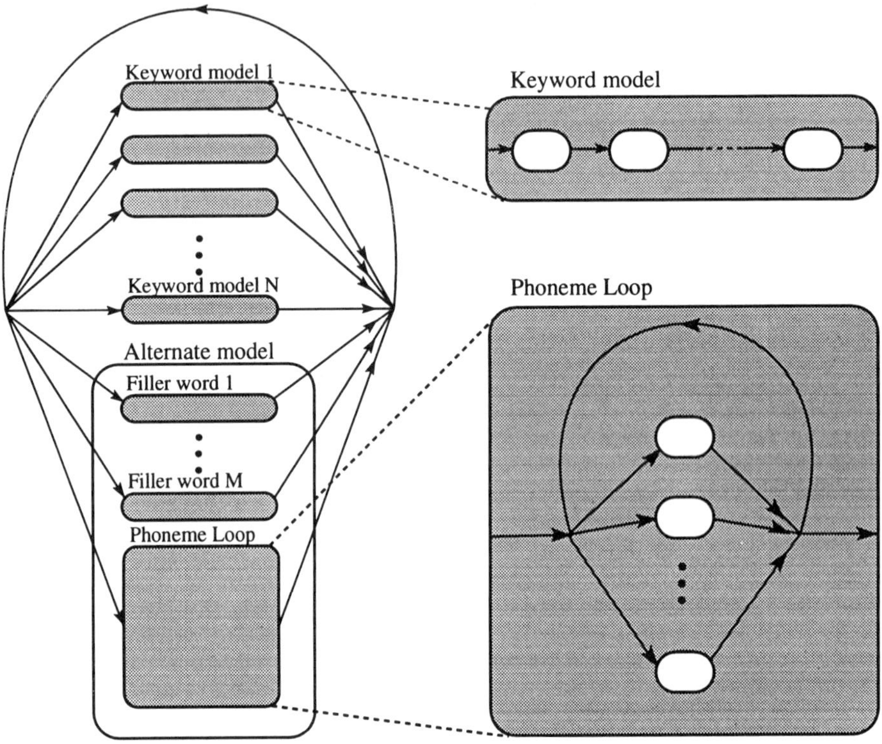

Figure 7 A large-vocabulary word spotting network presented in [29]. A parallel combination of a set of keywords and an alternate model are used. The keyword models are composed of linear sequences of context-dependent phonetic models. The alternate model is composed of a parallel set of non-keyword models and a context-independent phoneme loop as shown in the figure.

a much larger fraction of the vocabulary. The HMM network considered in [29] are shown in Figure 7.

Phrase Spotting

Detection of structured phrases is a relatively recent application of word spotting techniques. Although a small set of phrases can be enumerated and treated like a set of keywords, a structured phrase with many alternatives cannot be dealt with in this way. For example, detection of occurrences of any spoken time (e.g., "three o'clock") is addressed in [18, 19]. Conceptually, the approach

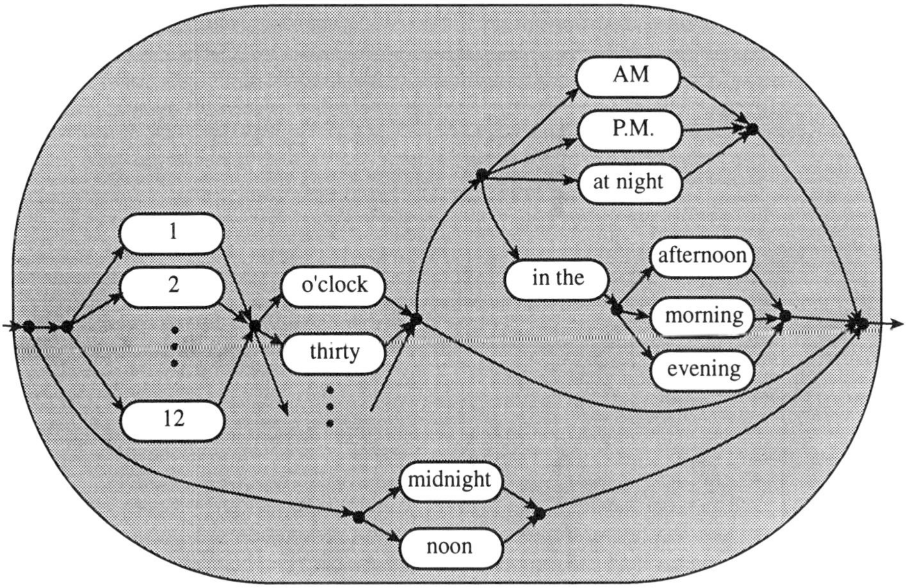

Figure 8 A portion of the HMM network for spotting occurrences of times [19]. This network is used in place of the simple linear keyword model used in previous work.

is to replace the linear left-to-right network with a possibly recurrent HMM network representing the phrase. For example, the network in Figure 8 shows the HMM network for detection of times. Note that computing the probability of exit from the phrase is slightly more complicated than in the linear keyword case. Once the presence of some phrase is detected, we haven't necessarily hypothesized the specific path through the phrase taken. Recognition techniques to recover the best path or N-best paths may still have to be employed [19].

Bottom-Up Approaches

In addition to the direct methods of keyword spotting described above, several "bottom-up" approaches have been presented. One approach [8] uses the phoneme dendrogram produced by the MIT Summit system [46]. A Viterbi-like search algorithm is used to find occurrences of a keyword's phoneme sequence in the dendrogram. Other approaches are also based on an initial phase of phoneme or consonant-vowel detection followed by a DP-based spotting of words [35, 26]. Finally, in a recent work, a dense phoneme lattice is computed using an HMM system followed again by detection of the keywords [17].

3.3 Secondary Classification

In order to improve the performance of word spotting systems, many attempts at "secondary classification" have been attempted. Hypothesized keyword occurrences (putative events) are either accepted or rejected based on a secondary test, or are scored based on additional information not incorporated into the primary word spotter. Several neural network, statistical, and regression-based approaches have been used.

Neural Network Classifiers

Several attempts to use Neural Networks as a means of secondary classification to reject putative keyword occurrences that are in fact false acceptances have been presented. In each approach, a fixed-length representation is first determined and then the result is presented to a classifier that has been trained on both true keyword occurrences and false acceptances obtained from a development data set.

In [24], various approaches to creating fixed-length representations that complemented a DTW-based primary word spotter are considered. The initial feature representation is a set of filterbank energies computed at regular intervals. In order to capture the "spatio-temporal" properties of the keyword two approaches are considered: time-normalized Fourier representations of each of the bands energies, and a fixed-size "grid" of frequency versus time. In [21], a NN is used for secondary classification after a primary HMM. A fixed-length segmentation into three equal-duration segments is used to create the fixed-length input needed by the NN classifier.

Segmental Modeling

In [14, 15] a secondary classification approach for an HMM word spotter using a statistical segment model is described. For each keyword, separate segmental models for the keyword and the set of false acceptances for that keyword obtained on a held out development set are trained. The model structure is based on a deterministic (i.e., using a segmenting preprocessor prior to segmental modeling) segmentation of the observation sequence for the hypothesized keyword. A segmental mixture model is used for each segment, where the distribution for each mixture component is based on the model for the sequence $x[1], \ldots, x[N]$

$$x[n] = \mu[n] + e[n]$$

where $\mu[n] = b_1 + b_2 n + b_3 n^2$ and $e[n] \sim N(0, \Sigma)$. For each mixture component, Σ and the b_i (or time-normalized coefficients) are estimated. Reranking the set of hypothesized keyword occurrences results in a substantial reduction in false acceptance rate. At a fixed detection probability, the false acceptance rate was reduced by over a factor of three.

Regression-Based Approaches

In [36] a regression-based rejection (scoring) approach is presented for the problem of embedded recognition. As in their previous work, [41, 42], the highest scoring model sequence, "filler"-keyword(i)-"filler" is determined using an HMM. A fixed-length representation, $y(i)$, of the complete observation sequence x is determined from the Viterbi alignment selecting only the subsequence aligned to the keyword, averaging the input for times aligned to each state. In addition, the data is aligned against a garbage model in place of the keyword model obtaining $y(g)$.

For each keyword i, as set of vectors $a_{i1}, \ldots, a_{ik}$ for the k keywords and a_{ig} are chosen by an iterative training procedure with the goal that the keyword j should be declared if

$$y(j) \cdot a_{ij} > y(g) \cdot a_{ig}$$

where i is the keyword with the highest HMM score. In addition, linear discriminant analysis is used to determine weights on several terms:

$$\begin{array}{ll} z_1 = \log \mathrm{P}(x|i) - \log \mathrm{P}(x|g) & z_3 = y(i) \cdot a_{ii} - y(g) \cdot a_{ig} \\ z_2 = \log \mathrm{P}(x|i) - \log \mathrm{P}(x|i') & z_4 = y(i) \cdot a_{ii} - y(i') \cdot a_{ii'} \end{array}$$

where keyword i' is the second best keyword based on the initial HMM scoring.

Using a large training corpus the false acceptance rate is reduced by a factor of four at a fixed detection probability.

A variety of other approaches to secondary classification and rejection have been reported. For example in [8], a Gaussian classifier based on another set of derived values including the top N match score, the $N-1$ score differences, and keyword durations, is used in a similar manner for rejection in an embedded recognition application.

A.	**Acoustic modeling**	1.	Whole word parametric model
		2.	Phonetically units
		3.	Signal normalization
B.	**Model structure**	1.	Acoustically-trained fillers
		2.	Keyword pieces in fillers
		3.	Phonetic models
		4.	Words models
		5.	Statistical language model
C.	**Search and scoring**	1.	Viterbi state sequence
		2.	Forward-backward scoring
		3.	Forward causal scoring
D.	**Secondary classification**	1.	Model-based
		2.	Regression-based

Table 1 Choices of approach to word spotting. Many combinations of approaches in each of the four areas are appropriate for particular applications.

4 DISCUSSION OF ALTERNATIVE APPROACHES

In the previous section, various reported approaches to word spotting were described. Each of the approaches is appropriate for some class of word spotting problems. Unfortunately there are few universally applicable approaches. There remains an art to word spotting that in large part corresponds to the appropriate choice of approach. Specifically, the choices fall into four primary areas as shown in Table 1. In the remainder of this section, the goal is to describe the alternatives available in each of these areas, referring to the survey in the previous section when possible, and to provide some guidance as to when the approaches are applicable. In practice, however, it is almost always an empirical question as to which approach provides the best overall performance in a particular application.

4.1 Acoustic Modeling

Issues in acoustic modeling for word spotting share most aspects with other areas of speech recognition. Phonetically-based sharing of model parameters (e.g., triphone models) has been shown to be effective when adequate amounts of transcribed training data are available. When little data is available for training, for example, when only a few instances of a keyword are available and no completely transcribed representative corpus is available, techniques that do not rely on phonetic structure remain useful. In particular, template-based approaches and whole-word parametric models may in fact be preferable.

Feature Normalization

An area of feature extraction and modeling that often does not receive adequate attention is that of signal "normalization." By this, we mean input feature transformations that attempt to compensate for overall variations in the input signal due, for example, to channel or speaker dependent effects. In speech recognition where the task is to classify an utterance based on the best matching model or model sequence, each model typically has similar characteristics such as amount of training, degree of constraint etc. When there is a mismatch between the trained models and the input data, the likelihood of the observations given each model is reduced. To the extent that the model mismatch is similar for all the hypotheses, the performance of a classifier is typically reduced but not catastrophically.

Word spotting can be thought of as a classification problem where the model of a word and the model of a competing filler do not have similar characteristics. For example, if there is a model mismatch due to a channel difference, the likelihood given the keyword model may reduced relatively more than that given the filler model. If keyword detection is based on a likelihood ratio test, the operating point may change drastically. In work with a whole-word HMM word spotter, the importance of channel equalization as a type of signal normalization was demonstrated [28]. Similarly, experience with word spotting in very noisy environments shows that although the shape of the ROC curve may not change dramatically, the operating point may be very sensitive to the model mismatch. One approach is to mitigate the effect of model mismatch by normalization or channel compensation [13]. Other approaches to this problem involve score normalization discussed below.

Generality of Keyword Models

A final area of acoustic modeling that may not be as significant in other areas of speech recognition has to do with the generality of the keyword model. An effect that we have observed is that it is not always desirable to make a more "accurate" model of the keyword or phrase one is attempting to detect. The

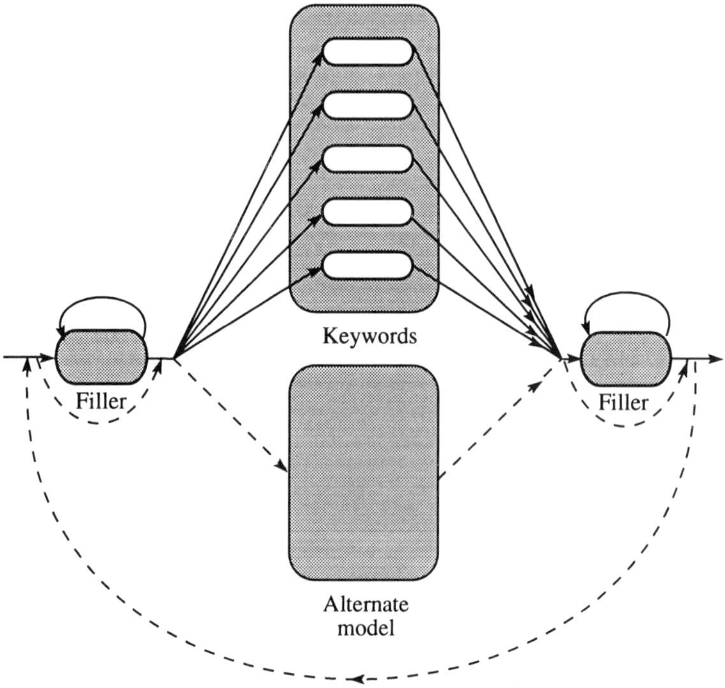

Figure 9 HMM structure for embedded recognition. Transitions to a competing Alternate Model and transitions allowing multiple keywords are optional depending on the applications.

reason for this is that there is always a tradeoff between detection probability and false acceptance probability. Although a more accurate model addresses increasing P_d, the result may be that a far greater increase in P_f actually results. In our whole-word HMM word spotter, we observed that a linear model with skips around each state is inferior to one with no skips, and further requiring a minimum duration of 20 ms rather than 10 ms in each state is even better. Similarly, in phrase spotting, it is tempting to introduce a rare form of a phrase, even with a small probability, in order to address a small number of misses with the result often being a large increase in false acceptances.

4.2 Model Structure

The model for non-keyword speech must serve two somewhat different functions. First, it should account for the non-keyword data. Second, as a competing model to the keywords, the filler model is used to discriminate between true keywords and false acceptances. A typical HMM structure for an embedded recognition problem is shown in Figure 9. If we assume that there is at most one occurrence of a keyword or phrase of interest is present in the utterance, then there is no transition back to allow multiple keywords. If there is always exactly one occurrence (i.e., there is no rejection of utterances containing no occurrences), then there is no competing alternative model.

Filler Structure

A variety of filler structures have been used successfully. These include:

1. A simple one-state HMM network,
2. A network of unsupervised units such as an ergodic HMM, or a parallel loop of clustered sequences,
3. A parallel network loop of subnetworks corresponding to keyword "pieces", phonetic models, or even models of whole words,
4. An explicit network characterizing typical word sequences.

The first two techniques do not require any knowledge of the typical vocabulary to be accounted for by the filler model. The models are acoustically-based relying on some characterization of the acoustic data to be modeled by the filler. In general, these types of filler are generally the least powerful due to two related effects. First, when used as filler models in an embedded recognition problem, there is often the tendency to "consume" part of the utterance that is actually a keyword or phrase. The second effect is that although the filler may model general speech well, in the role of a competing alternate model, a near "imposter" for a word may still score better with a keyword model than with the general alternate model.

The third technique has been found useful in several applications where little is known about the background speech. An interesting aspect of this type of model as an alternative model is that use of "pieces" of the keywords seems to provide good performance [5, 28]. One rationalization of this is that a near imposter will generally share much in common with a desired keyword. By including pieces of the keyword in the alternate model, there is no penalty introduced for those pieces. Use of word loops is the natural extension of this approach. Another rationalization of why these types of models are successful is that there is more similarity between the characteristics of the keyword model and the alternate model and therefore the system is less susceptible to the effects of model mismatch when channel of speaker conditions were not represented in training data.

In many applications, most notably in embedded recognition, it is often possible to characterize the filler surrounding the words of interest. A simple example of this is the presence of an optional "please" at the end of a response to a prompt. In the absence of acoustic training for the filler, a word-based approach is often quite successful.

Finally, an approach to filler modeling addresses the problem of fine-tuning an alternate model to compete with keyword models. Various approaches at discriminative training have been attempted. One difficulty with discriminative training is that a sufficient number and variety of false acceptances must be available both to characterize the ultimate application and to prevent over fitting of the training data.

Statistical Language Modeling

Another approach to overall HMM structure for word spotting including both embedded recognition and monitoring applications is based on statistical language modeling. The models are based on a training corpus of transcribed passages. The basic issue that comes up is the vocabulary that should be used in training the model. Also, for any incomplete vocabulary, a decision on how to treat the out-of-vocabulary word in the training data needs to be considered. The general result is that modeling a large vocabulary using an N-gram language model is an effective basis for word spotting either by simple CSR techniques or a probabilistic scoring technique. With a large but not complete vocabulary, there seems to be no need to make a model for the "unknown" word.

A statistical language model, such as a bigram model, has two main differences from use of a very large word loop as an alternative model. First, due to the bigram probabilities the alternate model should score somewhat better with typical word sequences than a model that assumes that all words can follow all other words with equal probability. A much more important effect is that the bigram probabilities provide a statistical context for the keywords of interest. For example, suppose we are spotting for city names in an utterance responding to "From which airport would you like to fly?" city names may be likely to be preceded by "from" and followed by "airport," "international," or "please." Therefore, by characterizing the statistical context of keywords of interest, the keyword detection problem is replaced by detection of the keyword ***in its typical context.*** In experiments using large vocabulary techniques, it has been shown that the vocabulary size can be reduced to include only the keywords generally adjacent to the keyword as well as typically confusable words [18]. A good model for the complete utterance does not seem to be needed.

Finally, in applications where there is often, but not always, a structured context, a combination of explicit specification of common carrier phrases and a statistical N-gram approach is useful [23].

4.3 Search and Scoring

Three basic search techniques have been used for word spotting. With any of these techniques, secondary classification or rejection can be performed. In addition, some CSR techniques can be adapted to the word spotting problem.

1. Viterbi search
2. Posterior probability scoring
3. Other recognition based approaches

Viterbi Search

The simplest adaptation of CSR techniques described above is simple automatic transcription with keywords being detected if they occur on the most likely state sequence. This is a very effective technique when the filler models can be well characterized in terms of allowable word sequences and there is either no requirement for rejection of utterances containing no keyword or utterances not containing keywords are typically very distinct for those containing the keywords.

The operating point for detection versus false acceptance is set implicitly based on the transition probabilities in the HMM. Scoring approaches include average observation likelihood along the interval containing the keyword as well as comparison of the likelihood of the best path through the keyword compared to the best path though any other word. Data following the keyword enters into the detection process by determining whether the keyword in the state traceback.

Posterior Probability Scoring

Based on the Forward-Backward algorithm, the probability of occupying a particular state or one in a set of states at some time, or traversing a particular transition after some time is straightforward to compute given the data in an entire passage or message. Using this type of scoring, the competing alternate model implicitly normalizes the score through its contribution to the denominator of the expression used to compute the posterior probability.

The probability of exiting a keyword can be computed as well, although there is a issue that a keyword occurrence with an uncertain endpoint will distribute the probability over a longer interval and therefore appear to have a lower score. Computing the probability that the keyword ends in a small interval is somewhat more reliable. A useful quantity used with a word spotter as a front-end for topic identification is the expected number of times a keyword occurs in a passage. This is simply computed summing the exit probability over all time in the entire passage.

If future data is not available, for example if one is monitoring an open microphone system for the occurrence of a trigger word, forward-only posterior probability scoring is effective. In fact, in the absence of a statistical language model, forward-only scoring provides nearly identical performance to forward-backward scoring. Note that it is not feasible to compute the probability of occupying a keyword based on only the forward path since a prefix of a word will make the scores within the word appear high.

Other CSR Techniques

Other approaches to the word spotting search are also feasible. One approach is based on phoneme recognition as a preprocessing of the input. Although it is feasible to consider a single recognized phoneme sequence as reduction of the data which is then used as input to a word spotter, alternatives such as generation of a phoneme lattice may be preferable [17].

Another technique for word spotting that overcomes some of the deficiencies of using the single Viterbi recognition sequence is to base word spotting on the N-best recognized sequences. If the probability of each sequence is computed, then the probability of a particular word occurring is simply the sum to the probabilities of the sequences containing the word. Although this is a straightforward application of the result of an N-best recognition system, the results could be obtained more directly without actually computing the N best sequences.

4.4 Secondary Classification

Secondary classification approaches generally use training data containing both the keywords of interest as well as false acceptances or potential false acceptances based on utterances not containing particular keywords. Some sort of classifier or model is constructed to distinguish between true occurrences of a keyword and false acceptances. The inputs to the classifier are often derived from the result of a primary word spotter. Classifiers can be based on probabilistic models, again forming a likelihood ratio as in [14]. Alternatively, scoring or classification can be based on a regression approach where based on a set of derived features of each hypothesized keyword occurrence, a parametric function is computed [36, 24].

The major consideration with these types of secondary classification schemes is whether adequate training data is available. Both true occurrences and a representative set of false acceptances are necessary. With adequate and representative data secondary classification can significantly reduce the false acceptance rate for a fixed detection probability. Due to the nature of typical ROC curves, the effect on detection probability at a fixed false acceptance rate may not be as dramatic.

5 RESEARCH PROBLEMS

Several aspects of word spotting remain active research areas. These include:

- Robust selection of operating point and prediction of expected performance
- Scoring that allows comparison of different keywords
- Secondary classification not requiring both keyword and imposter training

In addition, other speech recognition areas have very closely related open problems. These include

- Detection of unknown words
- Rejection of out-of-set utterances
- Rejection of utterances suspected to be misrecognized

5.1 Robust Selection of Operating Point

As discussed above, one difficulty with word spotting in general is that it is difficult to set reliably an operating point or predict performance of a word spotter based on training data. When representative data is available, cross-validation or bootstrapping techniques can be used [34]. However, if a new channel condition is encountered, it is not clear how the system should be adjusted to, for example, maintain the same detection probability possibly at the expense of more false acceptances. This problem of robustness to channel variability is particularly an issue in situations where variable amounts of noise may be present, for example in cellular telephone connections. Coping with channel variability while truly providing word spotting or utterance rejection capabilities is an open challenge.

5.2 Absolute Scoring

In all the word spotting scoring techniques described, the scores for one keyword are generally not comparable to those of another keyword. The nature of the score depends on many factors including the length of the keyword model (e.g., number of phonemes), the amount of training data used to construct the keyword model, the "uniqueness" of the keyword in general speech, and the duration of the occurrence. Just as it is difficult to predict the effect of channel differences, differences between keywords is also difficult to predict. Again, cross-validation or boot strapping techniques can be used if adequate occurrences of the keyword are available. However, using phonetically-based models, one does not necessarily have any keyword examples for cross-validation.

The problem of absolute scoring is particularly an issue with phrase spotting when often hypotheses of difference length must be compared. An example in spotting for a time involves score comparisons of "Any time after 12:10"

versus "Any time after 12 then". Techniques for reliably ordering the hypothesized occurrences of "after 12:10" and "after 12" in the example are not fully developed. Simple length normalization strategies are not generally adequate.

5.3 Secondary Classification Training

Secondary classification techniques currently require training containing both keyword occurrences as well as typical impostors. In addition, this training must be representative of the target channel conditions. This is often not available in applications where the channel is unpredictable and where there may be a large number of keywords or variants of a key phrase to be detected.

5.4 Related Areas of CSR

The CSR problem of detecting occurrences of words not in the known lexicon is very much the dual of the word spotting problem. Techniques which have been applied address the problem in a similar way to word spotting with an "unknown" word model competing with the known words [1]. The approach is very much like Viterbi-based word spotting. Unfortunately, all the robustness issues associated with channel variation and the implicit setting of operating points using transition probabilities are present.

Utterance rejection when the words spoken do not come from a predefined set of allowable sequences is another important problem in CSR applications. The basic approaches that have been taken are first, competing alternate models and second, explicit secondary classification. The issues of robustness and need for training are important here as they are in word spotting.

6 SUMMARY

Word spotting techniques originated in the areas of monitoring continuous input streams and in automatic end point detection for early recognition algorithms. Today, variants of word spotting including filler modeling, out-of-set utterance rejection, new word detection, and phrase detection have bridged the gap between simple word spotting and automated complete transcription. Speech recognition approaches in this range must be appropriately selected to match the characteristics of any particular application.

REFERENCES

[1] A. Asadi, R. Schwartz, and J. Makhoul, "Automatic detection of new words in a large vocabulary continuous speech recognition system," *Proc. International Conf. on Acoustics, Speech and Signal Processing (ICASSP) '90*, pp.125-129, 1990.

[2] R. Bakis, "Spoken word spotting via centisecond acoustic states," *IBM Technical Disclosure Bulletin*, 18(10), pp. 3479-81, March 1976. (Also described in IBM Tech. Report RC 4788, April 2, 1974).

[3] L. E. Baum, T. Petrie, G. Soules, and N. Weiss, "A maximization technique occurring the the statistical analysis of probabilistic functions of Markov chains," *Annals of Mathematical Statistics*, 41(1), pp. 164-171, 1970.

[4] H. Bourlard, B. D'hoore, and J.-M. Boite, "Optimizing recognition and rejection performance in wordspotting systems," *ICASSP'94*, pp. I-373–376, 1994.

[5] S. Boll, J. Porter, and L. Bahler, "Robust syntax free speech recognition," *ICASSP'88*, pp. 179-182, 1988

[6] J. S. Bridle, "An efficient elastic-template method for detecting given words in running speech." *Brit. Acoust. Soc Meeting*, pp. 1-4, April 1973.

[7] J. S. Bridle, M. D. Brown, and R. M. Chamberlain, "An algorithm for connected word recognition," *ICASSP'82*, pp. 899-902, 1982.

[8] B. Chigier, "Rejection and keyword spotting algorithms for a directory assistance city name recognition application," *ICASSP'92*, Vol. 2, pp. 93-6, 1992.

[9] R. W. Christiansen, and C. K. Rushforth, "Word spotting in continuous speech using linear predictive coding," *ICASSP'76*, pp. 557-6, 1976.

[10] R. W. Christiansen, C. K. Rushforth, and K. Craig, "Detecting and locating key words in continuous speech using linear predictive coding," *IEEE Trans. Acoustics Speech and Signal Proc. (ASSP)*, 25(5), pp. 361-367, 1977.

[11] G. D. Forney, "The Viterbi Algorithm," *Proc. IEEE*, Vol. 61, pp. 268-278, 1973.

[12] M. Franzini, "The TARGET architecture: a feature-oriented approach to connectionist word spotting," *Int. Joint Conf. on Neural Networks (IJCNN) '92*, Vol. 2, pp. 338-43, 1992.

[13] H. Gish, Y. Chow, and J. R. Rohlicek, "Probabilistic Vector Mapping of Noisy Speech Parameters for HMM Word Spotting," *ICASSP'90*, pp.117-120, 1990.

[14] H. Gish, K. Ng, and J. R. Rohlicek, "Secondary processing using speech segments for an HMM word spotting system," *Int. Conf. on Spoken Langauge Processing (ICSLP) '92*, pp. 17-20, 1992.

[15] H. Gish, and K. Ng, "Segmental speech model with applications to word spotting," *ICASSP'93*, pp. II-447-450, 1993.

[16] A. L. Higgins, and R. E. Wohlford, "Keyword recognition using template concatenation," *ICASSP'85*, pp. 1233-1236, 1985.

[17] D. A. James, and S. J. Young, "A fast lattice-based approach to vocabulary-independent wordspotting," *ICASSP'94*, pp. I-377–380 1994.

[18] P. Jeanrenaud, K. Ng, M. Siu, J. R. Rohlicek, and H. Gish, "Phonetic-based word spotter: various configurations and application to event spotting," *2nd Proc. European Conference on Speech Communication and Technology (EUROSPEECH) '93*, pp. 1057-1060, 1993.

[19] P. Jeanrenaud, M. Siu, J. R. Rohlicek, M. Meteer, and H. Gish, "Spotting events in continuous speech," *ICASSP'94*, pp. I-381–384.

[20] K. P. Li, J. A. Naylor, and M. L. Rossen, "A whole word recurrent neural network for keyword spotting," *ICASSP'92*, Vol. 2, pp. 81-84, 1992.

[21] R. P. Lippmann, and E. Singer, "Hybrid neural-network/HMM approaches to wordspotting," *ICASSP'93*, pp. I-565–568, 1993.

[22] J. McDonough, K. Ng, P. Jeanrenaud, H. Gish, and J. R. Rohlicek, "Approaches to topic identification on the Switchboard corpus," *ICASSP'94*, pp. I-385–388, 1994.

[23] M. Meteer, and J. R. Rohlicek, "Statistical language modeling combining N-gram and context-free grammars," *ICASSP'93*, vol. II, pp. 37-40, 1993.

[24] D. P. Morgan, C. L. Scofield, and J. E. Adcock, "Multiple neural network topologies applied to keyword spotting," *ICASSP'91*, Vol. 1, pp. 313-16, 1991.

[25] C. S. Myers, L. R. Rabiner, and A. E. Rosenberg, "An investigation of the use of dynamic time warping for word spotting and connected speech recognition," *ICASSP'80*, Vol. I, pp. 173-7, 1980.

[26] M. Okada, and M. Kohda, "Word spotting system based on stochastic models of phonemic segments," *Systems and Computing in Japan*, 22(9), pp.82-93, 1991.

[27] W. W. Peterson, T. G. Birdsall, and W. C. Fox, "The theory of signal detectability," *IRE Trans. Info. Theory*, PGIT-4, 171-212, Sept. 1954.

[28] J. R. Rohlicek, W. Russell, S. Roukos, and H. Gish, "Continuous hidden Markov modeling for speaker-independent word spotting," *ICASSP'89*, Vol. 1, pp. 627-30, 1989.

[29] J. R. Rohlicek, P. Jeanrenaud, K. Ng, H. Gish, B. Musicus, and M. Siu, "Phonetic training and language modeling for word spotting," *ICASSP'93*, pp. II-459-462, 1993.

[30] R. C. Rose, and D. B. Paul, "A hidden Markov model based keyword recognition system," *ICASSP'90*, pp. 129-132, 1990.

[31] R. C. Rose, E. I. Chang, and R. P. Lippmann, "Techniques for Information Retrieval from Voice Messages," *ICASSP'91*, Vol. 1, pp. 317-20, 1991.

[32] R. C. Rose, "Discriminant wordspotting techniques for rejecting non-vocabulary utterances in unconstrained speech", *ICASSP'92*, pp. II-105-8, 1992.

[33] H. Sakoe, and S. Chiba, "A dynamic programming approach to continuous speech recognition," *Seventh Int. Congress on Acoustics*, Budapest, pp. 65-68, 1971.

[34] M. Siu, H. Gish, and J. R. Rohlicek, "Predicting word spotting performance," *ICSLP'94*, pp. 2195-2198, 1994.

[35] N. Sugamura, "Continuous speech recognition using large vocabulary word spotting and CV syllable spotting," *ICASSP'90*, Vol. 1, pp. 121-4, 1990.

[36] R. A. Sukkar, and J. G. Wilpon, "Two pass classifier for utterance rejection in keyword spotting," *ICASSP'93*, pp. II-451-454, 1993.

[37] A. H. Waibel, T. Hanazawa, G. Hinton, K. Shikana, and K. Lang, "Phoneme recognition using time-delay neural networks," in *IEEE ASSP*, (37)3, pp. 328-339, 1989.

[38] M. Weintraub, "Keyword-spotting using SRI's DECIPHER large-vocabulary speech-recognition system," *ICASSP'93*, pp. II-463-466, 1993.

[39] L. D. Wilcox, and M. A. Bush, "HMM-based wordspotting for voice editing and indexing," *EUROSPEECH'91*, Vol. 1, pp. 25-8, 1991.

[40] J. G. Wilpon, and L. R. Rabiner, "Application of hidden Markov model to automatic speech endpoint detection," *Comput. Speech and Language*, 2(3-4), pp. 321-341, 1987.

[41] J. G. Wilpon, C. H. Lee, and L. R. Rabiner, "Application of hidden Markov models for recognition of a limited set of words in unconstrained speech," *ICASSP'89*, pp. 254-257, 1989.

[42] J. G. Wilpon, L. R. Rabiner, C.-H. Lee, and E. R. Goldman, "Automatic recognition of keywords in unconstrained speech using Hidden Markov Models." *IEEE ASSP* 38(11), pp. 1870-1878, 1990.

[43] J. G. Wilpon, L. G. Miller, and P. Modi, "Improvements and applications for key word recognition using hidden Markov modeling techniques," *ICASSP'91*, Vol. 1, pp. 309-12, 1991.

[44] T. Zeppenfeld, A. H. Waibel, "A hybrid neural network, dynamic programming word spotter," *ICASSP'92*, pp. II-77-80, 1992.

[45] T. Zeppenfeld, R. Houghton, and A. Waibel, "Improving the MS-TDNN for word spotting," *ICASSP'93*, pp. II-475-478, 1993.

[46] V. Zue, J. Glass, M. Phillips, and S. Seneff, "The MIT SUMMIT Speech Recognition System: A Progress Report," *Proc. of the First DARPA Speech and Natural Language Workshop*, pp. 178-189, 1989.

7

SPEECH RECOGNITION USING NEURAL NETWORKS

Stephen V. Kosonocky[1]

IBM T.J. Watson Research Center
Yorktown Heights, NY 10598, USA

1 INTRODUCTION

The field of artificial neural networks has grown rapidly in recent years. This has been accompanied by an insurgence of work in speech recognition. Most speech recognition research has centered on stochastic models, in particular the use of hidden Markov models (HMMs) [9][28][29][30][45][47]. Alternate techniques have focused on applying neural networks to classify speech signals [6][11][48]. The inspiration for using neural networks as a classifier stems from the fact that neural networks within the human brain are used for speech recognition. This analogy unfortunately falls short of being close to an actual model of the brain, but the modeling mechanism and the training procedures allow the possiblility of using a neural network as a stochastic model that can be discrimitively trained.

With most classification tasks, the input signal is first preprocessed to create a vector of representative features. Speech analysis techniques usually assume a time-varying filter model as the speech source. The important attributes of this model are not only the instantenous outputs of the model but also the nature of the time variation. To allow use of linear filter approximations, the speech signal is sampled in short time segments on the order of tens of milliseconds, and a frequency representation of that sample is used as the representative feature vector. The speech recognizer must then classify each of the individual feature vectors and combine this information with how they vary with time. Reconstruction of the time variation of these feature vectors has led to the use of HMM's using Viterbi recognition or other more general dynamic programming techniques. These techniques produce a locally optimum nonlinear alignment

[1] This work was done while the author was with the CAIP Center, Rutgers University

between two vectors. For the case of an HMM, the optimum alignment is chosen such that the outputs of the hidden states are maximized along connecting archs which include cost penalties for each path traversed [9][29][30]. The more general dynamic programming techniques maximize the match of the feature vectors to a reference template [1][4][5]. This reference template contains the information of how the feature vectors for a particular sound or word should vary.

Neural network approaches to classifying speech signals can be catagorized into three approaches; those which 1) directly incorporate the time information into the network, 2) time align the outputs of a series of static neural networks or 3) disregard the time information altogether. For the first case, these networks can generally be described as those which either incorporate a tapped-delay line for each of the inputs to a feed forward network, termed a Time-delayed Neural Network (TDNN) [40] or those which feed a tapped delay line by the output of a feed forward structure and feedback the the delayed outputs back to the inputs [22]. This is one version of a recurrent neural network. Comparisons of recurrent neural networks have been made in recent literature to HMMs [2][6][21][27]. These comparisons show the relationship between a recurrent neural network structure over time to an HMM network.

The second approach combines a static neural network with a time alignment procedure into a hybid network. These approaches can combine dynamic programming with conventional ANN structures [49][50] or alternative ANN structures [19][39]. Various approaches have been tried to construct hybrid neural networks and HMMs to exploit the temporal classification network in an HMM system. One example such a speech recognition system adds the discriminative ability of neural networks within the state models of an HMM speech recognizer [31], showing an increase in performance over conventional HMMs. Another example uses a neural network as a secondary processor [24][25], to correct errors from a conventional HMM system.

The third approach disregards the time information, and accumulates the statistical averages of each class occurring. This method is often used for speaker identification tasks [12][13], where the time information is not always a robust indicator of the speaker's identity, and can be neglected to minimize computational complexity.

This chapter will focus on the second approach and describe a classifier constructed by combining the discriminatory ability of a neural tree network (NTN) and a mixture of gaussians to produce a continuous density neural tree network (CDNTN), to model the posterior probability of a class occurring given a feature vector. A review of how an ANN can be used to directly model the posterior probability of a class will be given, followed by a brief review of an efficient tree implementation of a neural network (NTN). It will be shown how a mixture model can be added to the NTN to create the CDNTN classifier. A word spotting system based on the CDNTN will be described, by using the CDNTN to model the posterior probability of a subword segment within a keyword. The subword models can be connected together to form a Markov chain. Once a Markov chain is created for each keyword, state durations are extracted from the training data and clustered to form a state duration template to non-parametrically model the durations of each state. For testing, a dynamic time warping (DTW) algorithm is used to evaluate the state outputs for the test utterance against the state duration model extracted from the training data. The state duration template provides a temporal model for the state outputs during a keyword occurrence distinguishing between random state outputs and temporally aligned outputs during a keyword occurrence obviating the need for a recognition network used in traditional HMM implementations.

2 ARTIFICIAL NEURAL NETWORKS

Over the past decade Artificial Neural Networks (ANNs) have been applied to a multitude of classification applications. ANNs are an interconnection of simple computational elements or nodes connected by links with variable weights [23]. Each node or perceptron [35], computes a weighted sum of the outputs of the nodes feeding it, and sends the sum through an activation or squashing function (usually a non-decreasing sigmoid). The use of a sigmoidal squashing function insures that the output value of the perceptron is in the range [0,1]. A bias value is added to the sum in order to shift the threshold of the activation function. Using a single perceptron, it is possible to learn simple classifications formed by linear decision boundaries. Adding multiple layers of perceptrons (at least two) enables a network to learn arbitrary convex decision boundaries [23]. In general, multiple hidden layers can be constructed.

The method of learning most commonly used is known as supervised training, where inputs are presented to the ANN in succession with a target output for each output node. The use of a continuous differentiable activation function allows the computation of the gradient of the error between the MLP outputs and the target outputs, thus a gradient search algorithm for updating the weights of the MLP can be used. The most popular search algorithm is known as the Backpropagation algorithm [17]. Briefly, the backpropagation algorithm is a gradient search method that minimizes the expected sum-of-square error between a target and the outputs for each training pattern given. The gradient of the error at the outputs is used to adjust the weights of the output nodes and is propagated backwards in order to update the weights of the hidden layers. The error of the outputs of the hidden layers is approximated by the error of the output node multiplied by the connecting weight. A derivation of the standard backpropagation algorithm can be found in [35].

The functional capabilities of the MLP include Boolean logic functions [23], partitioning of a pattern space for classification and nonlinear functional approximations [8][18]. Neural networks implement nonlinear transformations of the form $y = f(x)$. Usually x is a member of R^n and y is a member of $[0,1]^m$ or R^m. The activation functions serve as basis functions, performing a nonlinear transformation of the input space. Neural Network implementations of nonlinear discriminate functions are possible by transforming the linearly inseparable input space, defined by the feature vectors and class labels, into a linearly separable space defined by the continuous valued outputs of the hidden layer. These outputs are then applied to a single output node which is capable of providing a linear discriminate function for the class associated with the output. What is not always known during classifier design is the number of basis functions necessary for optimal classification. It has been shown [8] [18] [16], that a minimal two layer MLP is capable of forming an arbitrarily close approximation to any continuous nonlinear mapping. These results are not specific to any single type of activation function, only that it is a nondecreasing continuous function. The most common type of activation function is given by a logistic function, but others such as $tanh()$, $sin()$, $cos()$ and radial basis functions have also been successfully used [8].

Training a neural network as a pattern classifier can be accomplished by the use of supervised training. Supervised training requires labeling each training vector according to class. A network is then created, associating each output of the neural network with a unique class. Commonly the training data is labeled with class labels of 0.0 for anti-class input vectors and a 1.0 for correct class labels for each output node. This type of labeling scheme allows the neural network to directly approximate the posterior probability of a class occurring. The network is trained using the backpropagation algorithm to minimize the sum of square error cost function given by [32]

$$\Delta = E\{\sum_{i=1}^{M}[y_i(x) - d_i(x)]^2\}, \tag{7.1}$$

where $y_i(x)$ is the output of the neural network for class i out of M classes, d_i is the label value for class i associated with the feature vector and $E\{\}$ is the expectation operator.

The pattern classification problem has now been transformed into a functional approximation of the target function given by $d_i(x)$. By using this type of labeling scheme, the target function becomes an impulse function which is expressed by

$$d_i(x) = \delta(x - x_i), \tag{7.2}$$

where x_i indicates the feature vector for class i and δ is the Dirac-delta function.

Care must be taken when designing the training set for a classification task. The classifier training paradigm can be defined in terms of a cost function based on the expectation operator of the joint probability of the input vector and the ith class, $P(x, C_i)$ [32]. The cost function used for training an MLP with the backpropagation algorithm, minimizes the error of the posterior probability for each output when each output node is trained with a target of 1.0 for a class occurring and 0.0 when the assigned class is not present i.e. eq. 7.2 [32]. This can be shown by expanding the cost function in eq. 7.1 and rearranging terms. It can be shown that the cost function can be written in a more statistically significant form [32] given by,

$$\Delta = E\{\sum_{i=1}^{M}[y_i(x) - E\{d_i|x\}]^2\} + E\{\sum_{i=1}^{M} var\{d_i|x\}\}. \tag{7.3}$$

The expected value of the of the label d_i given x, $E\{di|x\}$, is

$$E\{d_i|x\} = \sum_{j=1}^{M} d_i p(C_j|x). \tag{7.4}$$

It can be noted that the second summation in eq. 7.3 is not a function of the neural network output, so the weight updates will only affect the first expectation term. The expression in eq. 7.3 states that if each class has equal probability of occurring, the neural network learns to fit a surface to the function $d_i(x)$. If certain classes have higher probability of occurring then the cost function will be weighted in the direction of that particular class, causing the neural network to better approximate the more probable class at the expense of others. If x belongs to only one out of M classes, where d_i is the impulse function, eq. 7.4 reduces to $p(C_j|x)$ for $i \neq j$, which is the Bayesian posterior probability. Thus, by proper labeling, the cost function given in eq. 7.3 will minimize the expected value of the posterior probability of the class occurring.

The important point to note is that by setting up the classification problem in the typical supervised training scenario of training an MLP to choose one out of M classes, the MLP is learning to approximate the Bayesian posterior probability directly. It is apparent that manipulations of the training set to aid convergence or reduce computational complexity can directly affect the classification decision in a predictable way [32]. Segmenting the speech recognition task into two operations; phonetic recogntion based on extracted speech features, followed by word recognition based on recognized phonemes, allows a structured method to incoroporate a neural network to estimate the posterior probability of a phoneme occurring. The estimated phoneme probabilities can then be processed by a secondary temporal alignment procedure.

3 NEURAL TREE NETWORKS

An alternate computationally efficient description of a neural network is known as a Neural Tree Network (NTN). As described by Sankar and Mammone [37] a neural network can be embedded within a decision tree type classifier to create an NTN. The advantage of the NTN architecture is that a tree architecture allows an efficient implementation on a serial processor, combined with the fact that it is self organizing. The NTN is similar to a decision tree where an incoming unknown feature vector is tested at the root node of the tree, and depending on the outcome of the test, is passed to a child node, until it reaches a terminal leaf node. Class membership of the unknown feature vector is given as the class label assigned to the leaf. A directed graph can be used to describe the NTN architecture, and is shown in Fig. 1.

The tests at each node of the NTN are actually performed by "sub" neural networks, with the internal structure of these "sub" networks usually given as simple perceptrons, also known as a logistic regression model. The perceptrons performing a linear discrimination of the feature set presented to them. In general higher order discriminate functions, such as an MLP [38] at each node can be used. Class membership at each node is determined by testing if the activation level of the node is greater than or less then a threshold (typically 0.5). For the binary case, inclass vectors are labeled as a "1" if the activation level exceeds the threshold, and out-of-class vectors are labeled as a "0" if the activation level is less than the threshold. Associated with each of these outputs

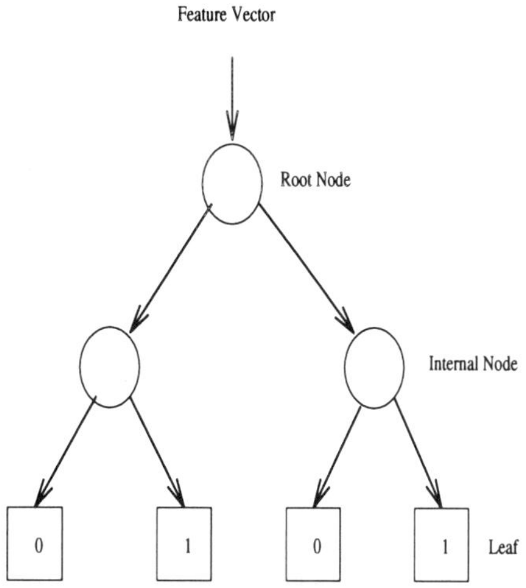

Figure 1 NTN classifier architecture

is a path to a child node where the unknown vector is directed depending on the outcome of the test at the current node. The result is the unknown feature vector is directed down the tree until it reaches a leaf node where it is classified by a label given to the leaf, determined at the time of training. The NTN shown in Fig. 1 performs a binary classification, with classes labeled as "0" and "1" at the leaves.

Obtaining the optimal size NTN is important for good generalization from the training data set to a test data set [36] [37]. Similar to the case using decision trees and the CART [3] algorithm used by Breiman et. al., creating fully grown NTNs, corresponding to correctly classifying the entire training data set, often leads to suboptimal performance on a test data set. This is similar to having too many hidden nodes in a neural network or over parameterization of a curve fitting algorithm. An MLP with more than the minimum number of hidden nodes can be thought of as a nonlinear parameterization using too many basis functions. The extra basis functions may cause the resultant function approximation to become unpredictable between training points. Breiman et. al. [3] describes an optimal pruning strategy for decision trees. Sankar and Mammone [36] [37] describe a similar optimal pruning strategy for NTN's based on minimization of a cost function relating tree complexity and error on a cross-validation test set. Both of these pruning strategies are backward pruning algorithms. Initially a maximal size tree is grown for a given training data set, and based on the outcome of tests on a cross-validation set, the algorithm deletes branches until the optimal tree is obtained satisfying the cost function. Other pruning strategies ased on forwbrd abrubt termination or percentage based stopping have also been described [12] [19].

It is possible to use a pruned NTN to define discrete regions within the feature space to provide a stochastic approximation for the feature vector. A discete region in feature space is defined by the hyperplane tests leading to a leaf j. The NTN is self organizing, producing regions which minimize the error of the underlying posterior probability of the training data within each region. Once the feature space is partitioned, probabilities of a class occuring in each of the regions can be approximated by counting the number of training feature vectors in each of the regions. The probability given by

$$p(C_i|x_j) = k_{ij}/(\sum_{l=1}^{M} k_{lj}). \tag{7.5}$$

specifies the probability of observing class i in leaf j, which is a quantized approximation of $p(x|C_i)$ within the leaf, where k_{ij} is the number of feature vectors of class i falling into leaf j. The posterior probability can be calculated directly by the class counts of the training feature vectors at each leaf by eq. 7.5 [3]. This posterior estimate includes the prior probability artificially introduced into the training data set. Expanding eq. 7.5 to include the prior probabilities of each class gives the posterior probability as [19],

$$p(C_i|x_j) = [(N_i/(\sum_{l=1}^{M} N_l))(k_{ij}/N_i)/[\sum_{m=1}^{M}(N_m/(\sum_{l=1}^{M} N_l))(k_{lj}/N_m)] \tag{7.6}$$

where N_i is the number of feature vectors of class i at the root of the tree. This allows the opportunity to normalize the prior probabilities when calculating the discrete probability estimates. New estimates can be used in place of the prior probabilities artificially introduced by the training data set.

4 CONTINUOUS DENSITY NEURAL TREE NETWORK

The process of building statistical models often requires assumptions that the vectors in the training data set are extracted from a single distribution. The components of the feature vectors are often created from a concatenation of discrete feature values relating to some physical process, or the vectors themselves can originate from independent processes, violating the assumptions on

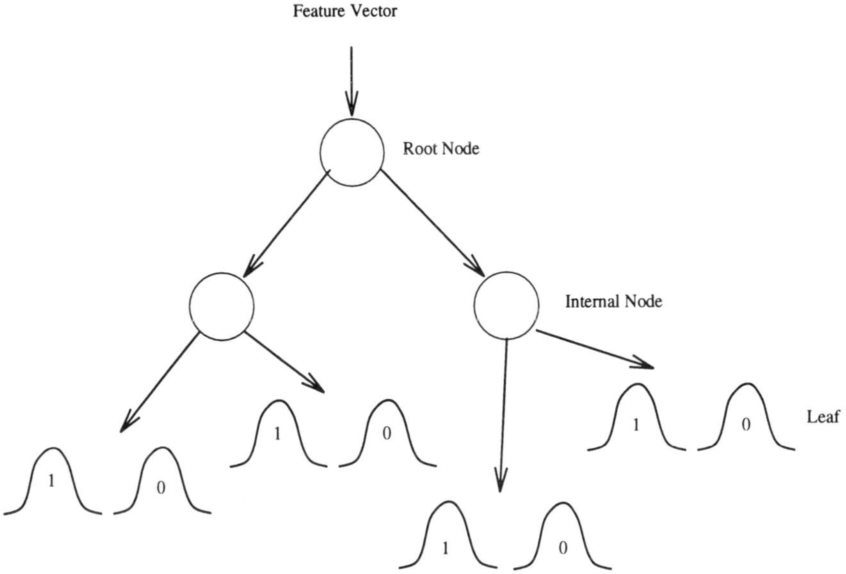

Figure 2 CDNTN graph model

the model. The NTN can be used to determine a locally independent statistical model, where the NTN determines the category within feature space of a given feature vector, then uses a leaf specific model to determine the probability distribution of the data in the leaf region. The posterior probability can then be found using Bayes' equation. This is based on the same assumptions made for using hidden states in a Markov model. As in more general graphical models [20][43], a mixture of gaussians can be used to model the probability distribution within the state of a location model, a model which uses different parametric models depending on the location within a directed graph.

Viewing the NTN as a graphical model, where each leaf represents an independent state, establishes the basis of using alternate models for modeling the probability distribution of the classes in the leaf region. The NTN can be used to partition the feature space discriminatively, then a parametric local mixture model can be used to model the distribution of data within the leaf region. This is similar to the graphical location models described in [43], except a non parametric neural tree network model is used to define transitions between nodes. This is known as a continuous density neural tree network (CDNTN) [19], formed by the addition of a continous parametric leaf model to the NTN.

Fig. 2 shows a graphical representation of the CDNTN, where a neural tree network is used to partition the data used to create local parametric models at the leaves. Unknown data vectors are first preprocessed by the NTN and then applied to the parametric leaf model. Once such parametric model is a mixture of Gaussian distributions, shown in the figure as single mixtures. The combined model provides a continuous estimate of the probability distribution within local regions of feature space. This new model can be constructed by modeling the distribution of vectors in each class at every leaf. A simple mixture of Gaussians can be used to approximate the distribution $p_j(x|C_i)$ of class C_i given feature vector x in the space defined by leaf j as,

$$p_j(x|C_i) = \sum_{m=1}^{M} c_{im} N(x, \mu_{im}, \Sigma_{im}) \tag{7.7}$$

and

$$N(x, \mu, \Sigma) = (2\pi)^{d/2|\Sigma|^{-1/2}} e^{-1/2(x-\mu)^T \Sigma^{-1}(x-\mu)}, \tag{7.8}$$

where c_{im} is the weight of the mth mixture and N indicates the normal operator given in eq. 7.8, with parameters μ_{im}, Σ_{im} expressing the mean vector and covariance matrix of mixture m for vectors of class i. Once the parameters for the distribution model for each class are found, the posterior probability $p_j(C_i|x)$ can be derived by Bayes' equation, and expressed as,

$$p_j(C_i|x_j) = [(n_i(\sum_{l=1}^{M} n_l))p_j(x|C_i)/[\sum_{m=1}^{M}(n_m/(\sum_{l=1}^{M} n_l))p_j(x|C_m)] \tag{7.9}$$

In eq. 7.9, the prior probability of each class within the space defined by leaf j is approximated by discrete sample counts and is shown in the bracketed expression multiplying the distribution term $p_j(x|C_i)$, where n_i is the number of samples of class i. The algorithm used for constructing the new CDNTN model for binary classified data can be described by the following procedures outlined in Fig. 3. The binary model can also be extended to the multi-class case using a multiple class NTN.

As described in the begin ing of the chapter, the statistical independence assumption of the data allows the use of a graphical model to partition the data into statistically independent groups. These groupings of data can then be used to define independent parametric models for each node in the graph model. The neural tree network partitions the data set according to a cost function defined in the training algorithm without any guarantee of statistical independence of the data arriving to different leaf regions during training. If the data arriving to each leaf were statistically independent from other leaf regions, then the

(1) Grow the NTN using algorithm described by Sankar & Mammone

(2) Stop tree growth at a predesignated maximum level

(3) Record the class population at each level

(4) Use K-Means clustering algorithm to find K clusters for each class for all vectors arriving at each leaf

(5) Determine the mean, variance and mixture coefficient for each cluster

Figure 3 CDNTN model training algorithm

approximation

$$p(C_i|x) \approx p_j(C_i|x) \tag{7.10}$$

can be made. For the case when the data within spaces defined by different leaf regions are dependent, the global probability of a class occurring can be approximated by multiplying the local probability $p_j(C_i|x)$ given by eq. 7.9 for region j with $p(C_i|x_j)$, the probability of class C_i given the a feature vector falling into region j. Discrete sample counts can be used to approximate the probability $p(C_i|x_j)$, given in eq. 7.6, and the global probability $p(C_i|x)$ can be found by,

$$p(C_i|x) \approx p(C_i|x_j)p_j(C_i|x). \tag{7.11}$$

The primary advantage of the CDNTN is to allow the modeling mechanisms embedded in the NTN to partition the feature space into distinct regions to allow a local parametric model to be developed on the data in that leaf region. This method is most useful when the subspace defined by each leaf is statistically independent. This is the same motivation for developing hidden states for a Markov chain. The different hidden states represent statistical models for independent processes that can describe the same class. Another advantage of partitioning the features space to construct local models is that it allows the opportunity to eliminate unlikely models by the hyperplanes defined in the NTN. These hyperplanes are placed to allow discrimination between classes. If a mixture model is used without any partitioning of the feature space, decisions can be made based on the tail-end of a distribution. The width of a distribution is determined by the variance of the data, which can be unreliable given a finite data set. The NTN allows a mechanism for disregarding broad regions of feature space by the discriminating hyperplanes.

5 CONTINUOUS DENSITY NEURAL TREE NETWORK WORD SPOTTING SYSTEM

The CDNTN model described earlier provides one example of combining a discriminate classifier with a mixture model. By using the CDNTN to model the posterior probability of a subword segment within a keyword, the subword models are connected together to form a Markov chain. For word spotting applications the added discriminate information in the CDNTN model allows better discrimination between subword models within a keyword. One such method is to apply discriminate training between subword state models. Feature vectors common to one subword difficult to separate from vectors assigned to other subwords, are grouped together by the NTN in a region with low confidence or low posterior probability. Feature vectors that are well separated from those assigned to other subwords, are grouped by the NTN into high confidence regions. This provides a natural partitioning of the subword data so that complex distributions are modeled more accurately. Once a Markov chain is created for each keyword, state durations are extracted from the training data and clustered to form a state duration template to non-parametrically model the durations of each state. For testing, a dynamic time warping algorithm is used to evaluate the state outputs for the test utterance against the state duration model extracted from the training data. The CDNTN based word spotting system is described in Fig. 4. A windowed segment of digitized speech signal is extracted and used to generate a series of spectrally based feature vectors. Each feature vector is applied to a CDNTN model trained to recognize a specific phoneme within a keyword. Each CDNTN model produces an estimate of the posterior probability of the feature vector occuring and the resulting probabilities are applied to a DTW routine. The DTW routine measures the distortion between the series of vectors formed by the CDNTN outputs over the time window, to a reference template. The reference template is created by applying actual keywords to the CDNTN models and averaging the results using a clustering method [44] to combine duration templates generated from multiple keyword utterances. The output word score in Fig. 4 is then compared to a keyword threshold to determine a putative keyword occurrence. The extraction window is incremented in time and the entire process is repeated, generating a sequence of keyword scores for every increment of the sliding window.

The Road Rally Speech Corpora [15] was used to train and test the system. This data base consists of two independently recorded sections of different speakers from different dialects. The training section, known as the Waterloo Corpus

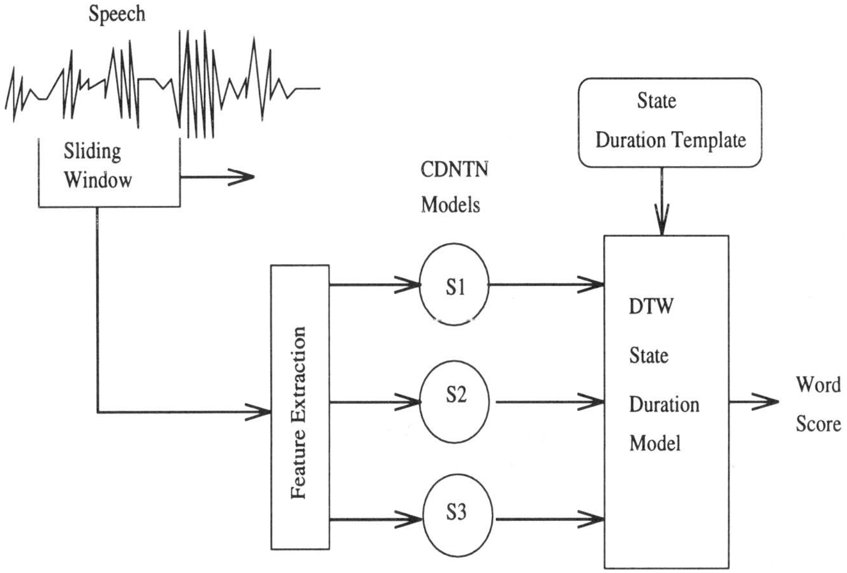

Figure 4 CDNTN word recognition system for a single word

is made up 56 speakers, 28 male and 28 female, reciting a paragraph about planning a road rally recorded through an actual telephone system sampled at 10KHz and filtered through a 300Hz to 3300Hz PCM FIR bandpass filter. The total duration of the Waterloo section consists of approximately two hours of read speech. Marking files for 20 keywords are provided, which specify locations of the 20 keywords within the speech files. A separate test corpus called the Stonehenge section, is made up of free unrestricted conversational speech independently recorded between two speakers planning a road rally. The Stonehenge speech was recorded on high quality microphones, and filtered using a 300Hz to 3300Hz PCM FIR bandpass filter to simulate telephone bandwidth quality.

Word marking files are provided for each test and training file to score the word spotting system. Word spotting systems are often reported by specifying hit rates as a function of false alarm rate/hour, known as a ROC curve. A ROC curve is generated by making a list of the putative keyword locations ordered by score from best to worst, across all talkers for each keyword. A tally is recorded for each correctly recognized keyword up to the 1st, 2nd, etc. false alarm encountered for each keyword and interpolated to false alarm rates/hour. A figure of merit (FOM) is defined as an average of hit rates allowing 0 to 10 false alarms per hour of speech to estimate the performance of the system for a variety of words at different false alarm rates.

Speech used for all word spotting experiments were parameterized into mel frequency scale cepstral coefficients (MFCC) using a 25.0 ms Hamming window every 10.0 msec, pre-emphasized by a first order difference with coefficient 0.97 [9]. The MFCC analysis used a 256 point FFT and an 18 channel triangular mel-scale filterbank. The first 8 MFCC coefficients were calculated from the filterbank outputs along with the normalized log energy added to make a 9 dimensional feature vector. The first and second differences were appended and the cepstral mean was subtracted, producing a 27 dimensional feature vector which was used for all experiments.

The Road Rally Speech Corpus was segmented using an HMM forced alignment. This was accomplished by training a serial network of three state HMM triphone models over all the utterances for each keyword. Once the HMM training was complete, the utterances were passed through the HMM networks and each feature vector was labeled according to the most likely phoneme model using a Viterbi search. This method was used to force a phonetically aligned segmentation for each keyword within the Waterloo section of the Road Rally Speech corpora. The forced alignment method can be replaced by one solely based on the CDNTN, and is described in [39]. A CDNTN segmental alignment procedure was implemented and experimental results showed a highly accurate segmentation is possible, when tested on a data base of phonetically segmented speech [39]. Once the speech data is phonetically segmented according to a phonetic dictionary, the phoneme segmentations are used to define subword states for discriminatively training a CDNTN to predict the posterior probability of a subword given a training vector. The anticlass data used for each subword CDNTN model consisted of the remaining subword vectors within the keyword labeled as not belonging to the subword being modeled. This amounts to training each subword model to predict the posterior output probability given a feature vector with respect to the other subwords within the keyword. CDNTN trees were grown for each phonetic subword occurring in each of the 20 keywords in the Road Rally Speech corpora. A total of 122 CDNTN trees were grown to model the subwords for the 20 keywords.

By restricting the training data used to develop the subword models to only phonetic data from that keyword, the CDNTN based keyword model can be used to define the most probable subword segmentations given a sequence of putative keyword feature vectors. Alternate strategies for constructing the training data set such as, adding confused words and randomly selecting alternate keyword data as anti-class data can be used. These methods were

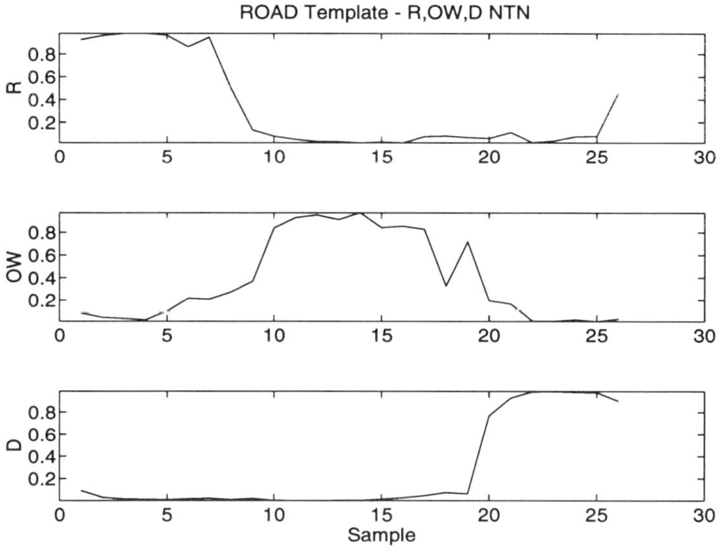

Figure 5 Averaged duration template for road state emissions

found to provide no substantial improvement in the keyword spotting system. A fundamental obstacle faced in using a discriminative classifier to construct subword models is that of defining an appropriate training set. A trade-off exists between trying to construct a global phonetic model using vast amounts of training data and one using locally appropriate data. Data sets using large amounts of anti-class data can mask the probability distribution of the specific phoneme model by artificially introducing a prior bias.

Subword duration templates were generated by applying the keyword utterances from the Waterloo section to the subword models for each keyword model. Fig. 5 shows the averaged template for the keyword *road* and Fig. 6 shows another template for the keyword *secondary*. As can be seen in Fig. 5 and Fig. 6, the CDNTN provides a relatively good model for the subwords across all the Waterloo speakers. A poor subword model will result in a blurred average template. The duration template can be thought of as a matched filter for the subword outputs during a keyword. If a non keyword is presented to the system, the subword models will output low random probabilities which will in turn produce high distortion values during the template match.

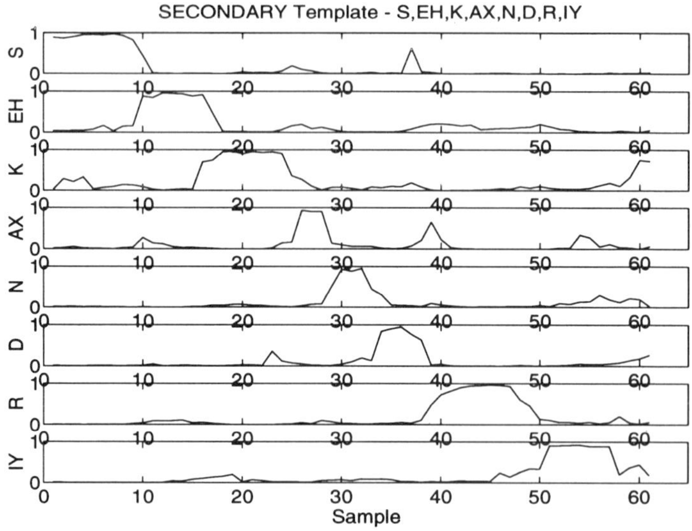

Figure 6 Averaged duration template for secondary state emissions

Fig. 7 shows a sample output of the duration distortion for the keyword *springfield* when the utterance "take the primary interstate west into *Springfield*" is spoken. The figure shows the distortion of the duration template algorithm as a function of feature sample. The keyword occurs in the utterance at the negative peak approximately at sample 325, and can be easily extracted by simply thresholding the output distortion.

Table 1 outlines the performance of the CDNTN based word spotting system for testing against 10 male speakers in the Stonehenge section of the Road Rally Speech Corpus. The features used were MFCC coefficients with added energy, and acceleration terms, with the mean removed. Each subword unit in this system was trained using a 5% percentage based forward pruning method with a maximum tree level set to seven. A maximum number of mixtures created for both in-class and anti-class vectors in each leaf was limited to six. An iterative k-means clustering method was used, starting with a single mixture and incrementally increasing the number dynamically until less than n were assigned to each cluster. The limiting constant n was chosen as the dimension of the feature vector, in this case 27. The training data for this system used only the keyword utterances from each of the 56 speakers in the Waterloo section of the Stonehenge speech corpora. Each recited paragraph from the Waterloo

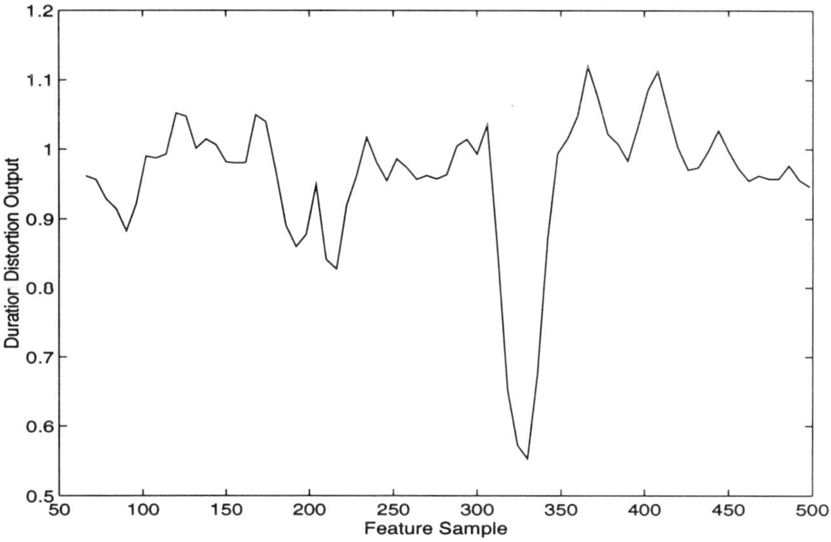

Figure 7 Sample output distortion of duration template match for springfield

Test Conditions	#Hits	#Actual	FOM	Hit Rate for 6FA's/hour
Cross Sex	750	900	36.55%	39.0%
Male Test	360	433	41.70%	44.0%
Female Test	390	467	33.28%	35.5%
Male Test(16 Keywords)	285	331	51.35%	53.9%

Table 1 CDNTN Keyword Spotting Performance

section provided 99 keyword tokens for training per speaker. Discriminative training data for the subword units was selected as the feature vectors assigned to the remaining subwords within the keyword. Subsequently, each keyword model can be trained using only data from the utterances of that keyword from the training data base. The total amount of keyword tokens used for training this system was 5,544 for all 20 keywords. This averages to 277.2 tokens/keyword and 5 tokens/keyword/speaker.

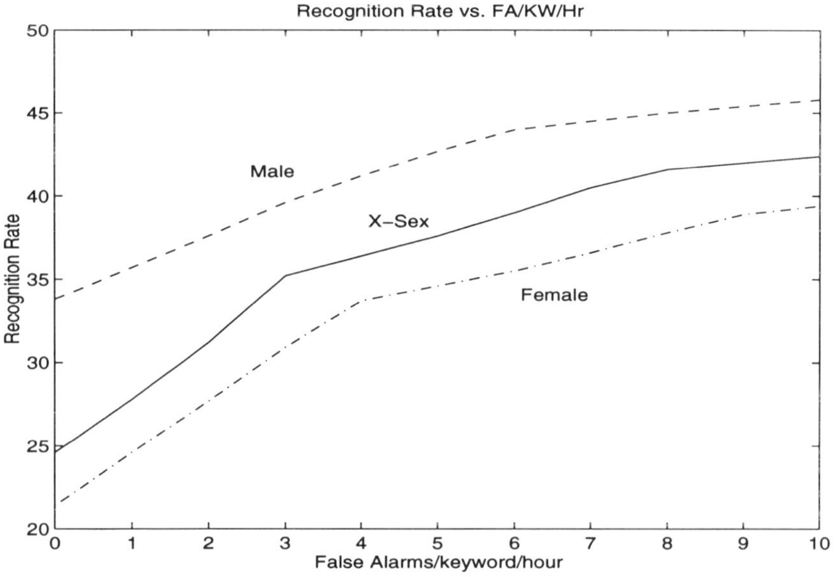

Figure 8 Overall performance for 20 keywords

Fig. 8 shows the overall performance of the system for all 20 keywords up to an error rate of 10 FA's/Keyword/Hour for the male, female and cross-sex tests. As can be seen in the figure, the male speakers scored better than the female speakers. Table 1 outlines the results for male, female and cross-sex tests. The high number of hits in Table 1 shows that this system can benefit greatly from a post processing method to further refine the keyword scores. A multi-level classifier system such as described by [24][25][46], has great potential for re-scoring false putative hits. Detailed investigation showed that a large majority of the keywords performed well, while a subset of keywords brought down the average score. Since no explicit background model is used, the shorter, simple keywords perform much worse that the longer keywords. This is a direct result from the fact that the longer keywords have more subword unit models which are more difficult to fit to random speech by the template matching duration model.

One of the primary advantages of using a discriminate classifier for modeling the state occupations within a HMM model is the use of the additional data provided by the anti-class feature vectors. Discriminative training maximizes the use of costly training data. The CDNTN model provides an effective means for blending the attributes of both the continuous mixture model and the discriminative neural network. The grouping action of the NTN allows a efficient parametric model to be made for the vectors in each region of the features space. A minimal number of exemplars are necessary to define the regions defined by the NTN leaves.

To measure the performance of the CDNTN system as a function of training tokens, a number of systems were trained varying the number of speakers in the training set. Each system was trained using the identical training parameters of 5% forward pruning, seven maximum NTN levels, and six maximum mixtures/class/leaf. The recited paragraph used in the Waterloo section contains 99 keyword tokens, which amounts to approximately 5 tokens/keyword/speaker. Since no background model is assumed, no extra tokens are needed to train the system. This considerably reduces the cost involved in training the system in terms of providing marked speech files for training the system. For many applications such as monitoring keywords from a non cooperative subject, large amounts of speech data maybe impossible to obtain. The CDNTN based word spotting system described maximizes the use of available tokens by the CDNTN state models to obtain superior performance to comparable conventional HMM systems [7][33][41][45][47]. Fig. 9 shows the system performance of the CDNTN word spotter compared to two HMM systems, as a function of average number of training tokens used per keyword. The HMM system in Fig. 9 shown as a dashed line, uses all the available training data from the Waterloo passage, which contains a total of 321 tokens for training both the keywords and the background model not including a silence model used. This translates to approximately 16 tokens/keyword/speaker. The CDNTN system requires only the keyword data, which translates to approximately 5 tokens/keyword/speaker for training, with no extra data for modeling background silence. Fig. 9 also shows the performance of the HMM system trained using pooled keyword data for both keyword triphone models and the background model (dot-dashed line), except for the silence model which uses background silence between keywords. Fig. 9 shows that the CDNTN based word spotting system provides superior performance to an HMM system when trained using limited amounts of training tokens. The neural network in the CDNTN combined with the internal mixture models created for both anti-class data and true-class data allows maximal use of the class information embedded in the limited amounts of training tokens. When sufficient quantities of data exist, the HMM model can produce better results. This is most likely due to the more complex recognition structure used in the HMM model combine with explicit background models trained to recognize non keyword speech.

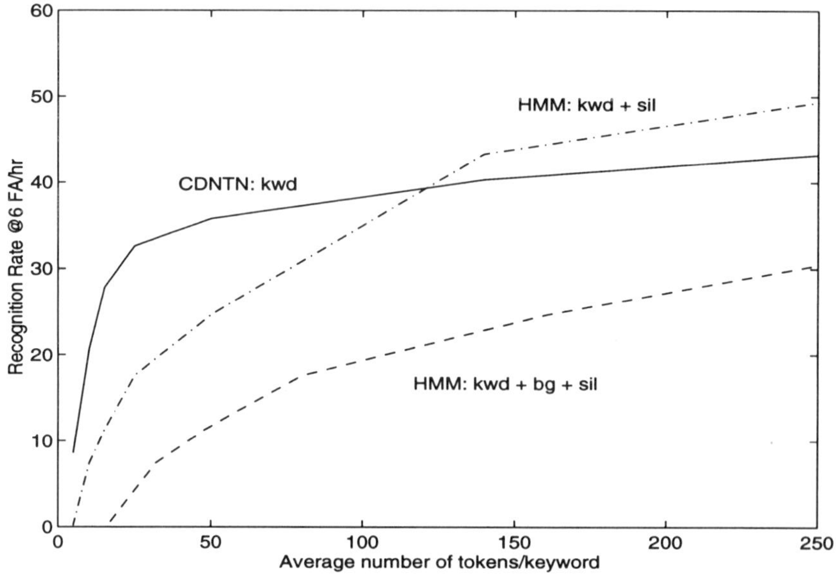

Figure 9 Comparison of CDNTN to HMM word spotting system

6 CONCLUDING REMARKS

This chapter examines the use of Arificial Neural Tree Networks for speech recognition. It was shown how ANN's can be used as a probablistic model. A CDNTN was described to show one specific implentation of an ANN, used to model the posterior probability of a class occurring. The CDNTN blends the non parametric discriminative training used by a conventional NTN model with a parametric Gaussian mixture model. A CDNTN based word spotting system was described. The system used a nonparametric state duration modeling method based on dynamic time warping template matching, as a method for obtaining non parametric duration models for a Markov chain using CDNTN state models. Results were given for a 20 keyword system for word spotting on the Road Rally Speech Corpora. The CDNTN was shown to out perform the best HMM system using small amounts of training data. The CDNTN system is most suitable for field applications of word spotting, which require fast training based on a limited number of exemplars. The CDNTN word spotting system allows each keyword to be trained independently to the number of keywords in the system. This allows convenient scaling for both large and small word spotting applications. No additional background data is needed and no background silence model is assumed for the system, allowing simplified training for different acoustic domains. The word spotting system described is not limited to using a CDNTN state model. Alternative ANN models can also be used to model the posterior probability of the subwords.

REFERENCES

[1] Bellman R, *Dynamic Programming*, Princeton University Press, Princeton, 1957.

[2] Bourlard H, Morgan N, "Connectionist Approaches to the Use of Markov Models for Speech Recognition", *Advances in Neural Information Processing Systems 3*, eds. R.P. Lippman, J.E. Moody & D.S. Touretzky, pp. 213-219, San Mateo, CA., Morgan Kaufmann, 1991.

[3] Breiman L, Friedman J.H, Olshen R.A, Stone C.J, *Classification and Regression Trees*, Wadsworth International Group, Belmont, CA. 1984.

[4] Bridle J.S, "An Efficient Elastic-Template Method for Detecting Given Words in Running Speech", British Acoustical Society Spring Meeting, Chelsea College, Paper No. 73SHC3, April 1973.

[5] Bridle J.S, "Pattern Recognition Techniques for Speech Recognition", J.C. Simon (ed.) *Spoken Language Generation and Understanding*, D. Reidel Publishing Co., pp. 129-145, 1980.

[6] Bridle J.S, "Neural Networks or Hidden Markov Models for Automatic Speech Recognition: Is there a choice?", Pietro Laface and Renato De Mori (ed.), *Speech Recognition and Understanding*, Springer-Verlag, Berlin, 1990.

[7] Carlson B, Chang E, Lippman R, Reynolds D, Zissman M. "Wordspotting on an HTK Foundation", Speech Research Symposium XIII 1993, June.

[8] Cybenko G, "Approximation by Superposition of a Sigmoidal Function", Math. of Control,Signals and Systems, 2:303-314, 1989.

[9] Deller J.R.Jr, Proakis J.G, Hansen J.H.L, *Descrete-Time Processing of Speech Signals*, Macmillan Publishing Co., New York, 1993.

[10] Duda R.O, Hart P.E, *Pattern Classification and Scene Analysis*, John Wiley & Sons, New York, 1973.

[11] Fallside F, "Neural Networks for Continuous Speech Recognition", Pietro Laface and Renato De Mori (ed.), *Speech Recognition and Understanding*, Springer-Verlag, Berlin, 1990.

[12] Farrell K, *Speaker Recognition using the Modified Neural Tree Network*, PhD. Dissertation, Rutgers University, New Brunswick, Oct. 1993.

[13] Farrell K, Mammone R, Assaleh K, "Speaker Recognition Using Neural Networks and Conventional Classifiers", IEEE Trans. on Speech and Audio Processing, Vol 2, No. 1, Part II, Jan. 1994.

[14] Finster H, "Automatic Speech Segmentation using Neural Network and Phonetic Transcription", Proc. of Int. Joint Conf. on Neural Networks, Baltimore, Vol. IV., pp. 734-736, June 1992.

[15] Fisher W.M, Doddington G.R, Goudie-Marhsal K.M, "The DARPA speech recognition research database: Specifications and status", Proceedings of the DARPA Speech Recognition Workshop, pp. 93-99, 1986.

[16] Funahashi K, "On the Approximate Realization of Continuous Mappings by Neural Networks", Neural Networks, 2:183-192, 1989.

[17] Hecht-Nielson R, *Neurocomputing*, Addison-Wesley, Reading, 1989.

[18] Hornik K, Stinchcombe M, White H, "Multilayer feedforward networks are universal approximators", Neual Networks, Vol. 2, pp. 359-366, 1989.

[19] Kosonocky S.V, *A Continuous Density Neural Tree Network Word Spotting System*, Ph.D Dissertation, Rutgers University, New Brunswick, NJ, Jan. 1995

[20] Krzanowski W.J, "The Location Model for Mixtures of Categorical and Continuous Variables", Journal of Classification, 10:25-49, 1993.

[21] Kung S. Y. *Digital Neural Networks*, Prentice Hall, Englewood Cliffs, 1993.

[22] Li K.P, Naylor J.A, "Whole Word Recurrent Neural Network for Keyword Spotting", Int. Conf. on Acoustics, Speech and Signal Processing, Vol. II, pp.81-84, 1992.

[23] Lippman R.P., "An Introduction to Computing with Neural Nets", IEEE Acoustics, Speech and Signal Processing Magazine, 4(2):4-22, April 1987.

[24] Morgan D.P, Scofield C.L, Lorenzo T.M, Real E.C, Loconto D.P, "A Keyword Spotter Which Incorporates Neural Netoworks for Secondary Processing", Int. Conf. on Acoustics, Speech and Signal Processing, New Mexico, Vol. 1, pp. 113-116, 1990.

[25] Morgan D.P, Scofield C.L, Adcock J.E, "Multiple Neural Network Topologies Applied to Keyword Spotting", International Conf. on Acoustics, Speech and Signal Proc., pp. 313-316, 1991.

[26] Myers C.S, Rabiner L.R, Rosenberg A.E, "An Investigation of the Use of Dynamic Time Warping for Word Spotting and Connected Speech Recognition", Int. Conf. on Acoustics, Speech and Signal Processing, Denver, pp. 173-177, 1980.

[27] Nadas A, "Some relationships between ANN's and HMMs",R. Mammmone (cd.), *Artificial Ncural Nctworks for Spccch and Vision*, Chapman & Hall, London, 1994.

[28] Paul D.B, "Speech Recognition Using Hidden Markov Models", The Lincoln Laboratory Journal, Vol. 3. No. 1, 1990.

[29] Rabiner L, "A Tutorial on Hidden Markov Models and Selected Applications in Speech Recognition", Proceedings of the IEEE, vol. 77, no. 2, Feb. 1989.

[30] Rabiner L, Juang B-H, *Fundamentals of Speech Recognition*, Prentice Hall, Englewood Cliffs, 1993.

[31] Renals S, Morgan N, Bourland H, Cohen M, Franco H, "Connectionis Probability Estimators in HMM Speech Recognition", IEEE Trans. on Speech and Audio Processing, Vol. 2, No. 1, Part II, Jan. 1994.

[32] Richard M.D, Lippman R.P, "Neural Network Classifiers Estimate Bayesian a posteriori Probabilities", Neural Computation 3, pp. 461 483, 1991.

[33] Rose R.C, Paul D.B, "A Hidden Markov Model Based Keyword Recognition System", Intr. Conf. on Acoustics Speech and Signal Processing, vol I, pp. 129-132, 1990.

[34] Rosenblatt F,"The perceptron: A probalistic model for information storage and organization in the brain", Psychological Review, 66:386-408, 1958.

[35] D.E. Rumelhart, G.E. Hinton, R.J. Williams, "Learning Internal Representations by Error Propagation" in Rumelhart and McClelland(Eds.), *Parallel Distributed Processing: Explorations in the Microstructure of Cognition*, Vol. 1, MIT Press, 1986.

[36] Sankar A, Mammone R.J, "Optimal Pruning of Neural Tree Networks for Improved Generalization", Proc. of the Int. Joint Conf. on Neural Networks, Seattle, 1991.

[37] Sankar A, Mammone R.J, "Growing and pruning neural tree networks", IEEE Trans. on Computers, C-42, pp. 221-229, March 1993.

[38] Samaddar S, Mammone R, "A neural tree network method for image segmentation", SPIE Conference, Neu- ral Networks for Image Processing, Innsbruck Austria, Oct. 1993.

[39] Sharma M, Mammone R.J, "Speech Recognition Using Sub-Word Neural Tree Network Models and Multiple Classifier Fusion", submmitted to the 1995 Int. Conf. on Acoustics, Speech and Signal Processing.

[40] Sivakumar S.C, Robertson W, Macleod K, "Improving Temporal Representation in TDNN Structure For Phoneme Recognition", Proc. the Int. Joint Conf. on Neural Networks, Baltimore, June 1992.

[41] Vrooman L.C, Narmandin Y, "Robust Speaker-Independent Hidden Markov Model Based Word Spotter", *Speech Recognition and Understanding, Recent Advances*, Edited by, P. Laface and R. DeMori, Springer Verlag, Berlin, 1992.

[42] Waibel A, "Connectionist Large Vocabulary Speech Recognition", Pietro Laface and Renato De Mori (ed.), *Speech Recognition and Understanding*, Springer-Verlag, Berlin, 1990.

[43] Whittaker J., *Graphical Models in Applied Mathematical Multivariate Statistics*, John Wiley & Sons, Chich- ester, 1990.

[44] Wilpon J.G, Rabiner L.R, "A Modified K-Means Clustering Algorithm for Use in Isolated Word Recognition", IEEE Trans. on Acoustics, Speech, and Signal Processing, Vol. ASSP-33, June, 1985.

[45] Wilpon J.G, Rabiner L.R, Lee C-H, Goldman E.R, "Automatic Recognition of Keywords in Unconstrained Speech Using Hidden Markov Models", IEEE Trans. on Acoustics, Speech and Signal Processing, Vol. 38. No. 11, Nov. 1990.

[46] Wu J, Chan C, "Isolated Word Recognition by Neural Network Models with Cross-Correlation Coefficients for Speech Dynamics", IEEE Transactions on Pattern Analysis and Machine Intelligence, Vol. 15, No. 11, pp. 1174-1185, Nov. 1993.

[47] Young S.J, Woodland P.C, Byrne W.J, "Spontaneous Speech Recognition for the Credit Card Corpus using the HTK Toolkit", submitted for publication to IEEE Transactions on Speech and Audio Processing Special Issue on Robust Methods, 1993.

[48] Zacks J, Thomas T.R,"A new neural network for articulatory speech recognition and its application to vowel identification",Computer Speech and Language, Vol. 8, pp. 189-209, 1994.

[49] Zeppenfeld T, Waibel A.H, "A Hybrid Neural Network, Dynamic Programming Word Spotter", International Conf. on Acoustics, Speech and Signal Proc., Vol II. pp. 77-80, 1992.

[50] Zeppenfeld T, Houghton R, Waibel A.H, "Improving The MS-TDNN for Word Spotting", International Conf. on Acoustics, Speech and Signal Proc., Vol. II, pp. 475-478, 1993.

8

CURRENT METHODS IN CONTINUOUS SPEECH RECOGNITION

P.S. Gopalakrishnan

IBM T.J. Watson Research Center, P.O. Box 704, Yorktown Heights, NY 10598

ABSTRACT

Several significant advances have been made in continuous speech recognition over the last few years. In this chapter, we will discuss some of the current techniques in feature extraction and modeling for large vocabulary continuous speech recognition.

1 INTRODUCTION

Large vocabulary continuous speech recognition has been the subject of intense research for several years. With the availability of significant computation power on the desktop in today's personal computers, the demand for versatile user interfaces and human computer interaction methodologies has brought renewed interest to the field. However, systems capable of providing high accuracy large vocabulary continuous speech recognition still remain largely in the realm of research. Error rates as well as computation time still remain high because of the complexity of the recognition problem. Several factors lead to this complexity. There is inherent variability in the speech signal produced by the same speaker at different times. In addition, there are common phenomena that result from the characteristics of the human vocal system which lead to variations in the way a sound is uttered in the context of other sounds. If the system is to be speaker independent, requiring no training data from a new speaker, then it has to handle the variability in speech between different speakers. The problem is more complicated if we have to deal with significant amounts of ambient noise. All these considerations lead to the use of more complex feature extraction and modeling methods to obtain higher accuracy recognition.

Significant progress has been made recently in this field, bringing reliable continuous speech recognition systems closer to reality. Recent progress has been driven by many factors, including the development of new analysis and modeling techniques, the availability of significant amounts of training data, as well as the availability of more powerful computers, which facilitate the use of complex models in recognition systems. In this chapter we will discuss some of the feature analysis and modeling methods that have yielded good results in continuous speech recognition.

In order to address the different issues in continuous speech recognition, we need to look at several areas. An accurate representation of the speech signal for recognition purposes must be obtained by designing a good feature set. The next issue to be considered is accurate models for these features. Finally, efficient search algorithms have to be designed in order to pick the best sequence of words given some acoustic input. Different sections of this chapter will deal with these different areas.

In this chapter, we will exclusively deal with transcription of speech and will not cover the area understanding and spoken language systems. Also, emphasis will be placed on current techniques specific to large vocabulary continuous speech recognition systems. Some of the more common modeling techniques that are fairly well-known by now will be discussed only briefly. We would like to warn the reader at the outset that a complete treatment of the work in this field cannot be attempted in a single chapter. A significant fraction of the work will be only briefly mentioned or omitted altogether. For instance, a large body of work dealing with segment-based approaches will not be covered here [32, 41]. We urge the reader to peruse the listed references for more details on different modeling techniques and algorithms.

1.1 Preliminaries

In order to understand the rest of the material in this chapter, let us examine the basic organization of a recognition system. A block diagram of a typical continuous speech recognition system is shown in Figure 1. The general approach to continuous speech recognition starts from a set of informative features extracted from the signal. Information about the distribution of energy in different frequency bands and the time varying nature of this distribution are usually encoded in the features. These form the acoustic observations for the recognition system.

The criterion used to select the word sequence to be output for a given acoustic input corresponds to a maximum *a posteriori* decoder. Given some acoustic observations Y the decoder chooses a sequence of words $W = w_1 w_2 \dots w_n$ such that

$$W = argmax_W Prob(W|Y) \quad (8.1)$$
$$= argmax_W Prob(Y|W) Prob(W) \quad (8.2)$$

We are assuming, implicitly, that all the information that is contained in the speech signal is encoded in the features. The recognizer does not know the probabilities required to find the maximum *a posteriori* word sequence, but estimates these from some models. The conditional probability $Prob(Y|W)$ is obtained from acoustic models. The *a priori* probability $Prob(W)$ is obtained from a model of the language. The language models typically tend to be some form of n-gram models for large vocabulary speech recognition. When dealing with large vocabularies, finding the word sequence that maximizes the *a posteriori* probability is a non-trivial task, requiring the use of sophisticated heuristic search algorithms.

The above discussion leads to the following organization of the material in this chapter. In the next section we will examine various feature extraction methods. Modeling techniques will be discussed in Section 3. In this chapter we will concentrate mostly on the acoustic models. Language models will be only briefly mentioned in Section 3. In Section 4 we will discuss some search techniques for finding the maximum probability word sequence. We close with some concluding remarks in Section 5.

2 FEATURE EXTRACTION

Many of the current large vocabulary speech recognition systems use frame-based features. Such features are extracted at equally spaced intervals from the speech signal and are designed to capture short term information such as the average energy in different frequency bands over that interval. Let us briefly examine one example of a frame-based feature set [10] to see what form of information is encoded in such features.

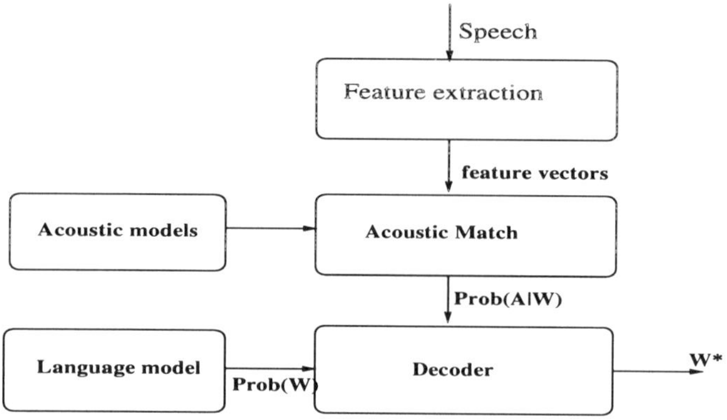

Figure 1 Recognition system organization

The main components of the feature extraction algorithm described in [10] are shown in Figure 2. The speech signal is acquired through a microphone, band-pass filtered, and digitized at a certain sampling frequency. For the sake of this discussion, let us assume that the sampling rate is 20kHz. A 512-point Hanning window is applied to this digitized signal. This is repeated every 10 milliseconds. In each window, a 512-point FFT is computed to get a 256-point power spectrum. This is then reduced to a smaller number of parameters by adding the power in each of 20 different frequency bands, approximating a critical-band filter bank. The center frequencies of these bands are distributed roughly linearly in the low frequency range and logarithmically in the higher frequency range. This yields a vector of 20 dimensions every 10 milliseconds. In the system described in [10] further processing is done using a model of the human auditory system to derive the feature vector from this 20-dimensional vector. The energy in each band is first converted to a loudness measure and then a short-term adaptation is performed, converting the loudness to a "neural" firing rate. The 20-dimensional vector thus obtained becomes the feature vector used in the recognition system. Each dimension in the feature vector is related to the signal power in the corresponding frequency band.

Several alternate feature extraction methods are prevalent in current recognition systems, including linear predictive analysis and related methods [21], cepstral parameters derived from LPC, and mel cepstral parameters [38, 26]. The reader is referred to [12] for a complete discussion of these methods.

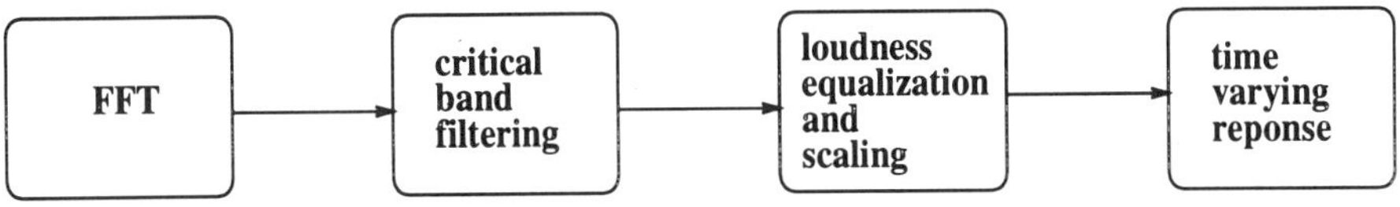

Figure 2 Components of a feature extraction algorithm

2.1 Long Term Information

The feature vectors derived using the methods discussed above do not incorporate any information about the way the signal changes over longer periods of time. It is well known that certain acoustic events are characterized by phenomena that take place over several time frames, and identification of such phenomena is crucial to the recognition of these events. For instance, consider the stop consonant K (as in ta*k*e). Examining the spectrogram of the signal from several utterances of this word, it can be observed that in the vowel preceding K the second and third formants move closer together just before the closure for K. Termed the *velar pinch*, this is often the signature of K. Thus, in order to distinguish between different stop consonants, one may have to look for evidence in the features adjacent to those that actually correspond to the stop consonant. There are several such characteristics one can point to in the speech signal that lead one to believe that incorporating information about such long term variations in the signal will be useful in recognition. This realization has lead to the development of several algorithms that incorporate long term information in addition to the short term spectral information in the features.

One possible approach to encoding long term information is to compute the time derivative of the frame based feature vectors over several frames. The derivative can be approximated by taking the difference $\delta_{t,i}$ at time t for feature dimension i which can be calculated easily as

$$\delta_{t,i} = f_{t+m,i} - f_{t-m,i} \tag{8.3}$$

for some value of m, typically 3 or 4. A better procedure is to estimate the slope of the feature value in each dimension using a linear regression fit over $2m$ frames centered around the current frame. The second order difference is then obtained from these first order differences using the same method. If $f_{t,1} f_{t,2} \dots f_{t,d}$ are the feature values at some time frame t, these are augmented by the first order differences δ_t and second order differences Δ_t to obtain an augmented $3d$ dimensional feature vector

$$\phi_t = [f_{t,1} \dots f_{t,d} \delta_{t,1} \dots \delta_{t,d} \Delta_{t,1} \dots \Delta_{t,d}]^T. \tag{8.4}$$

A more general approach results by viewing the above operation of taking differences as computing linear combinations of feature values from different time frames. Let $f_t = [f_{t,1} f_{t,2} \cdots f_{t,d}]^T$ be a d-dimensional feature vector for time frame t. Let us define a spliced vector s_t for time frame t to be the $((2m+1)d)$-dimensional vector obtained by simply adjoining the feature vectors from m frames preceding and following time frame t, including the vector at time t. That is,

$$s_t^T = f_{t-m}^T f_{t-m+1}^T \cdots f_t^T f_{t+1}^T \cdots f_{t+m}^T \tag{8.5}$$

The above operation of taking first and second order differences to get the augmented feature vector ϕ_t can then be viewed as the matrix vector product

$$\phi_t = D_\delta \times s_t \tag{8.6}$$

where D_δ is a $[3d \times (2m+1)d]$ matrix, and ϕ_t is the $3d$-dimensional feature vector given in (8.4). One can now attempt to generalize this by designing a different, hopefully better, linear combination of the feature values from different time frames. This may be accomplished by selecting a $[k \times (2m+1)d]$ matrix so as to capture information from the speech signal that will be useful in a recognition system. In essence, we are attempting to classify each time frame of speech into one of several classes. This classification problem can be tackled easily if we have a feature set that takes widely different values for these classes. We attempt to obtain such a feature set by formulating the following optimization problem.

Assume that a large amount of speech data is available for training and that a method exists for classifying each frame of speech in this training data to one of a set of L classes (we defer the discussion of how this is accomplished till Section 3). Let n be the total number of frames of speech present in the training data and assume that the spliced vector for each of these time frames are computed. The spliced vectors belonging to each class are collected together and the sample covariance matrix of the ith class is computed as W_i. Let n_i be the number of samples in class i. The average within-class covariance is then computed as

$$W = \frac{1}{n} \sum_i n_i W_i \tag{8.7}$$

where $n = \sum_i n_i$. Let O be the total covariance matrix of the n samples. The variance of the projection of the spliced vectors belonging to the ith class along a vector v is then given by $v^T W_i v$. The average within class variance along this direction can then be computed as $v^T W v$. We would like to determine the direction which maximizes the discrimination between different classes. Such a direction can be obtained by finding a vector v such that

$$v = argmax_u \frac{u^T O u}{u^T W u}. \tag{8.8}$$

This will result in the maximization of the overall variance while minimizing the average within class variance along v. This implies that this direction will be useful in providing discrimination between different classes while ensuring that the data from the same class does not vary widely along this direction. It can be shown that the solution of (8.8) is the eigen vector of $W^{-1}O$ with the largest eigen value. To obtain a set of k directions, the set of eigen vectors p_k of $W^{-1}O$ with the k largest eigen values are computed. The projection of the spliced vector s_t onto these eigenvectors yields a k-dimensional feature vector ϕ_t for each frame t. Described in [16], this technique has been successfully used in speech recognition to capture information about the variation of the frame-based features across several time frame [9, 7, 23, 20].

One drawback of this scheme, as with any supervised training scheme, is that any error in the initial assignment of frames to classes will result in non-informative directions being computed. This problem can be eliminated to some extent by iterating the whole modeling procedure - that is, re-classifying the training data after going through one training step and repeating the computations of the projections. Another consideration is that if the training data comes from several speakers, the inherent variability in the features for each speaker might bias the procedure. Hence the technique as discussed here may be weaker for speaker independent systems.

All these feature extraction procedures attempt to change the representation of the speech signal so as to make the modeling easier while attempting to capture as much information as possible to aid the recognition task. As the features incorporate more information the models have to be more complex to deal with them. We will examine these issues in more detail in the following section.

3 MODELING

Recall that acoustic models in the recognition system are used to compute the conditional probability $Prob(Y|W)$ of generating an observation sequence Y given the word sequence W. The most popular modeling method used in current systems makes use of hidden Markov models (HMM). We will assume that the reader is familiar with this modeling method and hence will not go into a detailed discussion of the basics of these models here. For the sake of completeness of the discussion and to establish some terminology, we sketch some essential concepts below. The reader is referred to [34, 35] for a more complete discussion.

A hidden Markov model (HMM) M is characterized by a set of states S, an $|S| \times |S|$ transition probability matrix A, where $a_{i,j}$ is the probability of moving from state i to state j, and a set of output probability distributions $B_{i,j}(y)$ of generating an observation y on the transition from state i to state j. For clarity, let us consider discrete observations at this time. In other words, the output generated on a transition is one out of a set of L discrete symbols. Then the output probability distribution $B_{i,j}$ is a discrete probability distribution. In addition, we are also given an initial probability distribution c_i over the states of the model, where c_i is the probability that the model starts in state i. Often a unique state is identified as the starting state $s \in S$ and the initial probability distribution assigns non-zero probability only to this starting state. Moreover, a state is also identified as the final state $f \in S$. Let us define a path p_n of length n to be a sequence of exactly n transitions. Given a sequence of observations $Y = y_1 y_2 \ldots y_n$ the probability of producing that sequence on a path p_n can be computed as a product of the transition and output probabilities along the path, and the initial state probability for the first state on the path. The probability $Prob_M(Y)$ of model M generating Y is computed as the sum

$$Prob_M(Y) = \sum_{p_n} Prob(Y|p), \tag{8.9}$$

where p is a path in the model which has a length of n. The conditional probability $Prob(Y|w)$ of generating Y given a word w is computed as above using the model M for the word. Often, we are interested in the probability that a model generates Y and is in the final state f at the end. The above sum is then computed for all paths that end in state f. Though the number of paths of length n in the model grows exponentially, there are efficient algorithms for computing these probabilities, whose computation time is bounded by $O(|S|^2 n)$.

Discrete observation HMMs are used as models in systems that use a vector quantizer (VQ) to label the feature vectors using a codebook of some size L. The output probability distribution on a transition of the HMM then gives the probability of generating any one of these VQ labels on that transition. Such models have the advantage that the distributions can be estimated without assuming any parametric form for them. However, such systems have to contend with the information loss in the VQ. More recent approaches eliminate the VQ and use the feature vectors directly as the observations generated on the HMM transitions. One can generalize the discrete HMMs to handle such multi-dimensional real valued (continuous) observations. In this case the output probability distributions are replaced by multi-variate density functions. Most of the above computations carry through to this situation with density values replacing probabilities. The drawback of using continuous observations is that one has to pick a parametric distribution such as a Gaussian as the density function, which makes the models more restrictive. As we shall see later, this problem can be alleviated to some extent by using mixtures of Gaussian distributions.

3.1 Estimating HMM Parameters

The parameters of the hidden Markov models are the transition and output probabilities, $\Theta = \{a_{i,j}, b_{i,j}(y)\}$. In the case of continuous observation HMMs, the parameters characterizing the output distributions are determined by the particular family of distributions that are used to represent the output probability. These parameters are estimated from training data. For speaker independent systems, they are estimated using training data from several speakers, whereas speaker dependent systems require these to be estimated using training data from each specific speaker. For both discrete and continuous observation HMMs, the parameters are estimated by selecting an optimization criterion and choosing the parameters to maximize this criterion.

Let Y be the acoustic observations obtained from the training data. Since we know the training script W, we can construct the HMMs for the training data as well. In maximum likelihood estimation, we select the parameter set Θ such that

$$\begin{aligned} \Theta &= argmax_{\Theta} Prob_{\Theta}(Y, W) \\ &= argmax_{\Theta} Prob_{\Theta}(Y|W) Prob_{\Theta}(W) \end{aligned} \tag{8.10}$$

Typically, the acoustic model parameters are estimated independent of the language model parameters. Hence, the second term in the above function $Prob_{\Theta}(W)$ is constant over Θ and can be ignored for acoustic parameter estimation. We are then left with the conditional probability as the maximization function. The maximization has to be done subject to certain linear constraints on the parameters Θ imposed by the fact that these for probability distributions. An efficient iterative maximization method, the Baum-Welch algorithm, exists for this formulation which is widely used to train HMMs [8]. The same technique can be applied for continuous observation HMMs for a large family of parametric distributions. An alternate optimization criterion that is sometimes used is the maximum mutual information function. Here we attempt to estimate parameters Θ such that

$$\Theta = argmax_{\Theta} log \frac{Prob_{\Theta}(Y, W)}{Prob_{\Theta}(Y) Prob(W)} \tag{8.11}$$

An extension of the Baum-Welch algorithm exists for maximizing this function [19]. The maximum mutual information estimation method has been successfully used in some systems [9, 30], but for large vocabulary systems it has not resulted in superior recognition results compared to maximum likelihood methods.

3.2 Constructing Models for Words

Given the basic HMM structure, one has to decide how to construct models for words. This is accomplished in most cases using a phonetic transcription of the word. The term *phonetic model* in the ensuing discussion is used to refer to word models obtained by concatenating the models corresponding to each phone in the phonetic transcription of the word. Each phone in the alphabet is assumed to be modeled using an HMM with a simple topology, for example as shown in Figure 3.2. Given a word w, its phonetic transcription is first retrieved to give the sequence $ph_1 ph_2 \ldots ph_k$ of phones, and the model for each phone ph_i in the transcription is obtained. These models are then concatenated by making the final state of the model for ph_i be the initial state of the model for ph_{i+1} (see Figure 3.2). This method of constructing word models would be adequate if not for the following complication that arises in continuous speech.

It is a well known phenomenon in continuous speech that the pronunciation of words changes in the proximity of other words. This occurs due to the characteristics of the articulatory system as well as due to speaking style and conventions. For instance, take the word *and.* When it is pronounced in the phrase *stop and go*, this word is often reduced to a single phone "*n*". The final stop *d* in *and* is dropped as it is impossible to pronounce the two stop consonants in sequence. Since this is a common phrase, the initial vowel is often reduced or dropped altogether. Such phenomena do not seem to affect the intelligibility of utterances for a human listener. However, in a speech recognition system we need to adjust the models to take into account these variations.

This phenomenon has been studied extensively by phoneticians who have constructed phonological rules that explain such context dependence [11, 31]. However, the use of such rules in recognition systems has not been very successful. Perhaps, a fundamental problem with this approach is that it relies on human perception rather than acoustic reality. Furthermore, this method only identifies gross changes, and the more subtle changes, which are generally unimportant to humans but may be of significant value in speech recognition by computers, are ignored. It is thus more appropriate to construct models for phones or words in context by using some algorithm on a large amount of speech data.

Some approaches attempt to model all distinct triphones (sequence of three consecutive phones) in an effort to capture phonological variations [26, 37]. Each triphone is modeled using a different HMM. The topology of the HMM (number of states and the connectivity between these states) is usually taken to be the same for all triphones, but the parameters are estimated separately. The drawback of this scheme is that as the vocabulary increases, the number of distinct triphones increases dramatically. Once the parameters are estimated, it may turn out that the parameters for several different triphones are close to being identical which leads to inefficient modeling. Also, triphones previously unseen in training data are also difficult to handle. There have been several approaches for tackling some of these drawbacks [22, 26]. A more general approach to constructing context dependent models is described in [4]. The method proposed in that paper has the advantage that it allows the models to take into account much longer contexts. It also has better powers of generalization, i.e. modeling contexts that do not occur in the training data.

Before we go on to describe this method, we need to examine a requirement underlying several methods discussed below, namely, the availability of a large set of speech data samples wherein each frame of acoustic observation has been labeled with a class identity. Typically, these classes are either phones or subphone units. Since these procedures require a large amount of training data, it is impractical to classify the data manually. Hence, the training procedure starts from an initial set of models that are trained using some unsupervised method. These models are used to classify each frame of acoustic observation to a phonetic class. Let us look at one possible scheme for labeling data with class labels. We start by deriving a vector quantizer (VQ) for the acoustic feature vectors generated from the training data. This can be done using an unsupervised method such as the "K-means" clustering algorithm [13]. The training data feature vectors are then labeled using this VQ, replacing each feature vector by the identity of the VQ class that is closest to it. We can now train the transition and output probabilities of a set of discrete observation phonetic HMMs using this data. This allows us to align the training data against the HMMs as follows. Since the word sequence that corresponds to the training data is known, we can construct phonetic models for the training sentences. In each sentence model, we find the path p such that the probability $Prob(Y|p)$

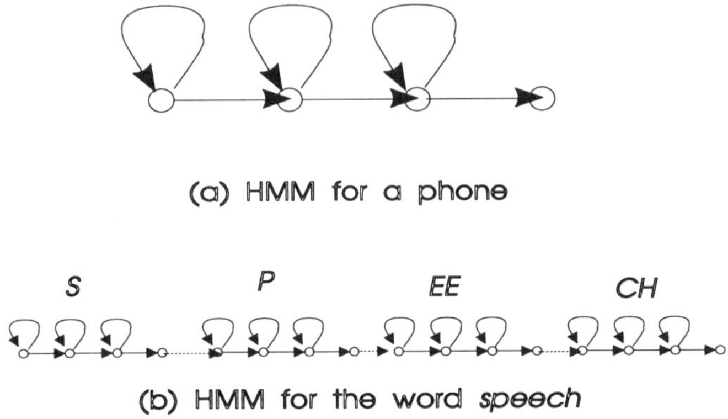

Figure 3 Phonetic HMMs for words

of the acoustic observations Y (the VQ label sequence) given the path p is maximum. This can be computed efficiently using a dynamic programming algorithm [39]. The maximum probability path specifies the transition within the model that is traversed for each time frame. This then allows us to classify each frame of speech using the identity of the phone or the transition within the phone. For the purpose of section 3.3, we label each frame by the identity of the phone. For the purpose of deriving models in section 3.4 we may label each frame by the identity of the transition that is taken at that time frame in the maximum probability path.

Now let us return to the issue of constructing models for phones and words in context. The procedure described in [4], which is summarized below, attempts to identify variations of each phone in different phonetic contexts by constructing decision networks.

3.3 Context Dependent Modeling Using Decision Networks

The procedure starts by going through an unsupervised training step to compute some initial models and using these models to align a large amount of training data against phonetic HMMs. The maximum probability path through these models corresponding to the VQ label sequence obtained for a training utterance will specify, for each phone, the sequence of VQ labels that is aligned against it. For each aligned phone a data record is constructed which contains the identity of the current phone, denoted as P_0, the context, i.e. the identities of the K previous phones and K following phones in the phone sequence, denoted as $P_{-K}, \ldots P_{-1}, P_1, \ldots P_K$, and the VQ label sequence aligned against the current phone, denoted as y. This collection of data is partitioned on the basis of P_0. Based on this annotated data a decision network is constructed for each phone.

To construct a binary decision network, the algorithm begins with a collection of data, which in this case consists of all the annotated samples for a particular phone. This is split into two subsets, and then each of these two subsets is split into two smaller subsets, and so on. The splitting is done on the basis of binary questions about the context P_i, for $i = \pm 1, \ldots \pm K$. A goodness-of-split evaluation function is used to select the question at each step. The goodness-of-split evaluation function is based on a probabilistic measure that is related to the homogeneity of a set of label strings. Finally, some stopping criterion is used to limit the size of the network that is generated. Splitting is terminated when the number of samples at a node falls below a threshold, or if the goodness of the best split falls below a threshold. The result is a binary network in which each terminal node represents one equivalence class of contexts. During recognition, given a phone and its context, the decision network of that phone is identified. By answering the questions about the context at the nodes of the network, the appropriate terminal node is identified, which specifies the model to be used.

Let Q denote a set of binary questions about the context. Let n denote a node in the network, and $m(q, n)$ the goodness of the split induced by question $q \in Q$ at node n. We will need to distinguish between tested and untested nodes. A tested node is one on which $m(q, n)$ has been evaluated for all questions $q \in Q$ and either the node has been split or designated as a terminal node. A sub-optimal greedy algorithm is used to construct the network, selecting the best question from the set Q at each node. In outline, the decision network construction algorithm works as follows. We start with all samples at the root node. In each iteration some untested node n is selected and $m(q, n)$ is evaluated for all possible questions $q \in Q$ at this node. If a stopping criterion is met, the node n is declared as a terminal node. Otherwise the question q with the highest value of $m(n, q)$ is associated with this node. Two new successor nodes are created. All samples that answer positively to the question q are transferred to the left successor and all other samples are transferred to the right successor. These steps are repeated till all nodes have been tested.

The most important aspects of this algorithm are the set of questions Q, the goodness-of-split evaluation function $m(q, n)$, and the stopping criteria. The question set used in [4] is obtained from common classification of phones. Let P denote the alphabet of phones, and N_P the size of this alphabet. The question set Q consists of questions of the form [Is $P_i \in S$] where $S \subset P$. We start with singleton subsets of P, e.g. $S = \{p\}$, $S = \{t\}$, etc. In addition, subsets corresponding to phonologically meaningful classes of phones commonly used in the analysis of speech [15], e.g., $S = \{p, t, k\}$ (all unvoiced stops), $S = \{p, t, k, b, d, g\}$ (all stops), etc. are added as questions. Each question is applied to each element P_i for $i = \pm 1, \ldots \pm K$, of the context. If there are N_S subsets in all, the number of questions N_Q is given by $N_Q = 2KN_S$. Thus there will be N_Q splits to be evaluated at each node of the network.

The goodness-of-split evaluation function is based on a probabilistic model of collections of label strings. Let Ω denote a particular class of parametric models that assign probabilities to label strings. For any model $\omega \in \Omega$ let $Pr_\omega(y)$ denote the probability assigned to label string y. Let Y_n be the set of label strings associated with node n. $Pr_\omega(Y_n) = \prod_{y \in Y_n} Pr_\omega(y)$ is a measure of how well the model ω fits the data at node n. Let $\omega_n \in Omega$ be the best model for Y_n, i.e. $Pr_{\omega_n}(Y_n) \geq Pr_\omega(Y_n)$ for all ω. $Pr_{\omega_n}(Y_n)$ is a measure of the purity of Y_n. If the label strings in Y_n are similar to each other, then $Pr_{\omega_n}(Y_n)$ will be large. A question q will split the data at node n into two subsets based on the outcome of question q. The goal is to pick q so as to make the successor nodes as pure as possible. Let Y_l and Y_r denote the subsets of label strings at the left and right successor nodes, respectively. Obviously, $Y_l \cup Y_r = Y_n$. Let ω_l and ω_r be the corresponding best models for the two subsets. Then

$$m(q, n) = log\left(\left(Pr_{\omega_l}(Y_l) Pr_{\omega_r}(Y_r)\right) / Pr_{\omega_n}(Y_n)\right) \tag{8.12}$$

is a measure of the improvement in purity as a result of the split. Since the goal is to divide the strings into subsets containing similar strings, this quantity serves as the goodness-of-split evaluation function. For computational reasons, a simple class of models Ω is chosen, for instance Poisson models as in [4].

Recall that y is a sequence of acoustic labels. Let us make the simplifying assumption that the labels in the sequence are independent of each other. A result of this assumption is that the order in which the labels occur is of no consequence. Now, a string y can be fully characterized by its histogram, i.e. the number of times each label in the acoustic label alphabet occurs in that string. Let us now represent the string y by its histogram $y_1 y_2 \ldots y_F$, a vector of length F where F is the size of the acoustic label alphabet and each y_i is the number of times label i occurs in string y. Each component y_i of the histogram is now modeled by an independent Poisson model with mean rate μ_i. Then, the probability assigned to y by ω is

$$Pr_{\omega}(y) = \prod_{i=1}^{F} \frac{\mu_i^{y_i} e^{-\mu_i}}{y_i!} \tag{8.13}$$

The joint probability of all the strings in the set Y_n is then

$$Pr_{\omega}(Y_n) = \prod_{y \in Y_n} \prod_{i=1}^{F} \frac{\mu_i^{y_i} e^{-\mu_i}}{y_i!} \tag{8.14}$$

It can be easily shown that $Pr_{\omega}(Y_n)$ is maximized by choosing the mean rate to be the sample average, i.e., the best model for Y_n has as its mean rate

$$\mu_{ni} = \frac{1}{N_n} \sum_{y \in Y_n} y_i \quad \text{for } i = 1, 2 \ldots F \tag{8.15}$$

Let μ_{li} and μ_{ri} for $i = 1, 2 \ldots F$, denote the optimal mean rates for Y_l and Y_r respectively. Then $m(q, n)$ can be written as

$$\begin{aligned} m(q,n) \quad = \quad & \sum_{i=1}^{F} \Bigg[\sum_{y \in Y_l} (y_i \log \mu_{li} - \mu_{li}) - \sum_{y \in Y_l} \log(y_i!) \\ & + \sum_{y \in Y_r} (y_i \log \mu_{ri} - \mu_{ri}) - \sum_{y \in Y_r} \log(y_i!) \\ & - \sum_{y \in Y_n} (y_i \log \mu_{ni} - \mu_{ni}) + \sum_{y \in Y_n} \log(y_i!) \Bigg] \end{aligned} \tag{8.16}$$

Now, it is easily seen that for $i = 1, 2, \ldots F$

$$\sum_{y \in Y_l} \log(y_i!) + \sum_{y \in Y_r} \log(y_i!) = \sum_{y \in Y_n} \log(y_i!) \tag{8.17}$$

and

$$\sum_{y \in Y_l} \mu_{li} + \sum_{y \in Y_r} \mu_{ri} = \sum_{y \in Y_n} \mu_{ni} \tag{8.18}$$

Eliminating these terms and using equation (8.15) we have the evaluation function

$$\begin{aligned} m(q,n) \quad = \quad & \sum_{i=1}^{F} \{ N_l \mu_{li} \log \mu_{li} + N_r \mu_{ri} \log \mu_{ri} \\ & - N_n \mu_{ni} \log \mu_{ni} \} \end{aligned} \tag{8.19}$$

where N_l is the total number of strings at the left node and N_r is the total number of strings at the right node resulting from split q. At each node, the question q that maximizes the evaluation function (8.19) is selected. The result of this procedure is a binary decision tree. In order to be more parsimonious in the number of models it is convenient to allow nodes at different parts of the tree to merge if they contain similar strings. This results in a binary network being constructed, where each node may have more than one predecessor but has exactly two successors. The criterion used to decide whether two nodes can be merged is exactly the same as that given in (8.19).

Two very simple stopping criteria are used to limit the size of the network. If the value $m(q, n)$ of the best split at a node n is less than a threshold T_m the node is designated to be a terminal node. Also, if the number of samples at a node falls below a threshold T_s then it is designated to be a terminal node, where T_m and T_s are empirically selected thresholds.

The terminal nodes of the network for a phone correspond to the different allophones of the phone. Separate models may be constructed for each allophone. This procedure thus succeeds in identifying significant variations of each phone and also encoding the phonetic contexts which lead to these different allophones. During recognition, a sequence of phones is constructed by concatenating the phonetic baseforms of the words. For each phone in this sequence, the appropriate decision is used to trace the path corresponding to the context provided by the phone sequence. This leads to a terminal node, and the model associated with this node is used for that phone. By concatenating the model for each such allophone the Markov model for the entire word sequence can be obtained.

The procedure described in this section allows us to model variations at a phone level. However, in some cases it may be more desirable to model these phenomena at the level of sub-phonetic units, for instance at the level of the transitions or states within a phone model. We will examine such approaches in the next section.

3.4 Continuous Parameter Models

Let us now examine some techniques for constructing continuous parameter HMMs for recognition. Many of the current large vocabulary speech recognition systems [2, 14, 17, 40] make use of continuous parameter models (or minor variations [7]) for large vocabulary continuous speech recognition. The main issues to be addressed are what family of models should be used to represent each class, how to estimate the parameters of these distributions for each class, and what techniques can be used to model all the variations exhibited by the feature vectors belonging to each class.

The general approach is to make use of some family of mixture distributions such as mixtures of Gaussian densities to model the features belonging to each class. Let us assume that classes are defined by assigning a unique identity to each of the transitions within the HMM for a phone. Thus, if the HMM for each phone has c transitions and there are P phones in the alphabet, the number of classes is cP. Let us denote the set of sub-phonetic classes as L. If we examine the feature vectors belonging to some class in this set they will populate disjoint regions of the feature space. This arises due to many reasons. The variability exhibited by speech from the same speaker is one factor. If the models are being constructed for a speaker independent system, we will be dealing with speech data from different speaker who may give rise to widely varying features for the same class. In addition, features may be influenced by the pronunciation changes resulting from phonological effects. Thus the output probability distribution associated with a class has to be multi-modal and has to assign significant probabilities to several disjoint regions in the feature space. On the other hand, we need to select a family of distributions such that the parameters of the distributions can be estimated from training data using some tractable estimation algorithms. The latter fact forces us to use model families such as Gaussian or Laplacian distributions. We attempt to meet the former requirement by making a mixture model. A mixture of Gaussian densities, for instance, is given by the density

$$M(f) = \sum_{g} c_g N(f; \mu_g, C_g), \tag{8.20}$$

where c_g is the mixture weight and $N(.)$ denotes a Gaussian density function with mean vector μ_g and covariance matrix C_g. These are the parameters of the distributions which have to be estimated using techniques such as the Baum-Welch algorithm.

One way of simplifying the distribution representing each class is to increase the class alphabet L so as to make each class cover a more compact region in the feature space. To this end, let us define a class alphabet L_a where each of the original classes $l \in L$ gives rise to several members in L_a in different phonetic contexts. The assumption here is that we can identify the regions corresponding to the phonology induced variations in the features by representing these regions using different classes in L_a. This would leave us with the problem of modeling the variations due to different speaker or the changes in the features produced by the same speaker at different instances. The problem of defining the appropriate set of classes L_a is now similar to that addressed in Section 3.2 and the same type of techniques can be used here as well. Thus, one approach is to consider each triphone separately and to define a new class l_a for each of the original classes l occurring in different triphones [14, 17, 25, 36]. A method similar to the one discussed in Section 3.3 is described in [7] which defines classes L_a by constructing decision networks associated with each class $l \in L$. The procedure is very similar to that describe in Section 3.3 except for the fact that here we attempt to separate feature vectors into similar clusters depending on the phonetic context unlike in the earlier procedure where we were dealing with label strings. The interested reader is referred to [7] for details on how this is achieved. Several systems currently use such techniques to obtain high accuracy recognition [22, 40].

Once the class alphabet L_a is defined, we need to estimate the parameters of the mixture distributions for each of the classes. Typically initial estimates of the mixtures are obtained using a clustering algorithm to arrive at a certain number of Gaussian densities. In outline, a possible clustering algorithm works as follows. We first divide the set of feature vectors that correspond to class l_a into some number of subsets randomly. For each subset we then estimate the sample means and covariances. With this set of distributions in hand, we can compute the likelihood assigned each feature vector in the class by each distribution. We could then assign feature vectors to the distribution that gives the highest likelihood. This generates a new partition of the feature vectors for class l_a. The distributions can now be reestimated and the procedure iterated until some convergence criterion is met.

The parameters are then further trained using the Baum-Welch algorithm. The number of parameters may be reduced by assuming that the covariance matrix for each Gaussian is diagonal. Modeling accuracy is then improved by increasing the number of different densities that are used in the mixtures. Typically the number of density functions used in speaker independent systems runs into several tens of thousands. With large amounts fo speech data currently available, it has become feasible to estimate such large numbers of parameters reliably.

3.5 Language Models

A language model is used in the recognition system to compute the *a priori* probability of a word sequence. The most popular language models used in current large vocabulary speech recognition systems are n-gram models which make the assumption that the *a priori* probability of a word sequence can be decomposed into conditional probabilities of each word given the n words preceding it. More precisely, the *a priori* probability is written as

$$\begin{aligned} Prob(w_1 w_2 \ldots w_k) &= Prob(w_1) * Prob(w_2|w_1) * \ldots * \\ &\quad Prob(w_n|w_{n-1} \ldots w_1) * \\ &\quad \prod_{i=n+1}^{k} Prob(w_i|w_{i-1} \ldots w_{i-n}) \end{aligned} \tag{8.21}$$

Typically n is one or two, giving a *bigram* or *trigram* model. The parameters of these models are estimated by processing a large amount of text, preferably from the same domain for which the recognition system is designed, such as business news articles or medical reports. If the vocabulary is of the order of several tens of thousands of words, the number of unique word triplets are astronomical and we will not observe all of these when estimating a trigram model from several million words of text. Hence the models that are usually used fall back to using the probability from the bigram when the trigram has never been observed in the training data, or use a mixture of the trigram, bigram, and unigram probabilities by replacing the conditional probability $Prob(w_3|w_2 w_1)$ by

$$\lambda_1 Prob(w_3|w_2 w_1) + \lambda_2 Prob(w_3|w_2) + \lambda_3 Prob(w_3). \tag{8.22}$$

The parameters $\lambda_1, \lambda_2, \lambda_3$ are estimated from training data [24] and depend on the number of instances of the trigram and bigram observed in the training data. The interested reader should refer to [24] for more details on such models.

Having discussed several methods for acoustic models and language models, let us now recap the procedure involved in building a continuous speech recognition system. We first select a vocabulary that the system will work with, perhaps a few tens of thousands. We then collect together a table of phonetic transcriptions for the words. These may be generated by hand or by some automatic methods [3]. We assume that a large quantity of speech data is available for training the models. We design an appropriate feature analysis algorithm. We then select a method for modeling co-articulation, either by assigning different models for different triphones or by constructing a decision network as discussed in Section 3.3. For continuous observation HMMs an appropriate class alphabet at the sub-phonetic level is chosen as well as a desirable model family (Section 3.4). The parameters of these models are then estimated using the Baum-Welch algorithm. This sets up the acoustic models. The language model has to be constructed from a large amount of text data as described in Section 3.5. The recognition system can then use these models to compute the conditional probability of the acoustic observations given a word as well as the prior probability of a word sequence. All that remains is to have a method for selecting the maximum probability word sequence given some new acoustic observations. This is often a fairly complex task in large vocabulary recognition systems. Let us examine this in the next section.

4 SEARCH

Let us recall the search problem that results from the maximum *a posteriori* decoding paradigm. Given some acoustic data Y we attempt to select the word sequence W that maximizes the probability $Prob(W|Y)$. This maximization has to be done over all possible word sequences in the language. When dealing with a large vocabulary system and $n - gram$ language models as described in Section 3.5 which assign non-zero probability to every word in the vocabulary at any position in a sequence, the number of word sequences over which this maximization has to be done is astronomical. Moreover, this is a discrete optimization problem for which no systematic method for finding the global optimum is known. Hence we have to take recourse to heuristic search algorithms to find the most likely word sequence.

There are two main search paradigms currently being used - time-synchronous Viterbi beam search and asynchronous search such as the envelope search algorithm described in [18]. Both rely on identifying word sequences that are unlikely to lead to the optimal word sequence early in the processing and thus eliminating large parts of the search space from consideration. The main difference between the two is that in the former all word sequences that cover the same amount of acoustic data are evaluated at the same time whereas the latter evaluates word sequences of different lengths covering different amounts of acoustic data in successive steps. Both of these are more or less comparable in performance but offer different trade-offs. It is easier to incorporate models that take into account information from a large amount of word context into the envelope search. On the other hand, the results of the decoding are available as soon as the last frame of speech has been processed in the time-synchronous search, whereas one has to wait for all iterations to end in the envelope search before the final result in known.

Many of the current large vocabulary continuous speech systems that make use of time-synchronous search schemes tend to use more sophisticated techniques to reduce the computation time. One of the complications that arises in current systems is the context dependent nature of the models. Since models may depend on the phonetic context across word boundaries, the number of distinct HMM states to be considered at each time frame increases dramatically. Hence, more sophisticated methods are needed to reduce this number to a manageable size. Several algorithms have been developed recently which use a multi-pass method to compute the most probable word sequence [1, 27, 28]. The essential idea here is to use a fast initial pass to identify a subset of states for each time frame that are likely to be on the maximum probability path. Subsequent passes are then used to re-evaluate these states using more sophisticated acoustic and language models.

As a way of illustrating some of the issues relevant to search, let us examine in more detail the envelope search technique.

4.1 Envelope Search

A popular search algorithm used in speech recognition systems is the envelope search [18, 33]. This is an asynchronous search scheme and is based on heuristic search algorithms such as the A^* algorithm [29]. In brief, this algorithm makes use of a heuristic evaluation function which is related to the likelihood of a word sequence to order all hypothesized word sequences on a priority queue. In each iteration the best hypothesized word sequence based on this heuristic evaluation function is selected and it is extended, generating more word sequence hypotheses. The algorithm described in [18] uses a *fast acoustic match* to reduce computation, a *detailed acoustic match* to compute the probability $Prob(Y|w)$ for some segment of acoustics Y and a word w, and a trigram language model to evaluate the prior probability of a word sequence.

The *detailed acoustic match* component computes the probability of the acoustics given a word w by constructing an HMM for the word and using it to compute the probability of some segment of input acoustic observations given that word. The computation is initialized by assuming a certain probability for the starting state of the model for w. This starting probability is obtained as a result of the detailed match on the preceding word. The detailed match computes the probability $Prob(Y_i^j|w)$ (where Y_i^j is the acoustic observations covered by word w) and also determines the probable ending time of word w. This is specified in terms of an *end-time distribution* (see Figure 4) $e(w, t)$ which is a probability distribution over time. The value $e(w, t)$ gives the probability of word w ending at time t which is used as the initial state probability for the word following w when a detailed match is done for that word. The detailed

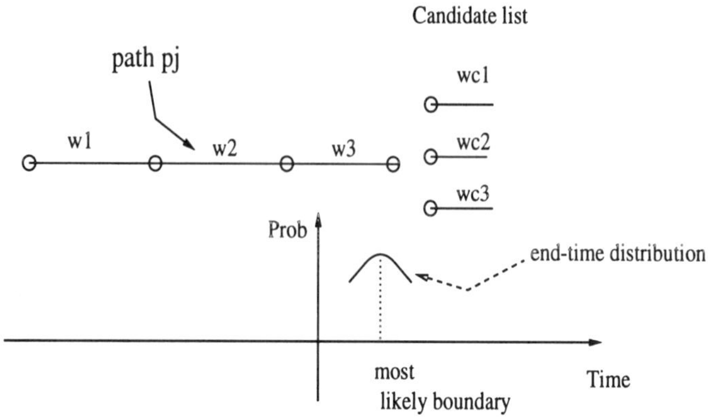

Figure 4 One search path and its extensions

match is computationally expensive and cannot be carried out in practice for all words in the vocabulary for a given starting time. The detailed match is restricted to a set of candidate words provided by the *fast match.*

The fast match is given the acoustic observations as input, as well as an end-time distribution $e(w, t)$ obtained from the detailed match of the preceding word w. The fast match first determines a set of words that are acoustically similar to the word following w by matching some part of the acoustic observations against models for words in the vocabulary. An efficient algorithm for this is described in detail in [6]. The essential characteristic of the fast match is that it requires much less computation compared to a detailed acoustic match. The result of this acoustic fast match computation is a list of candidate words $w_{c_1} \ldots w_{c_j}$ with associated acoustic fast match scores $a(w_c)$. This list is further pruned to a much smaller size by combining these scores with the *a priori* probability for each word obtained from the language model.

The decoder uses the above modules to search for the most likely sentence. The state of the decoder at any time is characterized by a set of ordered paths. Each path is a partial sentence that covers a certain amount of input speech. Let $w_1 w_2 \ldots w_i$ be the sequence of words corresponding to a path p_j (see Figure 4). The paths are ordered using their likelihood obtained as

$$l(p_j) = Prob(Y_1^k | w_1 \ldots w_i) * Prob(w_1 \ldots w_i), \tag{8.23}$$

where the first term is obtained from the acoustic match, Y_1^k being the acoustic evidence covered by this partial sentence, and the second term is obtained from the language model. Any path that covers all input speech is said to be a complete path. The decoding algorithm starts with the null path on the priority queue. In each iteration of the algorithm the best incomplete path is selected for expansion using certain criteria [18]. If there are no paths available for expansion the algorithm terminates and returns the best complete path. If, on the other hand, there is an incomplete path available for expansion then this path is expanded as follows.

Let the path p_j selected for expansion correspond to the word sequence $w_1\ w_2 \ldots w_i$. This path is associated with an end-time distribution $e(w_i, t)$. A fast match is first computed to determine a set of candidates for the word following w_i. The fast match uses the end-time distribution $e(w_i, t)$ and produces a set of candidate words $w_{c_1} w_{c_2} \ldots w_{c_m}$. A detailed acoustic match is then performed on each of these candidates to obtain the probability $Prob(Y_1^l|\ w_1\ w_2\ \ldots\ w_i w_{c_j})$. Now a set of m new paths are created which correspond to the word sequences obtained by adding each of the candidates w_{c_i} to the end of the sequence $w_1 \ldots w_i$. The detailed acoustic match gives the incremental acoustic probability associated with each of these words. The *a priori* probabilities are obtained from the language model. The likelihood scores given in (8.23) can then be computed for all these paths. These new paths are inserted in the queue with the associated scores and the current iteration of the algorithm then ends.

One advantage of this algorithm is that it is easier to incorporate language models that use a longer history to produce the prior probability for a word sequence. Also, it is easier to use acoustic models that take into account longer phonetic contexts without having to use multi-pass implementations. The implementation can also be designed to make efficient use of memory. On the negative side, since the algorithm has to compare word sequences that cover different length of acoustic observations, the design of the heuristic function and the selection of the best path to extend in each iteration becomes more complicated. Moreover, the result of decoding is not available as soon as the last frame of speech has been observed, since there may still be paths left that have to be extended.

The search space that large vocabulary continuous speech recognition systems have to contend with is large enough so that most of these algorithms strain the resources on currently available computers. As a result, these systems tend to take several seconds to process each second of speech. More advances in efficient search algorithms as well as more powerful computing resources will be required before such systems can be used in real time.

5 CONCLUSIONS

In this chapter we have attempted to discuss several current techniques in continuous speech recognition. Emphasis has been placed on material that is relevant to large vocabulary continuous speech recognition systems. We described several techniques in feature extraction and modeling that yield good recognition accuracy in current systems. This chapter is clearly not intended as a survey of the field or a detailed description of one system. Several excellent methods in this research area have been only briefly mentioned in the paper or not discussed at all. It is hoped that the reader will be able to use this as a starting point for further study in the field.

6 ACKNOWLEDGEMENT

Several modeling methods discussed in detail here have been developed jointly with members of the speech recognition group at IBM and I would like to thank all the members of the group for their help and encouragement in writing this paper.

REFERENCES

[1] F. Alleva, X. Huang, M. Hwang, "An improved search algorithm for continuous speech recognition," Proceedings of the 1993 International Conference on Acoustics, Speech, and Signal Processing, ICASSP-93, Minneapolis, MN, May 1993.

[2] X. Aubert, R. Haeb-Umbach, H. Ney, "Continuous mixture densities and linear discriminant analysis for improved context-dependent acoustic models," Proceedings of the 1993 International Conference on Acoustics, Speech, and Signal Processing, ICASSP-93, Minneapolis, MN, May 1993.

[3] L.R. Bahl, S. Das, P.V. de Souza, M. Epstein, R.L. Mercer, B. Merialdo, D. Nahamoo, M.A. Picheny, J. Powell, "Automatic phonetic baseform determination," Proceedings of the DARPA Speech and Natural Language Workshop, Hidden Valley, PA, June 1990, pp.179-184.

[4] L.R. Bahl, P.V. de Souza, P.S. Gopalakrishnan, D. Nahamoo, M.A. Picheny, "Decision Trees for Phonological Rules in Continuous Speech,"

Proceedings of the 1991 International Conference on Acoustics, Speech, and Signal Processing, ICASSP-91, Toronto, Canada, May 1991.

[5] L.R. Bahl, P.V. de Souza, P. S. Gopalakrishnan, D. Nahamoo, M. Picheny, "Word Lookahead Scheme for Cross-Word Right Context Models in a Stack Decoder," Proceddings of Eurospeech-93, Berlin, Germany, Sep.1993.

[6] L.R. Bahl, S.V. De Gennaro, P.S. Gopalakrishnan, R.L. Mercer, "A Fast Approximate Acoustic Match for Large Vocabulary Speech Recognition", *IEEE Transactions on Speech and Audio*, Jan 1993.

[7] L. Bahl, P. de Souza, P. S. Gopalakrishnan, D. Nahamoo, M. Picheny, "Robust methods for using context dependent features and models in a continuous speech recognizer," Proceedings of the 1994 International Conference on Acoustics, Speech, and Signal Processing, ICASSP-94, Adelaide, Australia, Apr. 1994.

[8] L.E. Baum, T. Petrie, G. Soules, N. Weiss, "A maximization technique occurring in the statistical analysis of probabilistic functions of Markov chains," *Annals of Mathematical Statistics*, 41, 1970, pp.164-171.

[9] P.F. Brown, "The acoustic modeling problem in automatic speech recognition," Ph. D. Thesis, Carnegie Mellon University, Pittsburg, PA, 1987.

[10] J.R. Cohen, "Application of an auditory model to speech recognition," *Journal of the Acoustical Society of America*, Vol.85, No.6, June 1989, pp.2623-2629.

[11] P.S. Cohen and R.L. Mercer, "The Phonological Component of an Automatic Speech Recognition System," in *Speech Recognition,* D.R Reddy, editor, Academic Press, New York, 1975, pp.275–320.

[12] J.R. Deller, J.G. Proakis, J.H.L. Hansen, *Discrete-time processing of speech signals*, McMillan Publishing, New York, 1993.

[13] E. Diday, J.C. Simon, "Cluster analysis," in *Digital Pattern Recognition*, K.S. Fu, ed, Springer-Verlag, NY, 1976.

[14] V. Digalakis, H. Murveit, "Genones: Optimizing the degree of mixture tying in a large vocabulary hidden Markov model based speech recognizer," Proceedings of the 1994 International Conference on Acoustics, Speech, and Signal Processing, ICASSP-94, Adelaide, South Australia, April 1994.

[15] G. Fant, *Speech Sounds and Features*, MIT Press, Cambridge, MA, 1973.

[16] H.P. Friedman, J. Rudin, "On some invariant criteria for grouping data," *Journal of American Statistical Association*, Dec 1967, pp.1159-1178.

[17] J.L. Gauvain, L.F. Lamel, G. Adda, M. Adda-Decker, "The LIMSI continuous speech dictation system: evaluation on the ARPA Wall Street Journal task," Proceedings of the 1994 International Conference on Acoustics, Speech, and Signal Processing, ICASSP-94, Adelaide, South Australia, April 1994.

[18] P.S. Gopalakrishnan, L. Bahl, R. Mercer, "A Tree Search Strategy for Large-Vocabulary Continuous Speech Recognition," Proceedings of the 1995 International Conference on Acoustics, Speech, and Signal Processing, ICASSP-95, Detroit, May 1995.

[19] P. Gopalakrishnan, D. Kanevsky, A. Nadas, D. Nahamoo, "An inequality for rational functions with applications to some statistical estimation problems," *IEEE Transactions on Information Theory*, Vol. 37, No. 1, January, 1991, pp.107-113.

[20] R. Haeb-Umback, H. Ney, "Linear discriminant analysis for improved large vocabulary continuous speech recognition," Proceedings of the 1992 International Conference on Acoustics, Speech, and Signal Processing, ICASSP-92, San Francisco, CA, March 1992.

[21] H. Hermansky, B.A. Hanson, H.J. Wakita, "Perceptually based linear predictive analysis of speech," Proceedings of the 1985 International Conference on Acoustics, Speech, and Signal Processing, ICASSP-85, Tampa, Florida, 1985, pp.509-512.

[22] M.-Y. Hwang, X. Huang, "Shared-distribution hidden Markov models for speech recognition," *IEEE Transactions on Speech and Audio Processing*, Vol. 1, No. 4, October 1993, pp.414-420.

[23] M.J. Hunt, D.C Bateman, S.M. Richardson, A. Piau, "An investigation of PLP and IMELDA acoustic representations and of their potential for combination," Proceedings of the 1991 International Conference on Acoustics, Speech, and Signal Processing, ICASSP-91, Toronto, Canada, May 1991.

[24] F. Jelinek, "Self-organized language modeling for speech recognition," in *Readings in Speech Recognition*, A. Waibel, K.-F. Lee, eds., Morgan-Kauffmann, Palo Alto, CA, 1990.

[25] F. Kubala, A. Anastasakos, J. Makhoul, L. Nguyen, R. Schwartz, G. Zavaliagkos, "Comparative experiments on large vocabulary speech recognition," Proceedings of the 1994 International Conference on Acoustics,

Speech, and Signal Processing, ICASSP-94, Adelaide, Australia, April, 1994.

[26] K.F. Lee, H.W. Hon, M.Y. Hwang, S. Mahajan, R. Reddy, "The Sphinx Speech Recognition System," Proceedings of the 1989 International Conference on Acoustics, Speech, and Signal Processing, ICASSP-89, Glasgow, Scotland, May 1989, pp.445-448.

[27] H. Murveit, J. Butzberger, V. Digalakis, M. Weintraub, "Progressive search algorithms for large vocabulary speech recognition," Proceedings of the ARPA workshop on Human Language Technology, Plainsboro, NJ, March 1993.

[28] L. Nguyen, R. Schwartz, F. Kubala, P. Placeway, "Search algorithms for software-only real-time recognition with very large vocabularies," Proceedings of the ARPA workshop on Human Language Technology, Plainsboro, NJ, March 1993.

[29] N. Nilsson, *Problem Solving Methods in Artificial Intelligence*, McGraw-Hill, New York, 1971.

[30] Y. Normandin, R. Cardin, R. De Mori, "High-performance connected digit recognition using maximum mutual information estimation," *IEEE Transactions on Speech and Audio Processing*, Vol. 2, No. 2, April 1994, pp.299-311.

[31] B.T. Oshika, V.W. Zue, R.V. Weeks, H. Nue and J. Auerbach, "The Role of Phonological Rules in Speech Understanding Research," *IEEE Transactions on Acoustics Speech and Signal Processing*, Vol. ASSP-23, 1975, pp. 104–112.

[32] M. Ostendorf, S. Roukos, "A stochastic segment model for phoneme-based continuous speech recognition," *IEEE Transactions on Acoustics Speech and Signal Processing*, Dec. 1989, pp.1857-1869.

[33] D. Paul, "An efficient A^* stack decoder algorithm for continuous speech recognition with a stochastic language model," Proceedings of the DARPA Workshop on Speech and Natural Language, Harriman, NY, Feb. 1992, pp.405-409.

[34] A.B. Poritz, "Hidden Markov models: A guided tour," Proceedings of the 1988 International Conference on Acoustics, Speech, and Signal Processing, ICASSP-88, New York, 1988, pp.7-13.

[35] L. Rabiner, B. Juang, "An introduction to hidden Markov models," *IEEE ASSP Magazine*, vol.3, 1986, pp.4-16.

[36] R. Roth, J. Baker, J. Baker, L. Gillick, M. Hunt, Y. Ito, S. Lowe, J. Orloff, B. Peskin, F. Scattone, "Large vocabulary continuous speech recognition of Wall Street Journal data," Proceedings of the 1993 International Conference on Acoustics, Speech, and Signal Processing, ICASSP-93, Minneapolis, April 1993.

[37] R. Schwartz, Y. Chow, O. Kimball, S. Roukos, M. Krasner, J. Makhoul, "Context-Dependent Modeling for Acoustic-Phonetic Recognition of Continuous Speech," Proceedings of the 1985 International Conference on Acoustics, Speech, and Signal Processing, ICASSP-85, April 1985.

[38] K. Shikano, "Evaluation of LPC spectral matching measures for phonetic unit recognition," (technical report), Computer Science Department, Carnegie-Mellon University, May 1985.

[39] A.J. Viterbi, "Error bounds for convolution codes and an asymptotically optimal decoding algorithm," *IEEE Transactions on Information Theory*, Vol. 13, No. 4, April 1967, pp.260-269.

[40] P.C Woodland, J.J. Odell, V. Valtchev, S.J. Young, "Large vocabulary continuous speech recognition using HTK," Proceedings of the 1994 International Conference on Acoustics, Speech, and Signal Processing, ICASSP-94, Adelaide, South Australia, April 1994.

[41] V. Zue, J. Glass, D. Goodine, M. Phillips, S. Seneff, "The Summit speech recognition system: phonological modeling and lexical access," Proceedings of the 1990 International Conference on Acoustics, Speech, and Signal Processing, ICASSP-90, Albuquerque, NM, April 1990.

9

LARGE VOCABULARY ISOLATED WORD RECOGNITION

Vishwa Gupta
Matthew Lennig

Bell-Northern Research
and INRS-Télécommunications,
Montreal, Quebec, Canada, H3E 1H6

1 INTRODUCTION

Many applications require recognition of spoken isolated words or phrases from a large vocabulary. For example, the goal of the 86000-word recognizer at INRS-Télécommunications [14] is to transcribe speech spoken as a sequence of isolated words. The sentences to be read are chosen arbitrarily from a variety of sources, including newspapers, books, magazines, etc. Another example is the StockTalk system running at BNR Montreal [24], which dispenses real time stock quotes by voice over the telephone for stocks traded in New York, Toronto and NASDAQ stock exchanges. The vocabulary for this system consists of words or phrases spoken in isolation. This system requires speaker-independent recognition over the telephone, while the first example requires speaker-dependent recognition over high quality microphones.

A typical flowchart for recognizing words or phrases in such applications is shown in Figure 1. There are many variations to this flow chart depending on the preferences of individual researchers, and many possibilities within each block. In describing these blocks, we will be guided by our own experiences, and to some extent will reflect our point of view.

At a higher level, we can divide the speech recognition algorithms into two categories: word-based or sub-word-based. Word-based recognition algorithms create a model for each word or phrase in the vocabulary. These algorithms are only suitable for small vocabularies. Each word-model has to be trained from several utterances of the word. For large vocabulary, word-based recognition requires unreasonable computing capabilities and a large effort in data collection in order to train the word models.

The alternative is to use sub-word based recognizers. The sub-words could correspond to phonemes, diphones, demi-syllables, speech segments etc. The idea in all these units is to reduce the total number of models required to transcribe any possible word in the vocabulary. Phonemes result in the smallest number of models. In phoneme based recognizers, each word is represented as a sequence of phonemes. Since there are roughly 40 phonemes in English, we can get away with training only 40 models instead of thousands for word-based systems. For this reason, phoneme-based recognizers require much less speech data to train the models. The computing complexity is also significantly less than that for word-based recognizers. In fact, the point where the computing for word-based recognizers becomes unreasonable is around a few hundred words.

Even though we are talking about models for phonemes, many researchers have shown that creating models for allophones [23] improves recognition accuracy. Allophones are phonemes in context. The acoustic realization of phonemes varies considerably depending on context (previous and following phonemes), on stress, on position in the syllable, on dialect etc. By representing the major variations as separate models, we can improve recognition accuracy significantly. However, there is some debate on the optimal number of allophones required to achieve good recognition accuracy. Various research groups have used anywhere from a few hundred to a few thousand allophones. Also, how to create these allophones is an issue. For example, Kai-Fu Lee [23] and Hon [15] have used phonemes in triphone contexts (previous and following phoneme) and a measure based on entropy reduction to choose their triphones. Bahl et al [6] at IBM have been recommending an approach based on data driven decision trees, and their allophone contexts can span up to 3 phonemes to the left and three phonemes to the right. The contextual variations due to co-articulation are a bigger problem in continuous speech than in words spokens in isolation, and Gopalakrishnan has probably addressed this issue in detail in the Chapter on continuous speech recognition.

Phonemes or allophones are not the only units used successfully in speech recognition. Bahl et al [5] use fenones as basic units. Fenones are data driven, and not driven by linguistic theory. A word in the vocabulary is represented as a sequence of fenones. Approximately 200 fenones are used to represent any word in the vocabulary. Fenones are probably described in detail in the article by Gopalakrishnan, since the speech recognition group at IBM has championed this approach. We will primarily stick to phonemes or allophones in our discussions. The search techniques or language models do not depend on the choice of phonemes or fenones, and all the algorithms which apply to phonemes also apply to other units as well.

Even though there are many ways to create phoneme or allophone models, the current best methods use Hidden Markov Models (HMMs). Many different styles of models have been used by various researchers. We can classify these styles into three broad categories based on the probability density function used for the observation vectors: discrete (or VQ-based) models, Gaussian mixture density based models, and tied mixture models. In VQ-based models, the observation vectors are mapped into discrete units by using one or more codebooks. For example, in Gupta et al [13], two separate codebooks are used to represent static and dynamic feature vectors. Kai-Fu Lee [23] uses three codebooks to represent the feature vector: one for the static parameters, one for the dynamic parameters, and one for the power and the dynamic power. Juang et al [18] use Gaussian mixture densities to represent the probability distribution of the observation vector. Tied mixture models also correspond to the Gaussian mixture models where the Gaussian mixtures are shared by all the transitions in one or more models.

Many different training algorithms have been used to train these models. These training algorithms include Viterbi training, forward-backward training, maximum mutual information (MMI) training, corrective training etc. In this Chapter we will focus on search algorithms, detailed matching strategies, and language modeling.

2 SEARCH ALGORITHMS

The goal of the recognizer is to transcribe words spoken as a sequence of isolated words. Since the words are spoken in isolated fashion, we can find the endpoints of each word, and transcribe each word separately. The task of the search algorithm is to find the most likely word hypothesis or to narrow this hypothesis to a few possible choices. Computing considerations dictate that we accomplish this with as little computing as possible without losing the correct choice from the hypothesis list.

A simplistic approach to large vocabulary recognition would compare the spoken word against all possible phonemic transcriptions for all the words in the vocabulary. To score all the words in the vocabulary in real-time would require hardware with computing capability on the order of hundreds of millions of floating point operations per second. The cost of the hardware to provide such computing capabilities would be prohibitive. To reduce the number of candidates for detailed comparison, we require an algorithm to perform fast

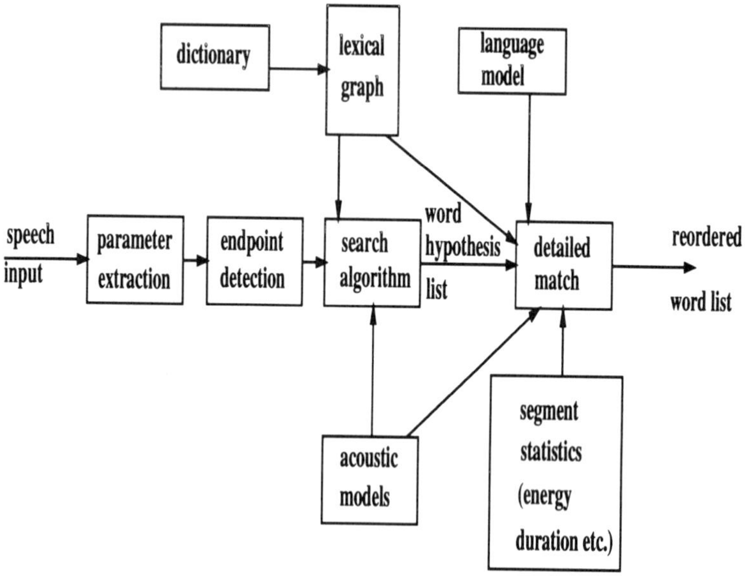

Figure 1 Block diagram for a typical large vocabulary isolated word speech recognition system.

search, reducing significantly the size of the list for subsequent detailed matching. A number of algorithms exist which reduce this search. These search algorithms can be classified into three general categories: search algorithms based on Viterbi decoding, those based on A^* heuristic search, and those based on fast match search.

2.1 Search Algorithms based on Viterbi decoding

The Viterbi decoding algorithm [31] provides an efficient way to get the most likely candidate. The earliest application of Viterbi decoding to speech recognition dates back to Vintsyuk [30]. The Viterbi algorithm provides the optimal frame synchronous search through a lexical graph whose branches represent allophonic or phonemic HMMs, and any path through the graph represents a lexical entry in the dictionary. Lee and Rabiner [22] give a good tutorial on the frame-synchronous search algorithms. Let us look at a rough estimate of computing involved in this frame-synchronous search algorithm. The computing complexity is proportional to the total number of transitions in all the HMMs associated with the lexical graph. For example, let us assume that the graph for a dictionary containing 20,000 phonemic transcriptions has a total of 100,000 branches. If each HMM has an average of 15 transitions, then the total computing per frame is proportional to 1,500,000. For a frame rate of 100 frames/second, we require computing capability in the order of 1500 million operations per second.

Another disadvantage of Viterbi algorithm is that it does not give us multiple choices economically. To generate multiple choices, we have to maintain multiple choices at each node in the lexical graph [22] resulting in a corresponding increase in computing. Soong [29] has shown how multiple choices can be obtained by carrying out a tree-trellis search after the Viterbi algorithm. His algorithm requires same computing complexity as the Viterbi algorithm for generating the top choice, but requires more memory for storing the intermediate values during Viterbi search. These values are then used during the tree-trellis search.

In order to reduce the computing even further, some people have resorted to beam search. Viterbi algorithm maintains all possible paths and extends these paths every frame, and the most likely path at the last frame corresponds to the optimal path. By extending only the most promising paths, the overall computing can be reduced significantly. For example, Ney [25] has shown that he can reduce the computing by an order of magnitude by using beam search together with proper structures to identify active partial paths. The beam search requires additional memory in order to track which paths are still alive and need to be extended during Viterbi search.

The tuning parameters which control beamwidth depend on the recognition task, and are usually set so as to minimize decoding errors while reducing computing as much as possible. These parameters may again have to be tuned even for the same task if the models or vocabulary or both are changed. If the graph represents a tree, then the beam search can also give multiple hypotheses. However, generating multiple hypotheses would require significant increase in computing, since the beam width would have to be increased significantly. To reduce the beamwidth, language model and any other sources of knowledge are incorporated during the frame-synchronous beam search [22] [25].

2.2 Search Algorithms based on A^* algorithm

A^* search algorithm [26] is a best-first optimal heuristic search as long as all its conditions are satisfied. The search keeps the partial paths on a stack and extends the partial path with the highest likelihood. The likelihood of the partial path includes the actual likelihood of the partial path and an estimate of the remaining path. The heuristic function which estimates the likelihood of the remaining path must provide a likelihood greater than or equal to the actual likelihood of the remaining path. As long as this condition is satisfied, A^* search ensures optimality. There are some trivial heuristic functions which satisfy the optimality condition (for example, likelihood of 1). However, such functions would lead to a slow breadth first search. The heuristic functions which yield fast and focussed search are the ones which estimate the likelihood of the remaining path close to the actual likelihood.

An early example of the A^* search algorithm is the stack decoding algorithm [16] [17]. In this algorithm, the likelihood of the remaining path was estimated as the likelihood of observing the remaining acoustic data independent of the path by collecting long term acoustic statistics. This likelihood was multiplied by a factor to ensure optimality. To avoid the search becoming breadth first search, the factor was controlled in order to ensure that the longest paths were slowly favored and extended.

Another example of the A^* search algorithm is the syllabic graph search algorithm used by Gupta, Lennig and Mermelstein [12]. In this algorithm, they first find the number of syllables n in the word using syllable models, and then carry out an A^* search through a network representing n syllables. This syllable network is much smaller than the network representing the entire lexicon (a few thousand branches versus a few hundred thousand branches). Every branch of the graph is scored independent of its phonetic context, allowing a fast computation of the branch likelihoods. The A^* search then generates multiple choices using these branch likelihoods. In this search strategy, the scoring of branch likelihoods is fast. The speed of search depends on number of choices necessary in order to include the correct choice among these hypotheses. For the 86,000 word recognizer, this algorithm required on an average 300 choices per word to keep the number of search errors to less than 2% (search error is when the correct hypothesis is not included in the hypothesis list).

Another example of A^* search is the A^*-admissible lexical search used by Kenny et al [20]. In this search algorithm, Kenny defines a class of heuristic functions which allow for a tradeoff between the efficiency of the search and the burden of evaluating the heuristic scores. The heuristic scores for the partial path are estimated using a reduced graph together with the Viterbi scoring algorithm. Two examples of these reduced graphs are two-phone and one-phone look ahead graphs. In two-phone look ahead graph, the reduced graph only ensures that the next two phones are identical to those in the lexical tree. This allows a much simpler graph to be constructed to estimate the heuristic scores. The heuristic scores by design turn out to be higher than the Viterbi scores. The two scores are merged by considering all possible segmentations corresponding to exact and heuristic scores. Only a few of these segmentation hypotheses have to be saved in order to achieve optimal decoding, resulting in significant savings in computing.

The advantage of this algorithm is that the actual score at the end of the search corresponds to Viterbi likelihood, and therefore, the search can be stopped as soon as the necessary n hypotheses have been obtained. While in the syllabic graph search, significantly more choices than n have to be obtained in order to get n best choices (the likelihoods in the first pass are rough likelihoods). Even though the search is fast in the first-pass, the second pass computing is significantly greater due to an average of 300 choices to be rescored in order to get the correct hypothesis as one of the choices. For this reason, the A^*-heuristic lexical search turns out to be faster since there is no rescoring involved (eliminating rescoring overhead) and fewer hypotheses have to be produced.

The disadvantage of this type of search algorithms is that they require large amount of storage. For example, in Kenny's A^* search algorithm, the heuristic scores have to be precomputed. Sufficient storage has to be allocated for the stack to store all possible active hypotheses at any given time. Since the total size of the stack cannot be predicted, the amount of memory necessary can only be estimated by trial and error. In any practical application, the bottom entries in the stack have to be pruned. Enough memory has to be allocated so that truncating the bottom hypotheses does not lead to suboptimal results.

2.3 Search Algorithms based on fastmatch algorithm

Another class of search algorithms called the fastmatch search algorithms rely on a fast method to evaluate all the words in the vocabulary in order to generate a short list of candidates. We then use detailed match to evaluate these candidates. The idea here is to get a short list as inexpensively as possible, and to evaluate this short list in detail. For example, in IBM's fast match algorithm [3], they collapse all the states in a hidden Markov model, reducing the model to a one state model. Just this modification reduces the computing by an order of magnitude. For example, for an HMM with 15 transitions, such a simplification reduces number of comparisons from 15 to one. They also use a beam search together with fast match in order to reduce the computing by as much as a factor of hundred. For the beam search, a threshold is used to truncate all the nodes which are unlikely to yield a valid hypothesis. This threshold is manipulated to allow paths of longer length to grow slowly. Such combination of simpler models and beam search reduces the computing by a factor of 100. Obviously, such reduced models require multiple choices to be produced in order to avoid decoding errors. In IBM's 20,000-word isolated-word speech recognizer, a maximum of 100 choices is sufficient in order to keep the percentage of decoding errors to a minimum.

2.4 Comparison of various Search Algorithms

The choice of the search algorithm depends on the requirements of the application and the hardware on which the application has to run. For example, Viterbi decoding using a beam search algorithm carries out all the computing frame synchronously, and provides the top choice result as soon as the input is finished. If the primary concern is the real-time delay in recognition, then this is probably the ideal algorithm to use. In systems with large memory, for example workstations, the computing complexity can be managed by tuning the beam search parameters to minimize computing while keeping the loss in recognition accuracy to a minimum. The drawback is that Viterbi decoding requires significant amount of computing and provides only one recognition choice. Post-processing to use more detailed models, segmental features, and language models is not possible with only one choice. These algorithms are generally embedded during the search algorithm, which makes the search algorithm even more expensive. Obviously, a tree-trellis search [29] or some other algorithm can be used after Viterbi decoding to provide multiple choices. The tree-trellis search requires even more memory since intermediate results during Viterbi decoding have to be saved for the tree-trellis search. Also, it introduces delay, since tree trellis search can only be performed after the complete utterance has been received.

A^*-search algorithms result in significant reduction in frame-synchronous computing. For example, Kenny et al [20] estimate the heuristic function using Viterbi decoding on a much smaller graph which they call a two-phone look ahead or a one-phone look ahead graph. Obviously, all the heuristic scores have to stored in order to be used during A^* search. For example, for a two-phone lookahead graph, they store the likelihoods for all possible two-phone sequences in the vocabulary for all the frames in the acoustic input. The A^* search also introduces decoding delays, since the search can only start after all the input is in. The delay depends on how rapidly the search proceeds, and this depends on the heuristic function. Even though the average delay in getting top few choices can be small, the maximum possible delay is unbounded. To avoid this maximum possible delay, either the search has to be aborted, or some short cuts have to be taken, resulting in sub-optimal decoding. Most of these parameters can be tuned to limit the loss in recognition accuracy. Because of the large memory requirements, these algorithms are well suited to hardware with large memory, for example workstations.

The fastmatch algorithm actually uses simpler models, in effect reducing memory requirements during frame synchronous search. The simpler models reduce the frame synchronous computing significantly. The beam search during fastmatch does bring some uncertainty to the amount of computing, but this can be controlled by adjusting tuning parameters to allow the longest paths to grow slowly. The fastmatch algorithm does introduce decoding delays, since all the hypotheses have to be rescored using detailed models, segment based features and language model etc. However, this delay can be controlled by limiting the number of choices at the expense of some loss in performance. All these facts show that fastmatch is ideally suited for hardware with limited memory resources.

3 DETAILED MATCHING ALGORITHMS

Once we have a short list of word hypotheses, we would like to rescore these hypotheses using best possible algorithms. Since only a few choices have to be rescored, we can use more complex algorithms to rescore them, without incurring heavy computational penalty. Let us look at some of the possible rescoring algorithms.

If the search algorithm used simpler models, then one obvious choice is to use the original models in their full complexity. Other possibilities include using segment based features. For example, imposing constraints based on segment durations and the energy profile of the segments. Where the spoken words are from sentences or phrases, a language model can be used to enhance recognition accuracy. Where a language model cannot be used, some a priori information about the word frequencies, etc. can possibly be used during rescoring.

Many researchers have used segment duration constraints to improve recognition accuracy. For example, Bush and Kopec [8] apply minimum duration constraints on digit-segments to improve digit-string recognition accuracy. They find that a minimum duration constraint of 50 ms for each acoustic segment is optimal for their digit-string recognizer. Soong [28] has used minimum durations to improve phoneme recognition accuracy in spoken Japanese text. He assigns minimum allophone durations to 2084 allophonic HMM's to improve phoneme recognition accuracy.

Many acoustic misrecognitions in the 86,000-word speaker-trained isolated word recognizer at INRS-Télécommunications [11] are due to phonemic hidden Markov models mapping to short segments of speech. When we force these models to map to longer segments corresponding to the observed minimum durations for the phonemes, then the likelihood of the incorrect phoneme sequences drops dramatically. This drop in the likelihood of the incorrect words results in significant reduction in the acoustic recognition error rate. Even in cases where acoustic recognition performance is unchanged, the likelihood of the correct word choice improves relative to the incorrect word choices, resulting in significant reduction in recognition error rate with the language model. On nine speakers, the error rate for acoustic recognition reduces from 18.6% to 17.3%, while the error rate with the language model reduces from 9.2% to 7.2%.

The minimum duration constraints we have found effective depend on the phoneme. For the sonorants and affricates, one duration minimum per phoneme is found to be effective everywhere, while for the remaining phonemes, the duration minimum depends on whether the phoneme occurs in initial, medial or final position in the word. These duration minima vary from 20 ms to 100 ms depending on the phoneme and the context in which it appears. Imposing duration constraints does increase computing requirements. An algorithm to impose these duration constraints efficiently is given in Gupta et al [11].

Segmental energy constraints have also been applied effectively to improve recognition accuracy. For example, Bush and Kopec [8], and Kopec and Bush [21] have applied energy constraints to improve isolated digit and digit-string recognition accuracy. The energy constraints have been imposed on either the peak energy in the speech segment or on the minimum energy in the speech segment. For example, imposing the constraint that the peak energy in a stressed vowel segment be above a certain threshold reduces certain digit insertion errors [8]. Forcing the minimum energy in voiceless fricative segments to be below a certain threshold reduces voiceless fricative confusion with sonorants [21]. These constraints have been incorporated in their recognition algorithm in order to improve recognition accuracy of their isolated digit and digit-string recognizers.

In the 86,000-word isolated-word recognizer at INRS-Télécommunications, we have succesfully applied energy constraints during training of the hidden Markov models for phonemes [11]. Many segmentation errors in the training data are between high energy and low energy phonemes. A number of these segmentation problems have been corrected by constraining the energy contours to prevent phonemes with high energy from mapping onto phonemes with lower energy. Segment boundaries between stops and sonorants, between fricatives and sonorants, between affricates and sonorants, and between breath and sonorants are most amenable to correction using energy constraints [11]. Segment boundaries between vowels, liquids, glides, and nasals can be corrected by using duration minima. Correction of the segment boundaries between phonemes in the training set leads to improved phoneme models, resulting in higher acoustic recognition accuracy and higher recognition accuracy with the language model.

Other successful strategies to improve recognition accuracy include combining results from two independent recognizers. For example, in a flexible vocabulary recognizer for common stocks listed in New York Stock Exchange [24], two separate feature parameters are used. The top choices using cepstral coefficients are rescored using line spectrum pair parameters. The log likelihoods from the two recognizers are then added. This method reduced the error rate from 5% to 4%. IBM has used a similar approach where they use two separate recognizers using different dictionaries. One recognizer uses phonemes to represents words in the dictionary, while another uses fenones to represent the words in the dictionary. Combining the likelihoods from the two recognizers results in a drop in word error rate from 2.5% (for fenonic models) to 1.8% for the combined fenonic and phonetic models [7].

4 LANGUAGE MODELS

For each word of spoken input, the acoustic recognizer generates a list of word hypotheses and their associated acoustic likelihoods. The language component takes this probabilistic word lattice as input and uses a statistical model of the syntactic, semantic, and pragmatic properties of English to generate the *a posteriori* most likely word string. A number of language models have been used previously for speech recognition. The trigram language model [17] has been used successfully for both a 5,000-word and a 20,000-word office correspondence task [1]. Derouault and Merialdo [9] use a tri-POS (parts-of-speech) model for a 250,000-word French recognizer. Application of a trigram language model to a recognition task with such a large vocabulary was considered infeasible by Derouault and Merialdo [9]. They also apply global syntactic constraints using a sentence parser. However, application of global syntactic constraints results in only a marginal improvement in the recognition accuracy of their system. The advantage of a tri-POS language model is that it requires significantly less memory for storage than a trigram language model and can be trained from a small training text corpus. However, the tri-POS language model estimates the probability of a word conditioned on the parts-of-speech of the previous two words, resulting in much weaker linguistic constraints than the trigram language model. Lee [23] and Rohlicek et al. [27] have used a simpler bigram or word-pair language model in a 997-word resource management task. To make best use of the available syntactic and semantic constraints, we have used a trigram language model in the 86,000-word vocabulary recognition system at INRS-Télécommunications.

The trigram language model parameters correspond to probabilities of words conditioned on the previous two words. The number of parameters to be estimated is enormous. For example, for the 86,000-word vocabulary recognition system, $86,000^3$ parameters have to be estimated. Even a training set consisting of 60 million words is too small to estimate these parameters reliably. Parameter estimates using relative frequencies would assign a value of zero to a large fraction of the parameters. A number of algorithms exist for estimating parameters for the trigram language model from sparse data. These algorithms include the deleted interpolation method [2], and the backoff method [19]. In the deleted interpolation method, the probability of word w_3 conditioned on the previous two words w_1 and w_2, $P(w_3|w_1w_2)$, is computed as a weighted average of the relative frequencies $f(w_3|w_1w_2)$, $f(w_3|w_2)$ and $f(w_3)$ in the training text corpus. (The relative frequency of w_3 conditioned on the context w_1w_2 is $f(w_3|w_1w_2) = \frac{C[w_1w_2w_3]}{C[w_1w_2]}$, where the function C counts the number of occurrences of its argument in the text.) The deleted interpolation method requires large amounts of storage for the parameters since both the weights and the relative frequencies are stored. The weights are estimated using the forward-backward algorithm which requires significant computing. The backoff method [19] is storage efficient as it does not require storage of weights. It uses Turing's estimate [10] to compute the probability mass of all the trigrams which do not occur in the training text corpus. This probability mass is then distributed among the unseen trigrams using the bigram and monogram counts. In the implementation of the trigram language model at INRS Télécommunications [14], we estimate the probabilities $P(w_3|w_1w_2)$ for the trigrams $w_1w_2w_3$ which do not occur in the training text corpus by direct application of Turing's estimate, without resorting to bigram and monogram counts to partition the probability mass.

We have found that the trigram language model is the most effective in improving recognition accuracy. Let us discuss some of the issues involved by analyzing our 86,000-word isolated word recognizer. The input to the acoustic recognizer consists of words separated by pauses of at least 150 ms. The spoken text consists of sentences read (without punctuation) from text which was selected randomly from magazines, books, and newspaper articles. An endpoint detector segments the acoustic data A for the spoken word string $W = w_1^n = w_1, w_2, \ldots, w_n$ into subsegments $A_1^n = A_1, \ldots, A_n$ corresponding to the words $w_1, w_2, \ldots, w_n$ in the word-string. For each segment A_i, the acoustic recognizer generates a list $\omega_{i1}, \ldots, \omega_{iN_i}$ of N_i most likely word choices, together with their likelihoods $(P(A_i|\omega_{ij}), j = 1, \ldots N_i)$. These likelihoods are smoothed by taking their seventh root before being passed to the second step of recognition. Such a normalization achieves a balance between likelihoods derived from the language model and the acoustic recognizer [4]. The probabilities

$$P(A|W_k) = P(A_1|\omega_{1k_1})P(A_2|\omega_{2k_2})\ldots P(A_n|\omega_{nk_n}),$$

(where ω_{ik_i} corresponds to one of the hypothesized words for the acoustic segment A_i) are used during search using the trigram language model. This search applies the trigram language model to find the most likely word string $\widehat{W}$ using

$$\widehat{W} = \operatorname*{argmax}_{W_k} P(W_k)P(A|W_k).$$

The acoustic recognizer generates the probabilities $P(A|W_k)$, while the language model provides the probabilities $P(W_k)$.

It is interesting to examine where the language model is most effective in correcting acoustic recognition errors. We analyzed the language model's ability to correct acoustic recognition errors as a function of the coverage of the word and its context by the training set of the language model. Each word in the test set is classified into one of six possible context coverage categories based on the neighboring words found in the training text corpus. To clarify these coverage categories, let us consider the word *sank* in the partial test sentence *every chair sank several inches*. The word *sank* is classified into one of the six categories as follows:

- **five-word context**: if $C[\textit{every chair sank}] > 0$ and $C[\textit{sank several inches}] > 0$ in the training text, then we consider word *sank* to have a five-word context. Note that, we are not implying that the sequence *every chair sank several inches* occurs in the training text nor that the sequence *chair sank several* occurs in the training text. We are only saying that both the sequences *every chair sank* and *sank several inches* are observed in the training text.
- **four-word context**: if $C[\textit{every chair sank}] > 0$ and $C[\textit{sank several}] > 0$, or $C[\textit{chair sank}] > 0$ and $C[\textit{sank several inches}] > 0$ in the training text, then the word *sank* has a four-word context.
- **three-word context**: if $C[\textit{every chair sank}] > 0$ and $C[\textit{sank several}] = 0$, or $C[\textit{chair sank}] = 0$ and $C[\textit{sank several inches}] > 0$, or $C[\textit{chair sank}] > 0$ and $C[\textit{sank several}] > 0$ in the training text, then *sank* has a three-word context.
- **two-word context**: if $C[\textit{chair sank}] > 0$ or $C[\textit{sank several}] > 0$ (but not both) in the training text, then *sank* has a two-word context.
- **one-word context**: if $C[\textit{chair sank}] = 0$ and $C[\textit{sank several}] = 0$, but $C[\textit{sank}] > 0$ in the training text, then *sank* has a one-word context.
- **zero-word context**: if $C[\textit{sank}] = 0$ in the training text, then *sank* has a zero-word context.

In the following text, when we refer to a test word having three-word context, we mean that for this word the conditions outlined are satisfied for the three-word context above, but not for any higher contexts (four-word or five-word contexts). In Table I, we have tabulated how effective the language model is in correcting acoustic recognition errors depending on these contexts. In compiling this table, we only consider words where the word hypothesis list includes the correct word, since only these words can be corrected by the language model. The search algorithm commits 3.4% search errors, therefore, the acoustic and language model recognition accuracies in Table I have been compiled from 96.6% of the test words. The effectiveness of the language model is found to be directly related to the context coverage observed for that word. For example, over 93% of the acoustic recognition errors are corrected for words having five-word context. Most of the remaining errors for these words are due to very low acoustic recognition likelihoods.

The language model is able to correct acoustic recognition errors even for words for which only two-word context is found. That is, the language model is

Context	Number of words	Acoustic recognition (%)	Recognition with language model(%)	Percentage of acoustic errors corrected
five-word	2000	88.3%	99.3%	93.6%
four-word	2156	85.3%	97.9%	85.8%
three-word	1646	82.0%	93.1%	61.6%
two-word	840	81.2%	85.1%	20.9%
one-word	306	80.4%	71.6%	-45.0%
zero-word	47	87.2%	59.6%	-216.0%
search errors	245			
average		84.7%	94.2%	62.3%

Table 1 Effectiveness of the trigram language model to correct words in the test set depending on contexts occurring in the training text corpus.

effective even when only one of the bigrams (with the word on the left or right of it) has a nonzero count in the training text. One example of a word from our test set having two-word context is *this* in the sequence *have weighed this thought with.* The 60-million word training text has $C[weighed\ this] = 0$, $C[this\ thought\ with] = 0$, and $C[this\ thought] > 0$. Another example is the word *wives* in the sequence *men without wives whereas the.* In this case, the training set has $C[without\ wives] = 0$, $C[wives\ whereas\ the] = 0$, but $C[wives\ whereas] > 0$.

The language model increases the recognition errors for words with neither the left nor the right context in the training text. In such cases, the language model does not have enough contextual information. Luckily, only 5% of the words fall in this category. Some examples of such words are *energy* and *wasters* in the sequence *most profligate energy wasters on.* Neither the bigram *profligate energy* nor *energy wasters* occurs in the training text. The word *energy* does occur in the training text. The word *wasters* does not occur even once in the 60-million word training text. Some other examples of words in the test set which do not occur in the training text are *barnburner*, *gondolier*, *heuristics*, *amoebas*, *sashaying*, *luminol*, *Marley*, *doorless*, etc.

REFERENCES

[1] A. Averbuch, L. Bahl, R. Bakis, P. Brown, G. Daggett, S. Das, K. Davies, S. De Gennaro, P. de Souza, E. Epstein, D. Fraleigh, F. Jelinek, B. Lewis, R. Mercer, J. Moorhead, A. Nádas, D. Nahamoo, M. Picheny, G. Shichman, P. Spinelli, D. Van Compernolle, and H. Wilkens. "Experiments with the Tangora 20,000 word speech recognizer," Proceedings of the 1987 IEEE International Conference on Acoustics, Speech, and Signal Processing, pp. 701–704.

[2] L.R. Bahl, F. Jelinek, and R.L. Mercer. "A Maximum Likelihood Approach to Continuous Speech Recognition," *IEEE Transactions on Pattern Analysis and Machine Intelligence*, **PAMI-5**(2), pp. 179–190, 1983.

[3] L.R. Bahl, S.V. De Gennaro, P.S. Gopalakrishnan, and R.L. Mercer. "A Fast Approximate Acoustic Match for Large Vocabulary Speech Recognition," *IEEE Transactions on Speech and Audio Processing*, vol. 1, pp. 59–67, January 1993.

[4] L.R. Bahl, R. Bakis, F. Jelinek, and R.L. Mercer. "Language-model/acoustic-channel-model balance mechanism," *IBM Technical Disclosure Bulletin* **23**(7B), pp. 3464–3465, 1980.

[5] L.R. Bahl, P.F. Brown, P. de Souza, R.L. Mercer, M.A. Picheny. "Acoustic Markov Models used in the Tangora Speech Recognition System," Proceedings of the 1988 IEEE International Conference on Acoustics, Speech, and Signal Processing, pp. 497–500.

[6] L.R. Bahl, P. de Souza, P.S. Gopalakrishnan, D. Nahamoo, M.A. Picheny. "Decision Trees for Phonological Rules in Continuous Speech," Proceedings of the 1991 IEEE International Conference on Acoustics, Speech, and Signal Processing, pp. 185–188.

[7] L.R. Bahl, P.F. Brown, P. de Souza, R.L. Mercer, M.A. Picheny. "A Method for the Construction of Acoustic Markov Models for Words," IBM Research Report RC 13099 (# 58580), IBM Thomas J. Watson Research Center, Distribution Services 73-F11, 1987, pp. 1–13.

[8] M.A. Bush, and G.E. Kopec. "Network-Based Connected Digit Recognition," IEEE Transactions on Acoustics, Speech, and Signal Processing, October 1987, pp. 1401–1413.

[9] A.-M. Derouault, and B. Merialdo. (1986). "Natural language modeling for phoneme-to-text transcription," *IEEE Transactions on Pattern Analysis and Machine Intelligence* **PAMI-8**(6), pp. 742–749, 1986.

[10] I.J. Good. "The population frequencies of species and the estimation of population parameters," *Biometrika* **40**(3 and 4), pp. 237–264, 1953.

[11] V. Gupta, M. Lennig, P. Mermelstein, P. Kenny, P.F. Seitz, and D. O'Shaughnessy. " Use of minimum duration and energy contour for phonemes to improve large vocabulary isolated-word recognition," *Computer, Speech and Language*, December 1992, pp. 345–359.

[12] V. Gupta, M. Lennig, and P. Mermelstein, "Fast search strategy in a large vocabulary word recognizer," *Journal of the Acoustical Society of America* **84**(6), 2007–2017, 1988.

[13] V. Gupta, M. Lennig, and P. Mermelstein. "Integration of acoustic information in a large vocabulary word recognizer," in Proceedings of the 1987 IEEE International Conference on Acoustics, Speech and Signal Processing, Dallas, pp. 697–700.

[14] V. Gupta, M. Lennig, and P. Mermelstein. " A language model for very large-vocabulary speech recognition," *Computer, Speech and Language*, December 1992, pp. 331–344.

[15] H.-W. Hon. *Vocabulary-Independent Speech Recognition: The VOCIND System.* Ph.D. Thesis, Computer Science Department, Carnegie Mellon University, March 1992.

[16] F. Jelinek. "Continuous speech recognition by statistical methods," *Proceedings of the IEEE* **64**(4), pp. 532–556, 1976.

[17] F. Jelinek. "The Development of an Experimental Discrete Dictation Recognizer," *Proceedings of the IEEE* **73**(11), pp. 1616–1624, 1985.

[18] B.H. Juang, S.E. Levinson, and M.M. Sondhi, "Maximum likelihood estimation for multivariate mixture observations of Markov chains," *IEEE Transactions on Information Theory*, **IT-32**(2), pp. 307–309, March 1986.

[19] S.M. Katz. "Estimation of probabilities from sparse data for the language model component of a speech recognizer," *IEEE Transactions on Acoustics, Speech, and Signal Processing* **ASSP-35**(3), pp. 400–401, 1987.

[20] P. Kenny, R. Hollan, V.N. Gupta, M. Lennig, P. Mermelstein, and D. O'Shaughnessy. "A^*-Admissible Heuristics for Rapid Lexical Access," *IEEE Transactions on Speech and audio Processing* **1**(1), pp. 49–58, 1993.

[21] G.E. Kopec, and M.A. Bush. "Network-Based Isolated Digit Recognition Using Vector Quantization," IEEE Transactions on Acoustics, Speech, and Signal Processing, 1985, pp. 850–867.

[22] C.-H. Lee, and L.R. Rabiner. "A Frame-Synchronous Network Search Algorithm for Connected Word Recognition," *IEEE Transactions on Acoustics, Speech, and Signal Processing* **ASSP-37**(11), pp. 1649–1658, November 1989.

[23] K.-F. Lee. *Large-vocabulary speaker-independent continuous speech recognition: The SPHINX System.* Ph.D. Thesis, Computer Science Department, Carnegie Mellon University, April 1988.

[24] M. Lennig, D. Sharp, P. Kenny, V. Gupta, and K. Precoda. "Flexible vocabulary recognition of speech," in Proceedings of the International Conference on Spoken Language Processing, Banff, October 1992, pp. 93–96.

[25] H. Ney, D. Mergel, A. Noll, and A. Paeseler. "Data Driven Search Organization for Continuous Speech Recognition," *IEEE Transactions on Acoustics, Speech, and Signal Processing*, **ASSP-40**, pp. 272–281, Feb. 1992.

[26] N.J. Nilsson. *Principles of Artificial Intelligence*, Tioga Publishing Co., Palo Alto, CA, 1980.

[27] J.R. Rohlicek, Y.-L. Chow, S. Roucos. "Statistical Language modeling using a small corpus from an application domain," in Proceedings of the 1988 IEEE International Conference on Acoustics, Speech and Signal Processing, pp. 267–270.

[28] F.K. Soong. "A phonetically labeled acoustic segment (PLAS) approach to speech analysis-synthesis" in Proceedings of the 1989 IEEE International Conference on Acoustics, Speech and Signal Processing, Glasgow, pp. 584–587.

[29] F.K. Soong, and E. Huang. "A Tree-Trellis based Fast Search for finding the N Best Sentence Hypotheses in Continuous Speech Recognition," in Proceedings of the DARPA Speech and Natural Language Workshop, Hidden Valley, June 1990.

[30] T.K. Vintsyuk. "Element-wise recognition of continuous speech composed of words from a specified dictionary," *Kibernetika*, vol. 7, pp. 133–143, March–April 1971.

[31] A.J. Viterbi. "Error bounds for convolutional codes and an asymptotically optimal decoding algorithm," *IEEE Transactions on Information Theory* **IT-13**, pp. 260–269, April 1967.

10

RECENT DEVELOPMENTS IN ROBUST SPEECH RECOGNITION

B. H. Juang

AT&T Bell Laboratories
Murray Hill, New Jersey 07974

ABSTRACT

Robust speech recognition refers to the problem of designing an automatic speech recognizer that works well in a wide range of unexpected or adverse environments. As the technology of automatic speech recognition moves out of the laboratories into field applications, the issue of robustness becomes a key element that distinguishes a successful deployment from a failed one.

In this paper, we present a review of several methods that were proposed in recent years to deal with the robustness problem. We classify these methods into two major areas, one aiming at features or distortion measures that are invariant to the operating environment and the other capitalizing on adaptive procedures in hopes of being able to modify either the speech signal or the reference speech patterns to reflect the influence of the operating environment for reliable recognition. This review may help point out directions for further algorithmic improvements.

1 INTRODUCTION

Recent advances in automatic speech recognition in the laboratory environment have triggered contemplation of a wide range of real world applications. As the technology moves out of the laboratory into field applications, however, the issue of robust speech recognition becomes extremely important. Robust speech recognition refers to the problem of designing an automatic speech recognition system which works equally well in various unknown or adverse operating environments. The reason that robustness is important is that the performance of existing speech recognition systems, whose designs are predicated on known or clean environments in the laboratory, often degrades rapidly under field conditions.

There are basically four categories of field conditions, namely, noise, distortion, articulation effects and pronunciation variations. These conditions are often not known a priori when the recognizer is trained in the laboratory and are often detrimental to the recognizer performance.

For example, the report by Dautrich et al. [1] gives a good account of how additive noise alone would cause performance degradation in isolated word recognition. It was shown that an isolated word recognizer trained in virtually noise free conditions and capable of delivering an accuracy of 95% had an order of magnitude increase in error rate when tested with noise-contaminated utterances at an SNR (signal-to-noise ratio) of 18 dB. Figure 1 (after [2]) summarizes the performance results and illustrates several key considerations in noisy speech recognition. First, as the curve (labeled MATCHING S/N) shows, the recognizer maintains its performance with moderate degradation if the SNR in the training data matches the SNR in the test data. Second, when the noise conditions in training and testing are mismatched, the recognition accuracy drops rapidly with SNR. (In the figure, the case shown involves training with clean speech data and testing with noisy data at various SNR.) Third, the degradation in recognition performance due to noise can be somewhat reduced if the reference patterns contain a moderate amount of noise. The curve labeled MISMATCHED S/N - TEST S/N = 18 dB represents the effects of adding various amounts of noise to the reference while keeping the SNR of the test data fixed at 18 dB. Although the results indicate certain ways to deal with the noise problem, the goal of robust speech recognition is to be able to achieve at least a performance similar to that under matched training and testing condition. Such a goal cannot be accomplished without proper processing of the speech signal and certain algorithmic improvements.

In the past few years, two broad categories of methods were proposed in the area of robust speech recognition. One aims at the speech features or the distortion measures (for pattern similarity evaluation) which are invariant under various field conditions. We shall name this class the invariant method. The other category capitalizes on adaptive procedures which adjust the speech parameters or the stored patterns according to certain estimates of the field condition under which the signal is received. We call this class the adaptive method. We review these two broad categories of techniques in this paper.

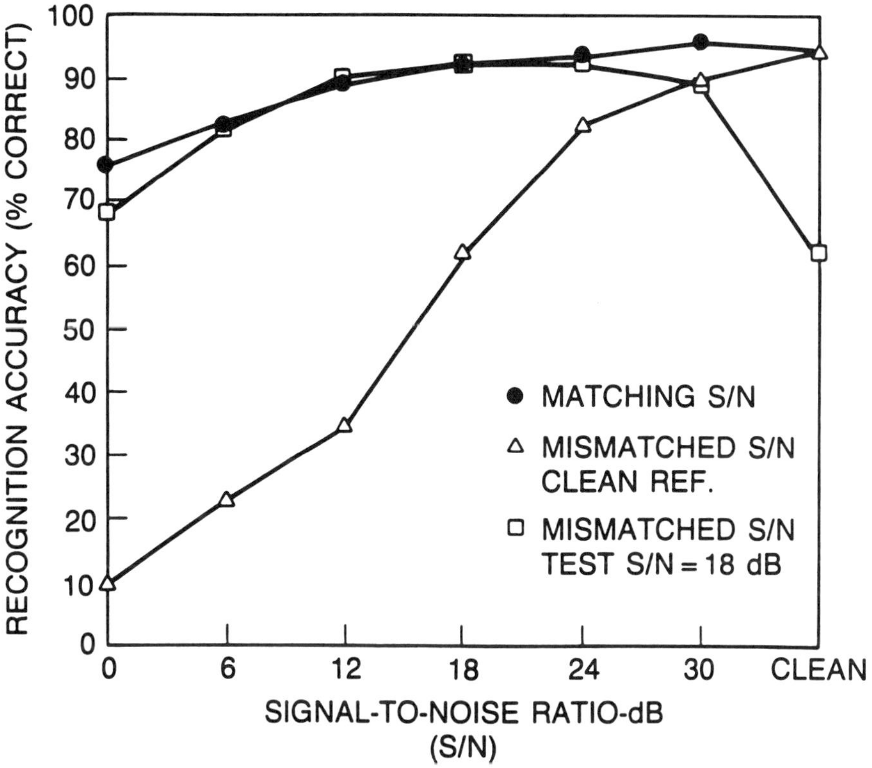

Figure 1 Speech recognition performance in noisy conditions; •: training and testing have matched S/N as indicated by the abscissa; △: only clean training reference is used and the abscissa indicates that the test S/N; o: training and testing S/N's are mismatched with test S/N fixed at 18 dB and the abscissa indicates the training S/N.

We begin by briefly discussing the general characteristics of field conditions under which a speech recognizer may be operated.

2 OPERATING CONDITIONS OF SPEECH RECOGNITION

Among the four conditions, three, namely noise, distortion and articulation effects are related to the signal environment and its effects on the signal being produced by the talker. The fourth condition, pronunciation variations, refers to the diversity the talker population demonstrates in pronouncing the words and is particularly important when the recognizer is deployed for speaker independent applications such as those found in telecommunications.

2.1 Noise

Acoustic ambient noise is usually considered additive, meaning the recorded signal is a sum of the intended speech signal and the ambient noise. The ambient noise is one of the primary concerns for a speech recognizer. Sources of acoustic ambient noise are abundant. For example, in the office environment, the office machinery (typewriter or printer), personal computers or workstations, telephone ringing and background conversation of people often emit enough acoustic noise to cause degradation in performance of a speech recognizer. Inside an automobile, the acoustic noise level due to engine, cooling fan, wind, tire and road noise is usually considerably high, particularly when the automobile is in motion. The spectrum of ambient noise is often not flat.

There exists also other types of noise such as electric noise and quantization noise. In digital mobile phone applications where the speech signal is encoded at low bit rates, substantial recognition performance degradation has been observed. Also in other applications involving wireless transmission of the speech signal, the electric noise (static, channel interference) may lead to undesirable results.

2.2 Distortion

Aside from the additive contamination due to noise like signals, the uttered signal inevitably undergoes a series of spectral distortions before being recorded and processed for speech recognition. The room in which the speech recognizer unit is deployed almost certainly has a varying degree of reverberation that can alter the signal spectrum. The microphone transducer, depending on its type and mounting position, also can significantly distort the speech spectrum. In the case where the transducer configuration used in testing is different from the one used during training of the reference patterns, the mismatch in spectral distortion becomes one of the major problems.

When the automatic recognizer is deployed in a telephone network, the telephone channel through which the speech signal travels can cause further distortion of the signal spectrum. Many microphones used in telephones such as the traditional carbon type also are known to have nonlinear effects on the speech signal. The nonlinearity while sometimes may be beneficial in reducing the noise effect usually introduces a distortion that may pose a severe problem in speech recognition.

2.3 Articulation Effects

Many factors affect the talker in his/her speaking manner. Even the psychological awareness of communicating with a speech recognition machine could make the talker produce a noticeable difference in his/her sound formants and rhythmic stability [3]. Characteristic changes in articulation due to environmental influence, known as the Lombard effect, can be dramatic [3, 4, 5]. Junqua et al. [3] estimated the increase in the first formant frequency due to an 85 dB SPL masking noise to be between 42 and 113 Hz. Furthermore, a significant change in spectral tilt was also observed; spectral tilt in the high-band decreases while that in the low-band increases for most of the vowels and liquids. These characteristic changes dramatically affect the performance of an automatic speech recognizer.

A major difficulty with the articulation effects is to quantify them. With acoustic noise or channel distortion which usually does not vary as rapidly as the speech itself in terms of spectral characteristics, we can to some extent measure or model them to help the recognizer design. Articulation effects, however, are an inherent product of the speech production process and are found to be context dependent [3]. The above qualitative characterization of the Lombard effect is not specific enough to provide an adequate guideline for enhancing the speech recognizer.

2.4 Pronunciation Variation

Talkers' pronunciation of a particular word often varies considerably, depending on their accents or regional dialects. Some of the variation may be due to physiological reasons, while others customary or cultural. For example, in North Ireland, the word "two" is spoken as "tooey" in which "oo" is pronounced as in "foot". The association between "two" and "tooey" can be considered translative to a significant extent, particularly from the signal modeling viewpoint. The traditional paradigm of signal modeling based upon the concept of acoustic isomorphism or similarity would find pronunciation variation a difficult phenomenon to characterize. It obviously entails a sensible, new definition of acoustic similarity with linguistic isomorphism explicitly taken into account. This subject is beyond the scope of this paper and will not be addressed in length.

3 DEALING WITH ADVERSE CONDITIONS

In this section, we review the two broad categories of techniques that were proposed for dealing with the general adverse conditions as laid out in the previous section. These two categories are the invariant method and the adaptive method, respectively.

To facilitate our discussion, Fig. 2 is a block diagram which depicts the framework we follow in the design of a recognizer. The speech signal is analyzed and the feature is extracted for comparison with stored references in the pattern matching block. The reference patterns which characterize individual speech classes to be recognized are obtained by a training procedure using a large set of known examples. A decision scheme determines the word or class of the unknown input based on the matching scores or evaluated similarity measures. Methods to be discussed for dealing with adverse conditions may take place in various blocks in the diagram.

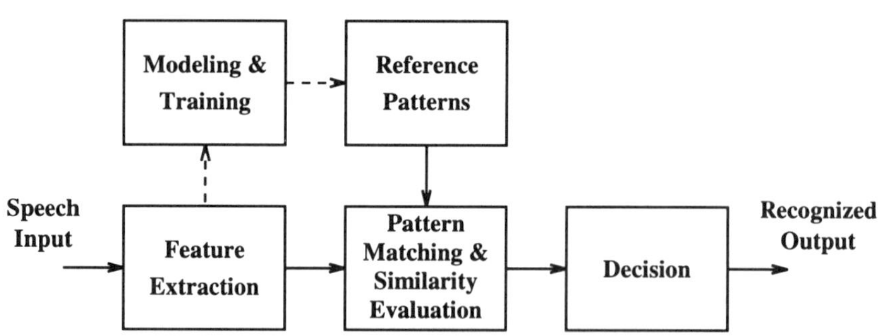

Figure 2 A block diagram of a speech recognition system based on the statistical pattern matching approach.

3.1 Invariant Methods

The category of invariant methods aims at using speech features or the associated similarity measures that are invariant under a wide range of speech signal conditions.

Robust Representations

One set of speech features that was shown to be robust to noise contamination is the liftered cepstrum. Let $c(n, m)$ be the cepstral sequence that represents a speech utterance. In the representation, n and m denote the (discrete) quefrency and the frame index respectively. A liftered cepstrum is a weighted cepstrum $w(n)\ c(n, m)$ where $w(n)$ normally takes the shape of a window; e.g. $w(n) = 1 + h \sin\left(\frac{n\pi}{L}\right)$, for $n = 1, 2, \ldots, L$, and $w(n) = 0$, otherwise, with h usually chosen to be $L/2$. When parameters in $w(n)$ are properly chosen, the liftered cepstrum has been shown to be advantageous in noisy speech recognition [6]. Indeed, one can argue that additive white noise tends to have substantial effects on the low quefrency terms (e.g. one that corresponds to the spectral tilt) and thus applying a cepstral lifter to deemphasize these terms would result in a representation less susceptible to the influence of additive noise than the unprocessed cepstrum.

The cepstral sequence $c(n, m)$ can also undergo filtering along the axis m to produce an enhanced representation. Techniques such as RASTA of Hermansky [7] and IIR filtering of Hanson and Applebaum [8] have shown advantages in performance. The key to improvements, however, may come mainly from suppression of artifacts in spectral measurements rather than the invariance itself in these methods.

The short-time modified coherence (SMC) representation of speech [9] takes advantage of the inherent coherence in adjacent segments of the speech signal to enhance the equivalent "signal-to-noise" ratio (SNR). The idea is that an unwindowed autocorrelation operation on the impulse response of an all-pole system does not alter its pole structure and estimation of the system parameters may be more reliably accomplished from the autocorrelation function than from the signal particularly when the signal is contaminated by noise. The autocorrelation sequence is defined by

$$\rho_i = \frac{1}{N} \sum_{j=1}^{N-1} x(j)x(j+1), \quad i = 0, 1, \ldots, N$$

where N is the window size, which for quasi-stationary signals like speech represents the coherence of adjacent signal segments. All-pole modeling of the autocorrelation sequence leads to a more robust representation than that of the signal itself. The SNR improvement is data dependent (i.e. a function of the speech impulse response) and was found to be about 10–12 dB for typical SNRs between 0 and 20 dB. For noisy speech recognition at 10 dB SNR, the SMC maintains an accuracy as high as that in a clean (low noise) condition.

Another type of robust speech representation is motivated by studies in the human auditory system. The human auditory system seems to perceive speech better than any machine processor when noise interference is present. Based on this premise, the Ensemble Interval Histogram (EIH) [10] was proposed to represent the auditory-nerve firing pattern which might be robust to noise corruption. The EIH model encompasses 1) the use of 85 simulated cochlear filters which split the speech signal in the frequency band of 200-3200 Hz, 2) measurements of level crossing for each of these cochlear filter outputs, and 3) accumulation of the histograms for the level-crossing intervals. The resultant histogram ensemble is reminiscent of a spectrum, but with built-in nonlinearities and non-uniform frequency resolution that are characteristic of the human auditory processing. Figure 3 shows a comparison of the traditional FFT spectrum and the EIH representation with their corresponding all-pole (linear prediction, LPC) spectral fit in clean as well as 0 dB SNR conditions for the vowel /e/. The robustness of the EIH to noise corruption is clearly displayed in the figure. When applied to noisy speech recognition in conjunction with a traditional dynamic time warping scheme, the EIH was reported to yield significant accuracy improvements for male speech [10]. Other aspects of the representation, particularly how the "brain" makes best use of the "auditory-nerve firing pattern", are not yet well understood.

Robust Distortion Measures

In discussing the invariance property of a representation, we often cannot separate it from the associated distortion measure which defines the dissimilarity between two observations. This is due to the fact that an ideal representation that is truly invariant to noise or other adverse conditions has not been found. As a result, the notion of robustness becomes relative and is usually referred to as the sensitivity the associated distortion measure displays. Obviously, when we speak of a representation in terms of the parameter vector in a space, the associated metric or distortion measure must be integrally considered.

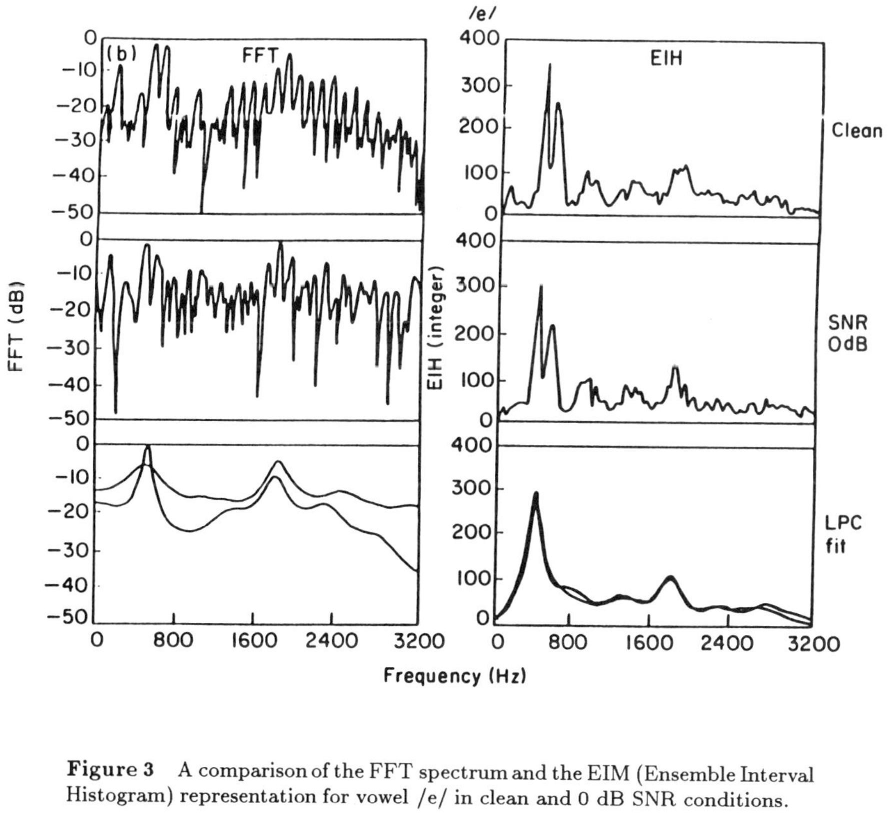

Figure 3 A comparison of the FFT spectrum and the EIM (Ensemble Interval Histogram) representation for vowel /e/ in clean and 0 dB SNR conditions.

The liftered cepstrum, in fact, implies a weighted Euclidean distance; i.e. $d(m) = \sum_n [w(n)c(n,m) - w(n)c'(n,m)]^2 = \sum_n w^2(n)[c(n,m) - c'(n,m)]^2$ where $c(n,m)$ and $c'(n,m)$ are the cepstral vectors of the two spectra being compared. The distance $d(m)$ defined with proper $w(n)$, as discussed above, is more robust than simply the Euclidean distance between $c(n,m)$ and $c'(n,m)$. A more extensive study of the weighted cepstral measures and their robustness to noise as well as channel conditions can be found in [11]–[12].

The pursuit of a robust distortion measure can be more effective if we have an analytical understanding of the effects of noise upon spectral parameters. Indeed, analytical as well as experimental evidence [13] indicates that additive white noise causes the cepstral vector norm (length of the cepstral vector, with the zeroth term excluded) to shrink but leaves the cepstral vector orientation more or less intact. The vector norm shrinkage is detrimental in the traditional distortion calculation where an Euclidean distance is employed. Furthermore, since the norm shrinkage was found to be a function of the noise level, the vector norm itself can be used to facilitate non-uniform weighting for each speech frame in the accumulative distance during dynamic sequence comparison. The result suggests the use of a projection operation to formulate several distortion measures to cope with the mismatched noisy condition where the reference patterns are noise-free and the test conditions are unknown. In [13], the following cepstral projection measure was found to be a good choice overall for clean as well as noisy speech:

$$d(\mathbb{C}_r, \mathbb{C}_t) = |\mathbb{C}_t| \left(1 - \frac{\mathbb{C}_r^* \mathbb{C}_t}{|\mathbb{C}_r|\,|\mathbb{C}_t|}\right)$$

where $\mathbb{C}_r$ and $\mathbb{C}_t$ are the reference and the test cepstral vector respectively and $*$ denotes matrix transpose. In a speaker dependent isolated word (39 words – 26 alphabet, 10 digits and 3 command words) recognition trial, the projection measure was shown to be superior to many other well known distortion measures. The above family of robust distortion measures were also found to lead to better recognition performance for Lombard speech [12].

3.2 Adaptive Methods

The adaptive method differs from the invariant method in the way the characteristics of the operating environment are taken into account. The invariant method assumes no explicit knowledge of the signal environment, while the adaptive method attempts to perform some estimation on the characteristics and adjusts the signals or the reference models accordingly in order to achieve a reliable pattern matching result. The key considerations in the adaptive method are the characteristic parameters of the operating environment to be estimated, the available amount of data for estimation (or speed of adaptation), and the techniques to accomplish such an estimation task. In the following, we discuss the adaptive method according to the characteristic parameters to be estimated in the process.

Waveform

When the adverse condition is due to additive noise alone, one can employ speech enhancement methods to suppress noise before recognition. One of the classical signal enhancement methods is adaptive noise cancellation using two signal sources [14]. Figure 4 illustrates the principle of the two-input noise cancelling approach. The technique adaptively adjusts the filter coefficients so that subtracting the filtered noise reference input from the primary (speech plus noise) input leads to an output signal with minimum energy. The main assumptions obviously are that the noise that contaminates the speech is a filtered version of the reference noise, and that the optimization method can produce an accurate noise estimate. The technique thus requires that the noise component in the corrupted signal and the noise reference have high coherence. This is often difficult to attain in a realistic operating environment. Only limited uses of such a technique were reported [15].

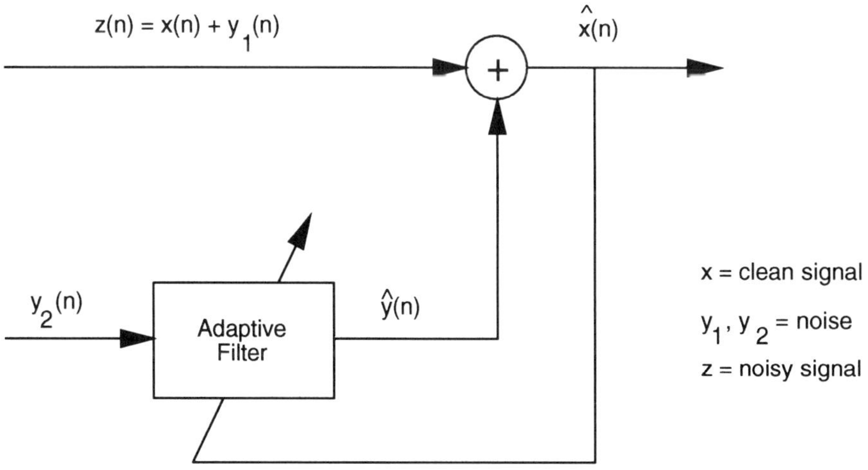

Figure 4 Schematic of the two-input noise cancelling approach.

Noise Spectrum

In many cases, the additive noise is assumed to be white. The characteristic parameter to be estimated is thus the noise power under the assumption. In a simplified setup [16], the noisy power spectrum $|Z|^2$ is modeled as a sum of the "clean" speech model $\sigma^2/|A|^2$ and the power spectrum of the noise, λ. An algorithm can then be devised to minimize the Itakura-Saito distortion between $|Z|^2$ and $\sigma^2/|A|^2 + \lambda$, by iteratively estimating the noise power level λ and the all-pole spectrum $\sigma^2/|A|^2$. The resultant all-pole model σ/A is considered an enhanced version and used in the recognizer. The method is applied to the test utterances as well as training utterances for best results. Obviously, the assumption in the method is rather strong and the increased computational complexity is a drawback.

Most noise is not strictly white and cannot be simply characterized by the noise level alone. In the presence of broadband noise, the fact that certain regions of the speech spectrum that are of lower level than noise will be more affected by the noise can be used advantageously in speech recognition. Recognizing this, Klatt [17] advocated use of noise masking in conjunction with a filter-bank analyzer. The key idea is to first choose, for each channel of the filter-bank output, the noise mask level as the higher of the noise level in the reference signal and that in the testing signal. Then, the channel output is replaced by the mask value if it is below the corresponding mask level. The resultant spectral pattern is thus "adapted" to the noisy environment. Many noise adaptation schemes [18, 19, 20] are based on this principle.

Channel Distortion and Characteristic Mismatch

The effect of the channel or signal path is often considered a form of distortion. This includes the situation where the reference patterns are established with the signal undergoing one path (from the talker's mouth to the machine and anything in between) and the speech signal is received during testing via another path. When the distortion is assumed to be linear, the received spectrum, $Y(\omega)$, is modeled as

$$Y(\omega) = H(\omega)X(\omega)$$

where $X(\omega)$ is the spectrum of the original signal (or the signal used in training) and $H(\omega)$ represents the channel effect. In the cepstral domain, the channel effect becomes additive; i.e.

$$\mathbb{C}_Y = \mathbb{C}_X + \mathbb{C}_H$$

where $\mathbb{C}_Y$, $\mathbb{C}_X$ and $\mathbb{C}_H$ are the cepstra corresponding to Y, X and H, respectively. The method of signal bias removal (SBR) [21] is to estimate $\mathbb{C}_H$ from the test utterance which is usually on the order of a couple of seconds long, and then subtract the estimated bias vector $\overline{\mathbb{C}}_H$ from each observed vector $\mathbb{C}_Y$, producing an improved $\overline{\mathbb{C}}_X = \mathbb{C}_Y - \overline{\mathbb{C}}_H$. The bias $\mathbb{C}_H$, representative of the channel effect or mismatch, is usually assumed to be constant during the period of the utterance.

Estimation of the bias vector can be easily accomplished in a maximum likelihood formulation [21] using an iterative algorithm, reminiscent of the EM (expectation-maximization) algorithm. In particular, the algorithm iterates between two estimates: $\overline{\mathbb{C}}_X$ and $\overline{\mathbb{C}}_H$. Given an estimate of $\mathbb{C}_X$, $\overline{\mathbb{C}}_X$, the bias vector $\overline{\mathbb{C}}_H$ can be estimated as an average of $\mathbb{C}_Y - \overline{\mathbb{C}}_X$ over the duration of the utterance. That is, with m denoting the frame index ranging from 1 to M,

$$\overline{\mathbb{C}}_H = \frac{1}{M} \sum_{m=1}^{M} \left(\mathbb{C}_y(m) - \overline{\mathbb{C}}_x(m) \right)$$

Then, given $\overline{\mathbb{C}}_H$, an improved $\tilde{\mathbb{C}}_x(m)$ is obtained as

$$\tilde{\mathbb{C}}_x(m) = \mathbb{C}_y(m) - \overline{\mathbb{C}}_H \ .$$

The improved $\tilde{\mathbb{C}}_x(m)$ is further used to refine the estimate $\overline{\mathbb{C}}_x(m)$. The process iterates until a certain criterion is met. In improving $\overline{\mathbb{C}}_x(m)$, one can use a vector quantization codebook containing undistorted, typical cepstral codewords, or for simplicity, an average of the undistorted cepstral vectors in the database. The latter case is equivalent to a codebook with one single codeword which is used as $\overline{\mathbb{C}}_x(m)$ throughout the utterance. The processing to produce $\overline{\mathbb{C}}_x(m)$ from $\tilde{\mathbb{C}}_x(m)$ is not required in this case and further iteration thus becomes unnecessary. This is a form of cepstral mean subtraction.

It was shown [21] that the method is able to recover 75% of the performance degradation (in terms of error rate increase) due to channel mismatch. Several algorithms of the similar nature can be found in Ref. [22].

Model Adaptation and Stochastic Matching

Noise masking can be implemented in a probabilistic modeling framework. The technique of Nadas et al. [23] uses the noise estimate to transform the probability density functions for an improved labeling result.

Let random variables X, V, and Z be the clean speech spectrum, the noise model spectrum, and the observed noisy spectrum respectively. (Note that $Z = z$ is the only observed value during test.) The masking procedure is to model Z as

$$Z = \max(X, V) \ .$$

Assume $F(X)$ and $G(Y)$ are the cumulative distribution functions for X and V respectively, with f and g denoting their corresponding probability density function (pdf). G is assumed to be obtained using independent noise data or estimated from the input. The pdf of Z, $h(\cdot)$, whose parameters are to be estimated, can be expressed as

$$h(z) = f(z)G(z) + F(z)g(z) \ .$$

The pdf h can be considered as the noise-compensated version of f, and is used to approximate the true noisy pdf, say $f'(z)$. Thus, similar to the design of a vector quantizer codebook, a set of $h(z)$'s can be designed as the noise-compensated prototypes of the speech given the noise estimate G. These prototypes are then used to facilitate labeling of each input noisy spectral vector for stochastic pattern matching. The technique is particularly suited for discrete HMM systems and has been shown to bring about substantial improvements in performance under noisy conditions.

Another model adaptation technique is in the broad category of composite source modeling and stochastic matching, of which the signal bias removal algorithm can be formulated as a special case. Let λ_x and λ_b be the speech model and the bias model, respectively, defined in the appropriate observation space. (For additive noise, the observation could be the spectral magnitude while for channel mismatch and distortion, it is usually defined in terms of cepstrum.) Also let N_x and N_b be the number of states of each model. The observed signal is assumed to be $y_t = x_t + b_t$, which is related to the composite model of an $(N_x \cdot N_b)$-state Markov chain as illustrated in Fig. 5. The model is defined as a joint measure $p_{\lambda_x \lambda_b}(X, Y) = p_{\lambda_x}(X) p_{\lambda_b}(Y - X) = \sum_{\{i_x\}} \sum_{\{i_b\}} p_{\lambda_x}(X, \{i_x\}) p_{\lambda_b}(B, \{i_b\})$ since we assume x to be independent of b. In the expression, X, Y, and B denote the random variable sequences for $\{x_t\}$, $\{y_t\}$ and $\{b_t\}$ respectively and $\{i_x\}$ and $\{i_b\}$ are the corresponding state sequences.

Note that λ_x is estimated a priori during training of the recognizer under a given condition. The bias process b_t, nevertheless, is to be estimated jointly with $\{x_t\}$ from $\{y_t\}$, the biased test signal, via a stochastic matching setup which aims at maximizing the likelihood $\mathcal{L}(\{y_t\}, \{x_t\} \mid \lambda_x, \lambda_b)$. This can be accomplished by an iterative procedure, again similar to the EM algorithm

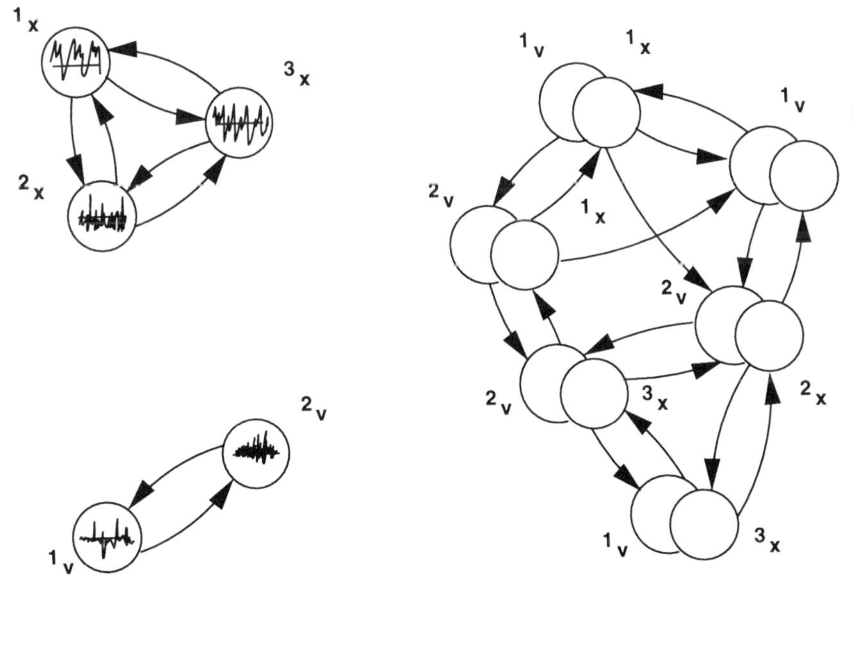

Figure 5 A composite hidden Markov model.

used in the formulation of the SBR method. It should be clear now that in SBR, the sequence $\{b_t\}$ is assumed to be a constant $b_t = b$ (which is to be estimated) rather than a random process. In other word, SBR is a special case of composite modeling in which one of the models represents a constant, deterministic sequence.

This method was used in speech enhancement by Ephraim, Malah and Juang [24] and became the basis of many robust recognition algorithms, such as those proposed by Rose [25], Cox and Bridle [26], Zhao [27] and Sankar and Lee [28].

4 CONCLUSIONS

In order for a speech recognizer to be useful, it has to be able to deliver high recognition accuracy over a wide range of operating conditions, many of which may not be known a priori. The issue of robustness becomes a critical one that distinguishes a successful deployment from a failed one. Many algorithms have been proposed to cope with ambient noise, distortions and other adverse problems in speech recognition. We group and review these algorithms in two broad categories: the invariant methods and the adaptive methods. Considerable success has been attained by these approaches. However, continued research on new models of the operating conditions as well as solutions to enhance the recognizer's robustness in performance is necessary to ensure maximal deployability and usefulness of a speech recognition system.

REFERENCES

[1] B. A. Dautrich, L. R. Rabiner, and T. B. Martin. On the effects of varying filter bank parameters on isolated word recognition. *IEEE Trans. Acoust. Speech & Signal Processing*, ASSP-31:793–806, August 1983.

[2] B. H. Juang. Speech recognition in adverse environments. *Computer Speech & Language*, 5:275–294, 1991.

[3] J.-C. Junqua and Yolanda Angelade. Acoustic and perceptual studies of Lomard speech: application to isolated-words automatic speech recognition. *Proc. ICASSP-90*, pages 841–844, Albuquerque, NM, April 1990.

[4] D. B. Pisoni, et al. Some acoustic-phonetic correlates of speech produced in noise. *Proc. ICASSP-85*, pages 1581–1584, Tampa, FL, March 1985.

[5] B. Stanton, et al. Acoustic-phonetic analysis of loud and Lombard speech in simulated cockpit conditions. *Proc. ICASSP-88*, pages 331–334, New York, NY, April 1988.

[6] B. A. Hanson and Hisashi Wakita. Spectral slope distance measures with linear prediction analysis for word recognition in noise. *Proc. ICASSP-86*, pages 757–760, Tokyo, JAPAN, April 1986.

[7] H. Hermansky, et al. RASTA-PLP speech analysis technique. *Proc. ICASSP-92*, pages 121–124, 1992.

[8] B. Hanson and Ted Applebaum. Subband or cepstral domain filtering for recognition of Lombard and channel distorted speech. *ICASSP-93*, pages 79–82, 1993.

[9] D. Mansour and B. H. Juang. The short-time modified coherence representation and noisy speech recognition. *IEEE Trans.*, ASSP-37(6):795–804, June 1989.

[10] O. Ghitza. Auditory nerve representation as a front-end for speech recognition in a noisy environment. *Computer Speech & Langauge*, 1(2):109–130, December 1986.

[11] F. Itakura and T. Umezaki. Distance measure for speech recognition based on the smoothed group delay spectrum. *Proc. ICASSP-87*, pages 1257–1260, Dallas, TX, April 1987.

[12] J.-C. Junqua and Hisashi Wakita. A comparative study of cepstral lifters and distance measures for all-pole models of speech in noise. *Proc. ICASSP-89*, pages 476–479, Glasgow, UK, May 1989.

[13] D. Mansour and B. H. Juang. A family of distortion measures based upon projection operation for robust speech recognition. *Proc. ICASSP-88*, New York, NY, April 1988. Also in *IEEE Trans.*, ASSP-37(11):1659–1671, November 1989.

[14] B. Widrow, et al. Adaptive noise cancelling: Principles and applications. *Proc. IEEE*, 63:1692–1716, 1975.

[15] P. Darlington, P. D. Wheeler, and G. A. Powell. Adaptive noise reduction in aircraft communication systems. *Proc. ICASSP-85*, pages 716–719, Tampa, FL, March 1985.

[16] Y. Ephraim, J. G. Wilpon, and L. R. Rabiner. A linear predictive front-end processor for speech recognition in noisy environments. *Proc. ICASSP-87*, pages 1324–1327, Dallas, TX, April 1987.

[17] D. H. Klatt. A digital filter bank for spectral matching. *Proc. ICASSP-76*, pages 573–576, Philadelphia, PA, 1976.

[18] J. S. Bridle, et al. A noise compensating spectrum distance measure applied to automatic speech recognition. *Proc. Institute of Acoustics (U.K.), Autumn Conf.*, Windermere, November 1984.

[19] John N. Holmes and Nigel C. Sedgwick. Noise compensation for speech recognition using probabilistic models. *Proc. ICASSP-86*, pages 741–744, Tokyo, JAPAN, April 1986.

[20] D. Van Compernolle. Noise adaptation in a hidden Markov model speech recognition system. *Computer Speech and Language*, 3:151–167, 1989.

[21] M. G. Rahim and B. H. Juang. Signal bias removal for robust telephone speech recognition in adverse environments. *Proc. ICASSP-94*, Adelaide, Australia, April 1994.

[22] A. Acero and R. M. Stern. Environmental robustness in automatic speech recognition. *Proc. ICASSP-90*, pages 849–952, Albuquerque, NM, April 1990.

[23] A. Nadas, et al. Speech recognition using noise-adaptive prototypes. *Proc. ICASSP-88*, pages 517–520, New York, NY, April 1988.

[24] Y. Ephraim, D. Malah, and B. H. Juang. On the application hidden Markov models for enhancing noisy speech. *IEEE Trans.*, ASSP-37:1846–1856, 1989.

[25] R. C. Rose, E. M. Hofstetter, and D. A. Reynolds. Integrated models of speech and background with application to speaker identification in noise. To be published in *IEEE Trans. Speech & Audio Processing*, 1994.

[26] S. J. Cox and J. S. Bridle. Unsupervised speaker adaptation by probabilistic spectrum filtering. *Proc. ICASSP-89*, pages 294–297, May 1989.

[27] Y. Zhao. Self-learning speaker adaptation based on spectral variation source decomposition. *Eurospeech'93*, pages 359–362, 1993.

[28] Sankar and Lee. Stochastic matching for robust speech recognition. *IEEE Signal Processing Letters*, 1(8):124–125, August 1994.

11

HOW DO HUMANS PROCESS AND RECOGNIZE SPEECH?

Jont B. Allen

Acoustics Research Dept., AT&T Bell Labs, Murray Hill, NJ 07974

ABSTRACT

Until the performance of automatic speech recognition (ASR) hardware surpasses human performance in accuracy and robustness, we stand to gain by understanding the basic principles behind human speech recognition (HSR). This problem was studied exhaustively at Bell Labs between the years of 1918 and 1950 by Harvey Fletcher and his colleagues. The motivation for these studies was to quantify the quality of speech sounds in the telephone plant to improve both speech intelligibility and preference. To do this he and his group studied the effects of filtering and noise on speech recognition accuracy for nonsense consonant-vowel-consonant (CVC) syllables, words, and sentences. Fletcher used the term *articulation* as the probability of correct recognition for *nonsense* speech sounds, and *intelligibility* as the probability of correction recognition for words (sounds having meaning). In 1919 Fletcher sought a nonlinear transformation $\mathcal{A}(s)$ of the articulation s for filtered and unfiltered speech defined to give an additive articulation density function $D(f)$ over frequency. The area under $D(f)$ is called the *Articulation Index*. The resulting transformation was shown to accurately predict the average articulation. Fletcher then went on to find relationships between the recognition errors for the nonsense speech sounds, words, and sentences. This work has recently been reviewed and partially replicated by Boothroyd and by Bronkhorst *et al.* Taken as a whole, these studies tell us a great deal about how humans process and recognize speech sounds.

Introduction

Speech recognition by machine is a critical core technology for the "information" age. Existing machine recognition systems do not work the way humans work. This is because automatic speech recognition (ASR) machines use spectral templates (e.g., cepstra, LPC, FFT, etc.), while humans work with *partial recognition* information across frequency, probably in the form of speech features that are local in frequency (e.g., formants). It has been shown, for example, that forcing partial phone recognition errors in one frequency region does not affect the partial recognition at other frequencies (i.e., the partial recognition errors across frequency are independent). To extract these frequency dependent features requires frequency–local signal processing, namely *independent feature–processing channels*. It seems to be this local feature–processing, uncoupled across frequency, that makes human speech recognition (HSR) robust to noise and reverberation.

The partially recognized speech units (i.e., extracted features) are then integrated into phones, and the phones are then grouped into syllables, then words, and so forth. This model of HSR was devised by Harvey Fletcher beginning in about 1921 in the form of an empirical probabilistic analysis of speech recognition scores obtained from a series of listening experiments. In these experiments, the context information in spoken speech was selectively and systematically removed, the speech was altered by filtering and additive noise, and the phone, syllable, word, and sentence recognition scores were experimentally estimated. In this paper we shall use the word *context* to refer to the supplementary information used to decode ambiguity (see Table 1) rather than the common usage in the speech field where context can refer to the sounds preceding and following a given sound. This analysis was first published in 1950, upon Fletcher's retirement from Bell Labs, 35 years after its conception [16]. I believe that very few people have attempted to understand this important publication.

In the machine case, when using spectral templates, errors across frequency are not independent. When presented with noise, filtering, reverberation, multiple speakers, and other degradations, machine systems are not robust because degradations at one frequency affect the entire template. There are also lessons to be learned about the effects of *context* and *redundancy* on the robustness of speech perception. These are difficult concepts to rigorously define. Thus, to reach the important goal of robust machine recognition, we need, *first*, a better understand of the signal processing of partial recognition, or feature recognition, in human speech recognition, and *second*, improved information processing models that describe how humans use *context* and *redundancy* to make speech robust.

In this paper I will describe articulation testing, the experimental results, and Fletcher's quantitative analysis methods (his model of HSR) developed at Bell Labs between 1918 and 1921. I will interpret and discuss the experimental data and Fletcher's model in terms of our present day knowledge of the auditory system. I then will relate this updated model to the problem of robust ASR.

Articulation Test Record

Date 3-16-28
Title of test Practice tests
Test no. 10
List Nos. 5-9-37

Syllable articulation 0.515 = S
Condition tested 1500∼ low pass filter
Observer W.H.S.
Caller E.B.

No.		Observed	Called	Observed	Called	Observed	Called
1	The first group is	ma'v	na'v	po'z	po't'h	kōb √	kōb
2	Can you hear	pōch √	pōch	nēz	nēzh	shēt'h	siz
3	I will now say	seng √	seng	jo'ch √	jo'ch	fūch √	fūch
4	As the fourth write	chūd √	chūd	t'ha'm √	t'ha'm	thōl √	thōl
5	Write down	run √	run	hab √	hab	pot'h √	pot'h
6	Did you understand	chiz	kiz	def	doth	wa'm √	wa'm
7	I continue with	foz	fozh	chech	chej	gūm	gūn
8	These sounds are	lo'l √	lo'l	lun	lon	nāsh	nāth
9	Try the combination	jās	zhāth	shāl √	shāl	vo'g √	vo'g
10	Please record	t'ha'th	t'ha'sh	muz √	muz	lung	long
11	Write the following	wūr √	wūr	lēd	bēd	diz	dizh
12	Now try	yāp √	yāp	wif √	wif	kak	tak
13	Thirteen will be	mad	maj	gōst √	gōst	t'ha'r	zha'r
14	You should observe	bēch	bēk	thāv	sāv	must √	must
15	Write clearly	gēm	dēm	kōf √	kōf	yo'd √	yo'd
16	Number 16 is	t'heb	veb	ra'g √	ra'g	jet √	jet
17	You may perceive	jok	jost	thip √	thip	rēp	rēj
18	I am about to say	gaf √	gaf	yar √	yar	t'hēp	hēp
19	Try to hear	hus √	hus	zhūt √	zhūt	—	chuv
20	Please write	hiv	thit'h	kūk	tūk	t'hef	t'hesh
21	Listen carefully to	tōg √	tōg	fung √	fung	bās √	bās
22	The last group is	sho't √	sho't	t'hev	vesh	t'ho'f	shaf

$cvc = 0.491$
$s^3 = 0.499$

$v = 0.909$
$c = 0.735$
$s = 0.793$

Figure 1 This shows an example of the articulation score sheet used in 1928. This example corresponds to a 1.5 kHz low pass filter applied to the speech. The first column is the carrier phrase, followed by three sets of "called" and "observed" CVCs. At the top is the overall CVC score ($S = 0.515$, or 51.5% correct). At the bottom the vowel score is shown as $v = 0.909$, namely 90.9% of the vowels were correctly received. Similarly for the consonants, $c = 0.735$ corresponding to 75.3% correct. The average phone score $s = (2c + v)/3$ was then computed (79.3%) along with an estimate of the CVC syllable articulation score S assuming independent C and V units, as $cvc = 0.491$. Finally $s^3 = 0.499$ corresponding to an average independent-phone model computed from the average phone articulation s.

1 THE THEORY OF HSR

History Beginning in about 1910, G. Campbell from AT&T, and later I.B. Crandall from Western Electric Engineering, initiated a series of experiments to explore the nature of human speech perception. Their goal was to improve speech communication over the telephone. After 1918 these experiments were continued by Harvey Fletcher and his colleagues at The Bell Telephone Laboratories (Western Electric Engineering until 1925) [16]. These studies led to a speech recognition measure called the *Articulation Index* which accurately characterizes speech intelligibility under conditions of filtering and noise. Fletcher used the word *articulation* in this perceptual context to mean the probability of correctly identifying nonsense speech sounds. The word *intelligibility* is defined as the probability of correctly identifying meaningful speech sounds, such as words.

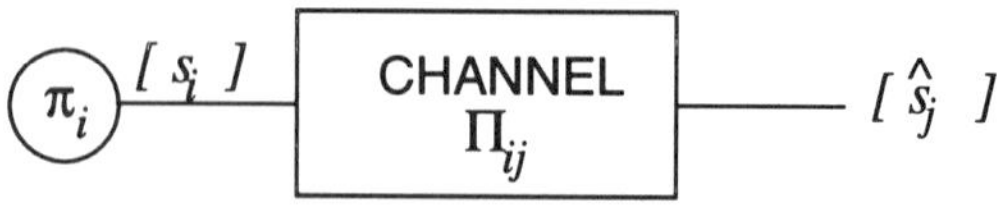

Figure 2 The human acoustic channel is represented here as a hypothetical information-theoretic M-symbol noisy channel. Input speech sounds spoken by an average speaker are received (heard) by an average listener. An input set of spoken phones $[s_i]$, labeled by i, are spoken with probability π_i. Errors occur in transmission as described by a transition probability matrix Π having elements Π_{ij}. The set of received phones is $[\hat{s}_j]$, indexed by j.

The original experiments in 1910 to 1918 began with normal conversational speech over a modified telephone channel. The subjects were asked to listen to distorted speech sounds and repeat what they heard. Fletcher refined the measurement methods, and by 1919 (see Fig. 1) they were using nonsense sounds (e.g., "yif," "vou," and "moush"), as described in a 1921 internal AT&T report by Fletcher [9]. These nonsense words were composed of three sounds (triphones): two consonants and one vowel, and two sounds (diphones): one consonant and one vowel. The class of triphones they used are referred to as nonsense "CVC" (consonant–vowel–consonant) syllables, which account for 34% of all syllables spoken on the telephone, while the CV and VC diphones each account for 20% of the syllables. The three types of syllables account for a total of 74% (34+20+20) [14, page 94]. This subset of nonsense sounds were viewed as an idealized subset of the language. The identification task on this subset was viewed as an idealized speech recognition task.

As shown in Fig. 1, the CVC's were scored for errors, both in terms of the C's and V's, and as syllable units. The errors were then described in terms of modifications to the telephone channel (channel distortion), which in turn was characterized in terms of channel bandwidth and signal to noise ratio. Human phone error rates for nonsense CVC's, under the best conditions are about 1.5% error (98.5% correct) [19, 16, 14].

Figure 2 shows an interpretation of the talker–listener experiment, modeled as an information–theoretic M–symbol noisy channel, where M is the number of phones in the language. During WWII while Fletcher was applying the Articulation Index theory to pilot–navigator communications as chairman of National Defense Research Committee,[1] Shannon was studying the entropy of language and applying it to cryptography. These studies were fundamental to the development of information theory.

[1] He received a citation signed by President Truman for this work.

Shannon came to Bell Labs in 1941. It is interesting that sometime before 1945, and the time that Fletcher retired in 1949, Fletcher was in Shannon's line of management.[2]

Context versus redundancy Context is a powerful force in speech recognition and has many *levels* [22]. As defined in Table 1, *context* is the setting surrounding a target sound. By setting we may mean the sounds adjacent to the target sound, the meaning of a sentence, a common situation of the speaker and listener, or in the most general sense, any *a priori* knowledge. We define the *contextual information* as the effect of context on the entropy of the recognition task. The context may be used by the listener to resolve the meaning of a word that is ambiguous, by the use of the contextual information derived from the context.

Redundancy is the *over specification of information.* Redundancy means that information may be removed and the meaning is still uniquely defined. When a message is coded in a manner that it is robust to errors, the message has redundancy. For the case of human listners listening to speech, if the redundancy is improperly interpreted by the listener, it can lead to confusion. Thus context can provide redundancy, resolve ambiguity, or create ambiguity.

Some examples The word "lead" is ambiguous and requires contextual information to resolve the ambiguity. The word "inflammable," on the other hand, is redundant since it has the same meaning as the word "flammable." Another famous example of this is demonstrated by the difference between the question "How do humans recognize speech?" and the question "How do humans wreck a nice beach?" These two sentences can be spoken so that only context can distinguish them, and in a situation with the proper context, the two will never be confused. Such a context could not be called redundant, since the information is only specified once. In the example "May we go to the store together?" the extra information is redundant because the "we" and "together" transmit the same information.

Context can be used across levels. Due to the meaning conveyed by a word or sentence, the listener can compensate for missing phone information. For example, the written sentence *Cn yu udrstnd ths sntnc?* is easily understood because of context even though all the vowels are missing. Thus for the same channel distortion, words and sentences have higher recognition scores than those of syllables, due to their contextual information.

---

[2] In 1945 the line of management was Fletcher, Director of Physical Research; Bode, Director; Schelkunoff, Head; and Shannon. During WWII Shannon worked on entropy of language, a field closely related to Fletcher's Articulation Index. However, I have not found any indication that Fletcher or his work influenced Shannon's thinking in any way.

NAME	DEFINITION
speech	the class of sounds generated by the human vocal tract
phone	the smallest classifiable speech sound unit
phoneme	the smallest classifiable speech information bearing unit
allophones	all the phone variants for a given phoneme
morpheme	the smallest speech unit having meaning
recognition	probability of correct unit identification
recognition error	1−recognition
recognition set	the set of all units being recognized
recognition set size	the number of members in the recognition set
feature	a binary unit of partial recognition
the number of features	$\log_2$ of the recognition set size
speech features	i.e., phone features
redundancy	the overspecification of source information
context	the overall situation or setting within which a sound occurs
contextual information	information derived from context
context entropy	reduction in entropy due to contextual information
context effect	the increase in recognition due to context,
intelligibility	recognition of sounds having context (e.g., words)
articulation	recognition in the absence of context effects
acoustic entropy	entropy derived from the articulation matrix

Table 1 Table of general definitions. The term *context* is used in the dictionary sense, with a meaning that differs from common usage by the speech community. We use *context* to refer to the setting of a sound or word, and *contextual information* as the implied information that may be inferred from that setting. The added contextual information may be either unique, thereby just resolving ambiguity, or it may be *redundant*. Nonsense sounds, used to measure articulation, are designed to have low redundancy while retaining their normal *coarticulations*.

Redundancy versus entropy We correct language and speech errors by the use of redundancy. What is the source of this redundancy in speech? It is not simply a matter of entropy. For example, assume that the probability of using one phone is different from another, leading to a lower entropy. If every time we heard a low probability phone we replaced it with one of high probability, the low–probability phone would have no utility. Thus the change in entropy due to phone probability does not seem to imply redundancy in speech.

Does redundancy play a role in the functional relationship between nonsense syllable articulation S and word intelligibility W? The increase recognition for words results from the lower *entropy* of words relative to nonsense syllables. If one could compute the entropy of nonsense syllables $\mathcal{H}_S$, and of words $\mathcal{H}_W$, they would find that $\mathcal{H}_S > \mathcal{H}_W$. For example, given 20 possible phones, then a triplet of CVC syllables would have a maximum entropy of $3\log_2(20) = 13$ bits (assuming a uniform distribution of the phones). It follows that the entropy of CVC words must be less than 13 bits, since CVC "words" (i.e., those CVC's that have meaning) are the subset of all CVC's. Does this imply redundancy? It certainly allows us to correct errors, as shown by Bronkhorst *et al.* [6]. Thus it passes the test given above for being redundant. When we fail to hear a phone, or we hear a very unlikely phone, we remap it to a likely word, using the context. For example: "I have a very furry *at." has the following possible solutions for "*" {b,c,h,m,n,p,r}. These solutions have different probabilities, but they all represent possible solutions. Other letters have a very small, or even a zero, probability.

Fletcher's methodology Fletcher clearly understood these concepts, when in about 1918 he introduced nonsense syllables into telephone speech testing to remove the strong influence of context entropy. In doing this he greatly simplified the problem he was working on. One may study many context questions by working with full text or phonetic transcriptions. However, to study the difficult problem of acoustic transcription, Fletcher (wisely) decided to systematically remove various levels of context, and its associated redundancy.

By 1929 Fletcher and Steinberg had found a functional relation between the nonsense CVC recognition rates and word and sentence recognition rates [2, 4, 3, 11, page 226]. These relations demonstrate that *word and sentence error rates can be predicted from the nonsense* $\{C,V\}$ *error rates for a given speech corpus.* Thus Fletcher, and later Steinberg, were the first to experimentally quantify context and redundancy in spoken language through their empirical articulation measurements [11], though they did not use the terms *context*, *redundancy*, or *entropy*.

Using sentences with context complicates the intelligibility testing, decreases the efficiency of the testing, and increases the variability of the results, because context greatly confounds the measurement of phone errors.

Intelligibility versus articulation Fletcher's basic approach (which he adapted from his colleague J.Q. Stewart) was to treat the speech units in terms of their empirical probabilities. As shown in Table 1, he defined *intelligibility* as the empirical probability of correct recognition when context is present (e.g., recognition of words) and *articulation* as the empirical probability of correct recognition when context is *not* present (e.g., recognition of nonsense syllables) [14]. The *articulation error* is 1–articulation. A *phone* is defined as a specific sound in the language. We use square brackets to specify the phones, such as $[s_i]$, where $i = 1, \ldots, M$, assuming M phones.

It is important to understand how Fletcher's nonsense syllables were generated. If one were to make a list of all the phones of English, and they then tried to make compound sounds by randomly drawing from the set with replacement ("zeroth–order" syllables), many of the resulting compound sounds would be unpronounceable. To get around this problem, Fletcher classified the phones into initial and final consonants and vowels. When he chose sounds conditioned on their placement ("first–order" syllables), all the sounds may be pronounced. The resulting phones he used are defined in [14, page 95].

SYMBOL	DEFINITION
α	gain applied to the speech
f_c	filter high–pass and low–pass cutoff frequency
$c(\alpha)$	consonant articulation
$v(\alpha)$	vowel articulation
$s(\alpha) = (2c + v)/3$	average phone articulation for CVCs
$e(\alpha) = 1 - s$	phone articulation error
$s_L(f_c, \alpha)$	s for low–pass filtered speech
$s_H(f_c, \alpha)$	s for high–pass filtered speech
$S(\alpha)$	nonsense syllable (CVC) articulation
$W(\alpha)$	word intelligibility
$I(\alpha)$	sentence intelligibly

Table 2 Table of specific symbol definitions.

2 THE ARTICULATION EXPERIMENT

After about 1918, the databases were formed from statistically balanced nonsense CVC, CV and VC syllable source lists. The syllable source lists were spoken and the listeners recorded what they heard. The list was chosen so that the set was closed, namely all sounds that were recorded were also included in the source list. Articulations c and v (see Table 2) were computed for the phones, and articulation S was computed for the syllables as shown in Fig. 1. Many crews of listeners were trained to do the task, and their performance was monitored, over years in some cases. Crew performance was found to stabilize within a few months.

We will assume that $[s_i]$ is the "called" phone, $[\hat{s}_j]$ is the "received" phone, M is the number of phones in the corpus, N_i the number of times $[s_i]$ is called, and $N_{ij}(\alpha)$ is the number of times that $[s_i]$ is called and $[\hat{s}_j]$ is received. The speech gain α was used to vary the speech signal to noise ratio (SNR), and thus the errors, in a natural manner. Using standard matrix notation, i is the row index, and j is the column index.

Each row corresponds to a different called phone, while each column corresponds to a different received phone. Because the set is closed, the frequency matrix N_{ij} is square. From these definitions the ith row sum $N_i = \sum_{j=1}^{M} N_{ij}(\alpha)$ is the number of times the ith phone is called. The empirical probability of calling each phone is given by $\pi_i = N_i / \sum_{i=1}^{M} N_i$. The lists were designed to make the called probabilities approximately equal, namely to make $\pi_i \approx 1/M$.

The empirical probability of hearing $[\hat{s}_j]$ after calling $[s_i]$ (e.g., $P([\hat{s}_j]|[s_i])$) is $\Pi_{ij}(\alpha) = N_{ij}(\alpha)/N_i$. The matrix $N(\alpha)$ defined by elements $N_{ij}(\alpha)$ is called the *frequency matrix.* The matrix $\Pi(\alpha)$ defined by elements $\Pi_{ij}(\alpha)$ is called the *confusion matrix*, or alternatively the *transition matrix.* If Π were measured for sounds having no context, we could also call it an *articulation matrix.* Since every called phone gives rise to a received phone, the row sum is unity, namely $\sum_{j=1}^{M} \Pi_{ij} = 1$. When the conditions are ideal ($\alpha = 1$), the matrix Π is close to diagonal, meaning that phone recognition is nearly perfect. When the signal to noise ratio is very poor ($\alpha = 0$), recognition goes to chance levels. In this case $\Pi_{ij} = 1/M$, since the observers understood that the sounds have equal probability, namely that $\pi_i = 1/M$.

Results Fletcher found that the CVC syllable articulation $S(\alpha)$ (the probability of correct identification of the CVC syllable) is accurately predicted from the phone articulations $c(\alpha)$ and $v(\alpha)$ by the relation

$$S(\alpha) = c^2 v \tag{11.1}$$

This formula reflects the fact that the three sound–units are heard as independent sounds, and that to correctly identify the syllable, all three sound–units must be correctly identified.

For example, suppose the CVC is *wif.* If a listener responds with *mif,* that would be one error of the first C sound–unit and the syllable would be scored as *incorrect.* To the extent that the above formula for S is accurate, it implies that coarticulations of the speech sounds are transformed by the auditory system into independent units at a early stage, before context is used, since context was not present during testing (see the discussion in section 3). From information theory we know that when quantities are independent, the have no mutual information [8, page 19]. Thus if Eq. 11.1 were exactly true, then the mutual information between the nonsense CVC syllables and the C and V consonants ($I(\text{CVC}; \{\text{C}, \text{V}\})$) would be zero.

Fletcher's measure of recognition To unify the computations across the various data bases, an average {C,V} phone articulation s was computed from the composition of {C,V} units in the data base. For example, for nonsense CVC's, Fletcher defined the empirical recognition probability of an average phone articulation as $s(\alpha) =$

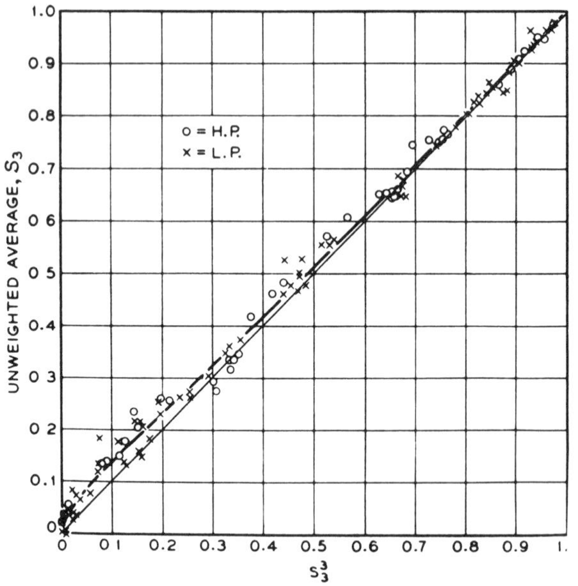

Figure 3 This figure, reproduced from [14, page 285], shows the CVC error S and the cube of the phone error s^3 for nonsense CVC's that have been low and high pass filtered. The error $\Delta S = S - s^3$ is the distance between the two solid lines. The maximum error is about 0.04. It is not clear if these results were corrected for guessing (i.e. $s(\mathcal{A} = 0) = 1/M \approx 0.05$). If the error ΔS were zero, it would mean that each CVC syllable was, on the average, perceived as three independent phone units. For further discussion see section 2.

$(2c + v)/3$. In terms of Π, $s(\alpha) = \frac{1}{M} \sum_{i=1}^{M} \Pi_{ii}(\alpha)$ which is the average of the diagonal elements of the transition matrix. The maximum of $s(\alpha) = s_{max} < 1$ occurs for a speech gain $\alpha \approx 1$. When $\alpha = 0$, $s(0) = 1/M$, because this is the case of proper (unbiased) guessing. Because the set is closed, the articulation error $e = 1 - s$ may also be computed from the sum of all the off–diagonal elements as $e(\alpha) = \frac{1}{M} \sum_{i=1}^{M} \sum_{\substack{j=1 \\ j \neq i}}^{M} \Pi_{ij}(\alpha)$, where the sum over j does not include the diagonal terms along $j = i$. For the case of guessing ($\alpha = 0$), $e = 1 - 1/M$.

After some manipulation it was shown that Eq. 11.1 may be approximated by

$$S \approx s^3 \tag{11.2}$$

with only a small error [17, 16, 14, page 283–5]. This approximation depends on the experimental variation of the relative consonant–vowel probability ratio $\lambda = c/v$. As shown in Fig. 3, when Fletcher plotted the syllable articulation S against the cube of the average phone articulation error s^3, he found an almost perfect agreement. A systematic error of less than 0.04 was found. It is not clear if or how they accounted for chance (guessing) as s becomes small (as $\alpha \longrightarrow 0$).

Other measures of recognition Other more precise measures of recognition besides s and S are interesting. When working with the {C,V} sounds, it is possible to compute the *acoustic entropy* of the speech corpus. For example we may define the *conditional entropy* given that phone $[s_i]$ was called, treating filtering, noise, and the listener as a "channel," as shown in Fig. 2. The idea here is to measure the quality of the channel using the estimated transition matrix probabilities, as a function of the physical attributes of the physical channel, namely filtering, the speech gain, and, in the case of hearing impaired listeners, the listener's hearing loss. This conditional entropy is given by the row sum $\mathcal{H}(\hat{s}|[s_i]) = -\sum_{j=1}^{M} \Pi_{ij} \log_2(\Pi_{ij})$. This measure is in *bits* and lies between 0 (perfect recognition) and $\log_2(M)$ (guessing). The conditional entropy of the received phone given a called phone is the expected value, namely the weighted column–sum over called phones $\mathcal{H}(\hat{s}|s) = \sum_{i=1}^{M} \pi_i \mathcal{H}(\hat{s}_j|[s_i])$. This quantity is between 0 (perfect recognition) and $\log_2(M)$ (guessing). The entropy is a more precise measure than s because it is sensitive to the difference between random guessing (maximum entropy) and specific errors between a few phones (e.g., [ba]–[da] confusions).

Coarticulation and HSR It has long been observed that the production of a speech sound is strongly dependent on the preceding and following sound (i.e., the production context). This interdependence is called *coarticulation*. Many speech researchers believe that coarticulation is what makes the speech recognition problem difficult.

Coarticulation is a "production" concept. From the results of the syllable articulation experiments (Eq. 11.1), one must conclude that humans decode syllables as *independent* phone units over time. For example, Fletcher [16], as shown in Fig. 3, and more recently Bronkhorst *et al.* [6] found that under conditions of low noise, *the phones are perceived independently*.

One might conclude that the problem of robust HSR may be split into two problems. First is the problem of decoding phones from the acoustic wave form. Second is the problem of utilizing contextual information (e.g., entropy) to edit corrections and fill in missing information.

This first problem leads us to the next obvious and important question, namely "How do we decode phones?" A second important insight, and a major simplification of this complex problem, comes from Fletcher's *articulation index*.

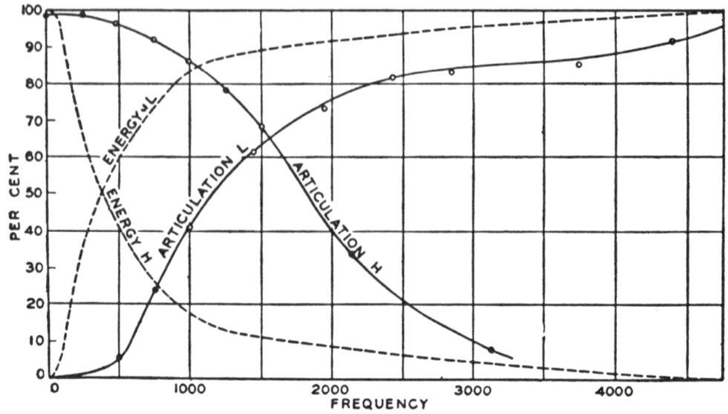

Figure 4 This figure is from [10, 11, p. 280]. It shows how the articulation and the energy of speech depend on low and high–pass filtering. The speech energy for the two filter outputs are shown by the dashed lines and the articulations are shown by the solid lines. Note how the energy curves cross at the 50% point, as they should for two sharp filters. This means that the two have equal energy at the corresponding cutoff frequency of 400 Hz. The curve labeled "ARTICULATION H" is the same as s_H and the curve labeled "ARTICULATION L" is the same as s_L. Note how the articulation curves are not equal at 50% but at 65%, at 1550 Hz (i.e., the articulation curves do not add to one at the cross over point). Thus he showed that articulation does not sum. Based on this data, Fletcher proposed a nonlinear transformation of the articulation bands that did sum. This transformation was called the articulation index.

3 THE ARTICULATION INDEX

The tight relation between the syllable and phone articulation (Eq. 11.1) emphasizes the fundamental importance of phone articulation to human speech recognition. Due to its basic nature, Fletcher was soon studying the phone articulation $s(\alpha)$ for various channel frequency responses and channel noise conditions [9, 16]. To do this he used low–pass and high–pass filters on the speech.[3] In these studies Fletcher soon found that the *partial articulations* (the articulations for each band) did not sum to the wide band articulation. He then showed that a nonlinear transformation of the partial articulations would make them additive [9, 10]. As described below, the nonlinearly transformed partial articulation defines an *articulation index density* $D(f)$ over frequency f. Integration, or summation, over this density gives the *articulation index* $\mathcal{A}$. The articulation index can be viewed as a fundamental internal variable of speech recognition. All recognition conditions are determined once $\mathcal{A}$ is determined for a given context.

As shown in Table 2, we designate the partial articulations of the low and high–pass filtered sounds as $s_L(f_c, \alpha)$ and $s_H(f_c, \alpha)$. The cut–off frequency of the filters is given by f_c, and the parameter α is the gain applied to the speech. By varying the

[3] It is interesting that George Campbell invented the lattice filter to do these experiments.

speech level, the signal to noise ratio of the speech was varied. As shown in Fig. 4, s_L approaches 0.985 for f_c above 8 kHz, and 0 for f_c below 100 Hz, and s_H is 0 for f_c above 8 kHz, and 0.985 for f_c below 100 Hz. Both functions are monotonic with frequency.

Fletcher [9, 10] showed that $s_L + s_H$ does not sum to s. (To motivate the discussion, we now know that for nonsense CVC's, $s = s_L + s_H - s_L s_H$, as will be shown below.) To resolve this "problem" he proposed finding an invertible nonlinear transformation $\mathcal{A}(s)$ of the articulation s, which he called the *Articulation Index*, that would make the two articulation bands add to one. In other words, he defined a transformation $\mathcal{A}(s)$ such that

$$\mathcal{A}(s_L(f_c, \alpha)) + \mathcal{A}(s_H(f_c, \alpha)) \equiv \mathcal{A}(s(\alpha)) \tag{11.3}$$

for all values of the filter cutoff frequency f_c and the speech gain α. There was, of course, no guarantee that such a transformation would actually exist, but his intuition suggested that it should. Since we are dealing with transformations of probabilities, the additivity condition is basically an argument of independence of the band probabilities.

He determined this transformation by finding the cutoff frequency $f_c = f_c^*$ such that

$$s_L(f_c^*, \alpha) = s_H(f_c^*, \alpha), \tag{11.4}$$

which is the frequency where the curves s_L and s_H cross in Fig. 4. He then argued that the two transformed articulations must be equal at f_c^*, and therefore must each be 1/2, namely

$$\mathcal{A}(s_L(f_c^*, \alpha)) = 0.5\mathcal{A}(s(\alpha)). \tag{11.5}$$

By repeating this procedure as a function of the speech gain α, he could empirically determine $\mathcal{A}(s)$, since the articulation $s(\alpha)$ is a function of the speech level.

What they found Under the conditions that the word corpus consisted of nonsense CVC's (the maximum entropy source condition), Fletcher found that the nonlinear transformation that gives articulation additivity is

$$\mathcal{A}(s) = \frac{\log_{10}(1 - s)}{\log_{10}(1 - s_{max})}. \tag{11.6}$$

The constant $s_{max} = 0.985$ is the maximum observed articulation and $e_{min} = 1 - s_{max} = 0.015$ is the corresponding minimum observed articulation error [14, page 282]. If we solve Eq. 11.6 for s we find $s(\mathcal{A}) = 1 - e_{min}^{\mathcal{A}}$. Note that when $\mathcal{A} = 0$, $s = 0$, and when $\mathcal{A} = 1$, $s = s_{max}$. This equation can also be written in terms of the articulation error $e = 1 - s$, which gives

$$e(\mathcal{A}) = e_{min}^{\mathcal{A}}, \tag{11.7}$$

The independent–channel model Fletcher (following Stewart) [9, 16, 14, 1] went on to show that the phones are processed in independent *articulation bands* (frequency channels), and that these independent estimates of the speech sounds in each frequency band are "optimally" merged, as given by the following two band example [16, 14]:

> If 10 errors out of 100 spoken sounds are made when only band 1 is used, and 20 errors are made when only band 2 is used, then when both bands 1 and 2 are used simultaneously, the error is $e = 0.1 \times 0.2 = 0.02$, or two errors will be made.

For the two band example, using Eqs. 11.3 and 11.6, we find $\log(1 - s) = \log(1 - s_L) + \log(1 - s_H)$, which becomes $1 - s = (1 - s_L)(1 - s_H)$, or in terms of the articulation error $e = 1 - s$

$$e = e_L e_H. \tag{11.8}$$

This equation is true for every value of f_c. The definition of $\Pi(\alpha)$ may be extended to the low and high–pass filtered speech case as $\Pi_L(f_c, \alpha)$ and $\Pi_H(f_c, \alpha)$.

Equation 11.8 says that the articulation errors due to low–pass filtering are independent of the articulation errors due to high–pass filtering. I interpret this equation to mean that *we are listening to independent sets of phone features in the two bands* and processing them independently, up to the point where they are fused to produce the phone estimates. The term *feature* implies the recognition of partial information. To the extent that the recognition in each band is independent, there is no mutual information (no redundancy) between the bands [8].

The articulation index density As a result of the additivity required by Eq. 11.3, the nonlinear transformation $\mathcal{A}(s)$ transforms $s(f_c, \alpha)$ into an integral over an articulation index density $D(f)$. This follows if we let each term of Eq. 11.3 correspond to an integral of $D(f)$ over frequency, namely

$$\mathcal{A}(s_L(f_c)) = \int_0^{f_c} D(f)df \tag{11.9}$$

$$\mathcal{A}(s_H(f_c)) = \int_{f_c}^{\infty} D(f)df \tag{11.10}$$

$$\mathcal{A}(s) = \int_0^{\infty} D(f)df. \tag{11.11}$$

The density $D(f)$ may then be uniquely determined from

$$D(f_c) = \frac{\partial}{\partial f_c} \mathcal{A}(s_L(f_c)). \tag{11.12}$$

From these studies Fletcher was able to derive the *density over frequency*, $D(f)$, of the phone articulation index. The distribution $D(f)$ indicates the density of speech recognition information over frequency [9]. Thus $D(f)$ is called the *importance function*; it is shown in Fig. 177 and is tabulated in Table 63 on page 333 in Fletcher 1953 book [14, 15].

The multichannel model Given the concept of the articulation index density, it follows that Eq. 11.8 may be generalized to a multichannel articulation band model, namely

$$e = e_1 e_2 \ldots e_K, \tag{11.13}$$

where K is the number of independent articulation bands.

Since the independent channel model was first proposed by J.Q. Stewart (i.e., Eq. 11.8), but was fully developed by Fletcher [14, page 281], it seems proper to call it the *Fletcher–Stewart multi–independent channel (MIC) model of phone perception.* Stewart worked for AT&T *Development and Research* while Fletcher worked for *Western Electric Engineering.*

It is easy to show that the relation between the kth band error e_k and the density is given by

$$e_k = e_{min}^{D_k} \tag{11.14}$$

where

$$D_k = \int_{f_k}^{f_{k+1}} D(f)df. \tag{11.15}$$

The frequency limits f_k were chosen so that all the D_k's were equal, which means that under optimum listening conditions (α near 1), $D_k = 1/K$.

It was first observed by Galt, working with Fletcher, that the equal spacing of the articulation index density function (i.e., $D_1 = D_2 = \cdots$) *approximately corresponds to equal spacing along the basilar membrane since Eq. 11.9 is very similar to the cochlear map function, which is the relation between normalized place X on the basilar membrane and characteristic frequency F (in Hz) along the basilar membrane.* The normalized place variable is defined as $X = (L - x)/L$, where x is in mm from the stapes, and $L = 35$ mm is the length of the basilar membrane. It is frequently expressed as a percent [14, page 293]. From the human cochlear map function

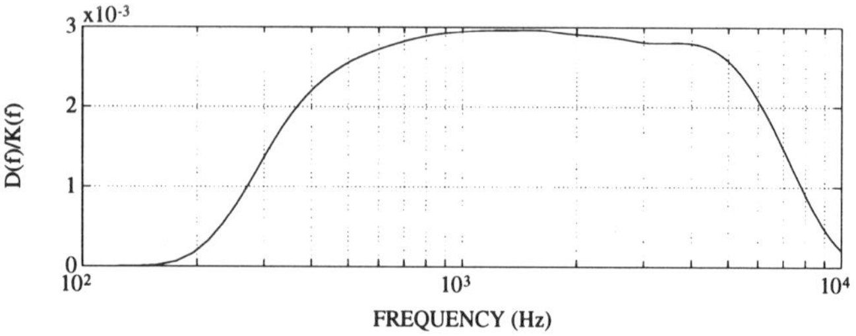

Figure 5 Figure showing the ratio of $D(f)/\kappa(f)$. This ratio is a measure of the phone articulation per critical band. The ratio has been normalized so that its area is 1.

[21], $F(X) = 165(10^{2.1X} - 0.88)$ (Hz), we know that the distance along the basilar membrane, between 300 Hz and 8 kHz, is 20 mm. Thus, for $K = 20$, there is about 1 articulation band/mm, corresponding to about 80 hair cells or about 800 neurons. Since Fletcher identified a critical band to be about 0.5 mm along the basilar membrane, one articulation band represents two critical bands. However the number K was chosen for convenience, and should not be taken as a particularly significant number. It has been reported for example that 10 bands is too few, and 30 bands gives no improvement in accuracy over 20 bands.

The slope of the cochlear map dF/dX was found to be proportional to the *critical ratio* $\kappa(f)$, which is the ratio of a tone probe level to a noise masker spectral level at the masked threshold [1, 14, 13, 12]. The critical ratio $\kappa(f)$ is an important psychophysical measure of the relative bandwidths of normal human cochlear filters. The ratio of $D(f)$, the articulation index density, and $\kappa(f)$, the critical ratio, is a measure of the relative speech articulation/mm or per critical band. As shown in Fig. 5 this ratio is approximately uniform over the speech band. These results have been scaled so that $\int D(f)/\kappa(f)df = 1$.

A physical measure of the articulation errors Steinberg, Galt, Anderson [23], and French and Steinberg [19] went on to show that the signal to noise ratio expressed in dB, in each critical band, normalized to 30 dB, determines the band articulation index D_k in band k corresponding to the band articulation error e_k, as given by Eq. 11.14. This relation is given by $D_k(\alpha) = \mathrm{SNR}_k(\alpha)/30K$. The band signal to noise ratio SNR_k, expressed in dB, is set to zero when it becomes negative, and is set to 30 when it is greater than 30. Thus the articulation index depends on the signal to noise ratios in each band rather than the speech energy spectrum.

There seems to be a slight but important error in this formulation. When the band SNR_k is poor (i.e., 0), Eq. 11.14 requires that $e_k = 1$. However we know that the total error must be less than $1 - 1/M$ due to guessing. Thus this formula needs to be modified to be realistic. In fact, if the independent speech features are binary, then the maximum error of the product of all the channels that detect a given feature must be 0.5 when the SNR for those channels is poor. Thus if we could more accurately measure and model the errors due to chance, it might lead to a better understanding of the feature space.

Articulation bands versus critical bands It is important to clarify the difference between articulation bands and critical bands. Critical bands represent filtered and neurally encoded signal intensity, and are related to the *partial loudness* of the speech as a function of the position along the basilar membrane X. The partial loudness is sometimes called the *neural excitation pattern* [1]. Articulation bands, on the other hand, are a measure of *partial phone recognition*, as a function along a tonotopic axis X_{AI} similar, but not equal, to X. Thus a very important transformation has taken place between the critical band signal and the articulation band signal, namely the *neural representation of signal intensity has been transformed into a measure of partial recognition*. We must not assume that this is a trivial transformation. If it were, robust ASR would have been achieved many years ago. It is worth remembering that Fletcher discovered both the critical band and the articulation band, and nowhere did he suggest equating them.

The intelligibility of words When words are used instead of nonsense syllables, the syllable errors must be further transformed to determine the word intelligibility $W(\mathcal{A})$. This case represents a decrease in the speech entropy due to word context. These methods were partially worked out by Fletcher and Steinberg [10, 18, 16]. Boothroyd described similar results using the empirical expression $W(\mathcal{A}) = 1 - (1 - S(\mathcal{A}))^j$, where the constant $j > 1$ depends on the entropy of the word corpus and may be empirically determined [2, 4, 3].

Summary For speech having a signal to noise ratio of SNR_k dB, where k labels the $K = 20$ frequency bands

$$D_k(\alpha) = \frac{1}{K}\text{SNR}_k(\alpha)/30 \tag{11.16}$$

$$\mathcal{A}(\alpha) = \sum_{k=1}^{K} D_k(\alpha) \tag{11.17}$$

$$s(\mathcal{A}) = 1 - e_{min}^{\mathcal{A}} \tag{11.18}$$

$$S(\mathcal{A}) = s^3 \tag{11.19}$$

$$W(\mathcal{A}) = 1 - (1 - S(\mathcal{A}))^j \tag{11.20}$$

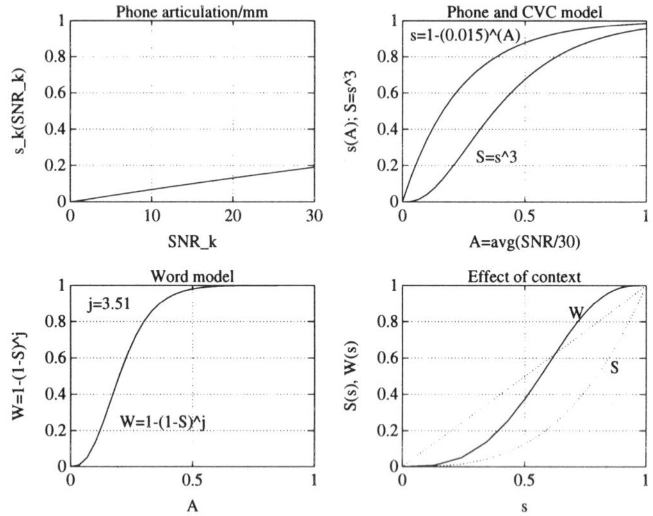

Figure 6 The partial phone articulation density $s_k = 1 - e_k$ at one point along the tonotopic axis is shown in the upper-left plot, as a function of the band signal to noise ratio SNR_k along the basilar membrane, expressed in dB. The partial articulation density is in units of probability of correct identification of nonsense phone units per millimeter along the basilar membrane. The phone articulation s is shown as a function of the average SNR/30 in the upper-right plot along with the nonsense CVC syllable articulation S. When meaningful sounds (words) are used, the effects of context must be taken into account. This effect is shown in the lower-left plot. One may also plot one variable against the other, as shown in the lower-right plot.

When $\mathrm{SNR}_k(\alpha)/30$ is less than 0 it is set to 0, and when it is greater than 1, it is set to 1. The constant j is greater than 1. The constant e_{min} is a measure of the subject's performance.

This model has been tested with hundreds of combinations of channel parameters and is impressively accurate over a large range of channel conditions [16, 19].

4 DISCUSSION

Thus partial recognition results from estimates of independent features across frequency (Eq. 11.13). For example if noise fills the upper band and clean speech fills the lower band, the speech articulation will be identical to the case of sharply low–pass filtered speech having no energy in the upper band. Estimates of the feature density give one feature per 5 mm, corresponding to 1 octave in frequency. One octave is about 5 to 10 critical bands since each critical band is between 0.5 and 1 mm. Only after the features are detected are they merged into estimates of the phones. This implies that the CNS is processing the cochlear channels over (across) time and extracting independent features across frequency. This analysis might be called *across–time processing*. The presence of noise in the upper band, however, confounds an ASR template–norm based approach which does not treat the frequency channels as independent during a feature–vector extraction phase. Such a template based analysis might be called *across–frequency processing*. Since $\mathcal{A}$ depends on the SNR rather than the energy spectrum, filtering the speech (e.g., a spectral tilt of 12 dB/oct) does not change the articulation unless the filtering reduces the SNR below 30 dB.

Band correlation and CMR The extraction of frequency–local independent features might be done by finding correlations between adjacent auditory (critical band) channels. For example, when noise is added to the speech and the SNR across frequency changes, a correlation measure between adjacent bands will decrease in a manner that is consistent with the phone articulation reduction. In such a model, a large number of neural signals would be grouped by the CNS to form one of the phone features. This view is most strongly supported by "comodulation release from masking" (CMR) experiments which demonstrate that correlations of neural envelope signals are computed by the auditory system.

The recognition chain Fletcher's analysis defines a heuristic model of human speech recognition as a layered hierarchy as shown in Fig. 7. The acoustic signal enters the cochlea and is broken into frequency bands (critical bands) which define the signal to noise ratios SNR_k, where k lables the cochlear frequency channel. There are about 3000 hair cells along the basilar membrane corresponding to a heavily overlapped set of cochlear filters. These outputs are then processed by the first "layer" which defines the phone features represented by the partial articulation errors e_k, as given by Eq. 11.16 and Eq. 11.14. Usually $K = 20$ bands are assumed, corresponding to 1 mm along the basilar membrane (e.g., one or two critical bands). The next layer defines a phone space, measured as articulations s found from Eq. 11.18. There are about 20 phones per {C,V} unit. The phones are then transformed into syllable units having articulation S (Eq. 11.19), and then into words with intelligibility (Eq. 11.20). The approximate number of nonsense CVCs is about 8000. The number of CVC words is much less, and could be estimated using data from the tables of Chapter 5 of [14]. A plot of typical values for these articulation measures is shown in Fig. 6.

Implications for modern machine speech recognition Fletcher's Articulation Index studies and models have important implications for ASR. Typical ASR systems start with a "front–end" that transforms the speech signal into a "feature vector" which is then processed by a "back–end" classifier. These systems frequently place a heavy emphasis on word and language models as a method of increasing the recognition scores.

Because of the confusion and misunderstanding based on coarticulation arguments, only a small amount of research has been done on the automatic recognition of nonsense CVC's. From the work of Fletcher and Steinberg, it should be clear that *the* real challenge in machine recognition today is human–like performance for phones and nonsense CVC under conditions of typical channel distortions. Since the human performance is well known under these conditions [16, 19], nonsense CVC's represent an excellent database. Decreasing the error rates for these elementary signals would have a major impact on overall system performance and robustness.

Average phone entropy If we use the phone frequencies given by Fletcher [14, page 95] and calculate the phone entropy, we find 4.3 bits for the initial consonant C_i, 4.5 bits for the V, and 4.1 bits for the final consonant C_f. The average phone entropy is then 4.3 bits/phone, or $M = 2^{4.3} = 19.7$ possible phones, on the average. This entropy

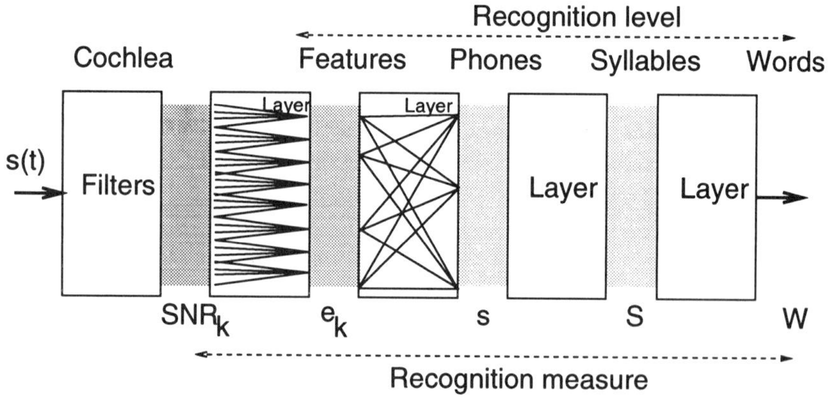

Figure 7 A hypothetical cascade of recognition layers, starting with the cochlea. The articulation measures shown at the bottom are defined in Table 2. The words along the top describe the physical correlate of the measure. No feedback is assumed between layers in this over simplified model of HSR. The first layer, the cochlea, determines the signal to noise ratio in about 3000 overlapping critical band channels. The next layer extracts features (i.e. partial recognition) from the speech in a local manner, as indicated by the network wiring. The output of this layer is measured in terms of the K=20 or so feature errors e_k. Next the features are mapped onto the $M = 20$ or so phones. This process necessarily integrates across the entire tonotopic axis. Then syllables and words are formed.

would be relevant for telephone speech when using a phone recognizer front–end that treats the phone string as context free. The actual entropy would drop as context effects are included.

What is the nature of the phone feature space In this section I would like to speculate on the possible nature of the phone feature space, given what we know about the auditory system. From this point on I shall use the term *feature* to mean 1 bit of independent partial recognition information. This definition seems consistent with the notion that a feature is a binary concept, namely it describes something that is, or is not, present [7]. If M objects are to be described with no redundancy, then it takes $\log_2(M)$ features to describe them. If we interpret the bits computed from the phone entropy as features, then there are 4.3 features/phone, on the average.

It is common practice to use 20 articulation bands (K=20), where each band corresponds to 1 mm along the basilar membrane. Thus, on the average, there are 4.3/20 = 0.215 features/mm. Perhaps a more intuitive way of expressing this is that there are 20/4.3=4.65 mm/feature. Since 1 octave corresponds to about 5 mm along the basilar membrane, the average phone feature density is about 1 feature/octave and the average feature length is 1 octave/feature.

We know that the cochlea (as well as the eye) breaks the external world down into a tonotopic (spatial) array of critical bands (pixels). It is then the job of the CNS to "reconstruct" the scene from these pieces. There is evidence that this is done by making a huge cascade of local binary decisions. Much of this decision making process is done at a subconscious level. For example, in a visual scene, the edge of a piece of paper is seen to be continuous, not because it is continuous, but because that is a rational decision to make given the local pixel inputs. Each time the elements are found to be correlated, they are fused together (e.g., the edge of the paper looks continuous; a musical chord sounds as one; a voice in noise is fused as a unit, independent of the noise).

From this point of view the actual number of articulation bands (K=20) is unimportant as long as they are not under-represented, just like the Nyquist sampling theorem. What *is* important is that the feature length along the tonotopic axis as defined by the local feature extractors. Assigning probabilities to the detection of (binary) features provides a natural and unique reduction of the data from 3000 narrow–band neural channels to 4.3 (on the average) tonotopic feature regions of various lengths, that depend on the specific input phone. The number of correlated regions along the tonotopic axis might determine the dimension of the space, for example, with the length of the correlation determining the coordinates (e.g., the probability of the feature being present). Such a model is a natural generalization of Fletcher's (and Munson's) model of tonal loudness which was the first model to propose the idea of a neural excitation pattern [1].

Across–time versus across–frequency processing The template based approach used in ASR could be called an *across–frequency* processing scheme. It appears that HSR is achieved using an *across–time* processing scheme, with only local coupling across frequency. There is considerable evidence for this. First, the articulation channels are independent. Second, the human listener is quite insensitive to dispersive (frequency dependent) delay, such as all–pass filters. This famous fact is frequently referred to as "Ohm's Law of acoustics," which claims that the ear is *phase–deaf* [24]. Room reverberation is an important form of such a degradation. The reverberation time in a room must reach at least 0.3 – 0.5 seconds before one is aware of its presence, and must be in the range of seconds before it becomes a serious degradation to speech communication. Reverberation is typically very frequency dependent, with only short relative delays at high frequencies, and long relative delays at low frequencies. With the feature extraction done across–time rather than across–frequency, the system is much less sensitive to this common type of frequency dependent degradation.

Coarticulation Across–time processing may also resolve the paradox of coarticulation which results from trying to assign each phone a spectral template. Associating a spectral template to a particular sound assumes (incorrectly) that the timing of the features over frequency must be synchronous. From psychophysical experiments we known that, under many conditions, our ability to perceive the magnitude (and even the relative order) of temporal events can be very poor. Phone recognition is most certainly not the synchronous timing of feature events, but some more abstract relation between the presence and absence of features, and their geometrical relations in a multidimensional feature space [5]. This transformation has to be viewed as a form of "categorical perception" [20].

5 SUMMARY

How do humans process and recognize speech? (Recall the rhetorical title of this paper.) We are still looking for the answer, but Fletcher's experiments and analysis tell us where to look. The most elementary auditory speech processing model (Fig. 7) is a cascade of the cochlea followed by the following recognition layers: features, phones, syllables, words, sentences, meaning, etc. The basis of this model is the articulation results for the various context levels.

To understand how speech is recognized it is necessary to systematically control context factors since context is very important. This was done by working with a database of balanced nonsense CVC, CV, and VC syllables. Syllables having context decreases the speech entropy. We know the relation between the phone and nonsense syllable error rates (Eq. 11.1). The phones may be represented as a multidimensional features space leading to the idea of the categorical perception of these units [20].

The most important problem in HSR for those interested in ASR is feature and phone perception. This is the part of the system that goes from an acoustic signal to the most basic speech recognition element. The speech sounds are divided into a time–frequency continuum of feature–bearing frequency bands by the cochlea. There are about 4.3 independent binary features represented along approximately 20 mm of basilar membrane. These feature channels form the basis of the articulation channel errors e_k. The bands are processed in such a way as to robustly extract and isolate the $\approx$ 20 possible elemental sounds for each phone time slot (each C or V) (i.e., 4.3 bits/phone).

Equations 11.3 and 11.6 indicate that the articulation error *information* $\mathcal{I}$ defined by $\mathcal{I}(e) = \log_2(e)$ is additive (non–redundant) and defines a tonotopic density, because the partial recognition errors are independent (Eq. 11.13).

The signal to noise ratio of each cochlear hair cell signal is important to the formation of the feature channels since e_k is known to depend directly on these SNR's rather than on the spectral energy. There are many more articulation bands than features, and we have estimated that each feature, on the average, occupies about 5 mm (1 octave) along the basilar membrane. The model is consistent with the idea of using correlations between neighboring cochlear channels to form the output of the feature layer. If two filter bands were not correlated due to a poor SNR, then the correlator output would be small.

At each layer the entropy is decreased as meaning is utilized and extracted. By removing meaning (context) from the speech, we may effectively disable the processing for that layer. This allows us to quantify the layer's utilization of the entropy of speech. Using this method, the recognition probability $W(S)$ between the words and nonsense syllables, due to the word intelligibility, has been empirically estimated. The same technique has been applied to quantify meaning in sentences.

Finally, the articulation formulas (and the data they are based on) provide an important knowledge database against which we may benchmark the machine recognizers, to drive their performance toward that of the human listener.

REFERENCES

[1] Jont B. Allen. Harvey fletcher 1884–1981. In Jont B. Allen, editor, *The ASA edition of Speech and Hearing in Communication*. Acoustical Society of America, New York, 1994.

[2] A. Boothroyd. Statistical theory of the speech discrimination score. *J. Acoust. Soc. Am.*, 43(2), 1968.

[3] A. Boothroyd. Speech preception, sensorineural hearing loss, and hearing aids. In G.A. Studebaker and I. Hochberg, editors, *Acoustical Factors affecting Hearing aid performance*, pages 277–299, Boston, 1993. Allyn and Bacon.

[4] A. Boothroyd and S. Nittrouer. Mathematical treatment of context effects in phoneme and word recognition. *J. Acoust. Soc. Am.*, 84(1), 1988.

[5] L.D. Braida. Integration models of speech intelligibility. In *Speech communication metrics and human performance*, pages 1–20, Washington DC, 1993. NAS–CHABA.

[6] A.W. Bronkhorst. A model for context effects in speech recognition. *Journal of the Acoustical Society of America*, 93(1):499–509, January 1993.

[7] N. Chomsky and M. Halle. *The sound pattern of English*. Harper & Row, New York, 1968.

[8] T.M. Cover and J.A. Thomas. *Elements of information theory*. John Wiley & Sons, Inc., New York, 1991.

[9] Harvey Fletcher. An empirical theory of telephone quality. *AT&T Internal Memorandum*, 101(6), October 1921. First document about the Articulation Index. Case 211031, report 21839. The only know copy is in Fletcher's archives which are held by Brigham Young University, Provo UT. (UA029, Box 19 #2, MSS1233).

[10] Harvey Fletcher. The nature of speech and its interpretation. *Journal of the Franklin Institute*, 193(6):729–747, June 1922.

[11] Harvey Fletcher. *Speech and Hearing*. D. Van Nostrand Company, Inc., New York, 1929.

[12] Harvey Fletcher. Loudness, masking and their relation to the hearing process and the problem of noise measurement. *Journal of the Acoustical Society of America*, 9:275–293, April 1938.

[13] Harvey Fletcher. The mechanism of hearing as revealed through experiments on the masking effect of thermal noise. *Proceedings National Academy Science*, 24:265–274, 1938.

[14] Harvey Fletcher. *Speech and Hearing in Communication*. Robert E. Krieger Publishing Company, 1953.

[15] Harvey Fletcher. Speech and hearing in communication. In Jont B. Allen, editor, *The ASA edition of Speech and Hearing in Communication*. Acoustical Society of Am., New York, 1994.

[16] Harvey Fletcher and R.H. Galt. Perception of speech and its relation to telephony. *Journal of the Acoustical Society of America*, 22:89–151, March 1950.

[17] Harvey Fletcher and J.C. Steinberg. The dependence of the loudness of a complex sound upon the energy in the various frequency regions of the sound. *Physical Review*, 24(3):306–317, September 1924.

[18] Harvey Fletcher and J.C. Steinberg. Articulation testing methods. *Bell System Technical Journal*, 8:806–854, October 1929.

[19] N.R. French and J.C. Steinberg. Factors governing the intelligibility of speech sounds. *Journal of the Acoustical Society of America*, 19:90–119, 1947.

[20] K.P. Green, K.K. Kuhl, A.N. Meltzoff, and E.B. Stevens. Integrating speech information across talkers, gender, and sensory modality: Female faces and male voices in the McGurk effect. *Perception and Psychophysics*, 50(6):524–536, 1991.

[21] D.D. Greenwood. A cochlear frequency-position function for several species – 29 years later. *Journal of the Acoustical Society of America*, 87:2592–2605, 1990.

[22] G.A. Miller. Speech and language. In S.S. Stevens, editor, *Handbook of Experimental Psychology*, chapter 21, pages 789–810. John Wiley & Sons, INC., New York, 1951.

[23] J.C. Steinberg, R.H. Galt, and A.B. Anderson. Study of speech and hearing - case 20871-3. *Bell System Technical Memorandum*, 328(67):1–36, I:1–17, II:1–16, III:1–9, Figs:1–115, 1937.

[24] S.S. Stevens and H. Davis. *Hearing, Its Psychology and Physiology*. Reprinted by the Am. Inst. Phy. for the Acoust. Soc. Am., 1983.

PART 3

SPEAKER RECOGNITION

12

DATA FUSION TECHNIQUES FOR SPEAKER RECOGNITION

Kevin R. Farrell
Richard J. Mammone

Dictaphone Corporation
Stratford, Connecticut, 06497, USA
Caip Center, Rutgers University
Piscataway, New Jersey, 08855-1390, USA

1 INTRODUCTION

Speaker recognition refers to the capability of recognizing a person based on his or her voice. Specifically, this consists of either speaker verification or speaker identification. The objective of speaker verification is to verify a person's claimed identity based on a sample of speech from that person. The objective of speaker identification is to use a person's voice to identify that person among a predetermined set of people.

Speaker recognition by machine is generally interpreted as a pattern recognition problem. There are two phases involved, namely feature extraction and classification. The feature extraction phase obtains parameters from the speech signal that are characteristic of the speaker. These parameters, or *features*, are then used by the classifier to determine the speaker identity or authenticity.

Two general approaches have been considered for constructing classifiers for speaker recognition systems. These can be categorized as those which use unsupervised training algorithms and those which use supervised training algorithms. Unsupervised training algorithms utilize unlabeled training data. Hence, the algorithm only considers the data for the speaker to be modeled. These approaches include vector quantization [1], hidden Markov models [2, 3], and Gaussian mixture models [4]. Supervised training algorithms use labeled training data and tend to be discriminant-based. These approaches include multilayer perceptrons [5], radial basis functions [6], and neural tree networks [7]. Speaker models based on supervised training capture the differences of the target speaker to other speakers (interspeaker variability), whereas models based on unsupervised training use a self-similarity measure (intraspeaker variability).

Numerous unsupervised and supervised training algorithms for generating speaker models have been evaluated and compared for speaker recognition tasks [5, 7]. However, there has been little research devoted to the *combination* of these two methods. Intuitively, this is appealing as both methods are modeling different speaker characteristics, i.e., interspeaker versus intraspeaker. This chapter evaluates methods based on data fusion for combining the results of classifiers based on unsupervised and supervised training algorithms. The following section provides a brief introduction to data fusion in addition to a description of the more popular data fusion techniques. These techniques are then evaluated for text-independent, closed-set speaker identification and text-dependent speaker verification. A summary of the observations is then provided.

2 DATA FUSION

It is often advantageous to combine the opinions of several experts when making a decision. For example, when one is obtaining a medical diagnosis, a decision for subsequent care may become easier after obtaining several opinions as opposed to just one. This concept has been exploited in the field of data fusion for tasks including handwriting recognition [8] and remote sensing [10].

The general form of a data fusion system is illustrated in Figure 1. Given a set of feature vectors, each expert outputs its own observation, which consists of a probability measure and/or class label. The combiner will then use one of many methods to collapse these observations into a single decision. The set of feature vectors can also be different from expert to expert, which would be a case of *sensor* fusion [9]. However, the work in this chapter only considers the case for different experts and not different features.

There are numerous ways to combine the opinions of multiple experts. For example, if the outputs of all experts are probabilities then a simple combination method would be to take a weighted sum of the probabilities or of the logs of the probabilities. These methods are known as the linear opinion pool and log opinion pools [10], respectively. If the outputs of the experts are class labels, then methods such as voting [8] or ranking [11] can be used. For fuzzy decisions, Dempster-Shafer theory can also be used for the combination of experts

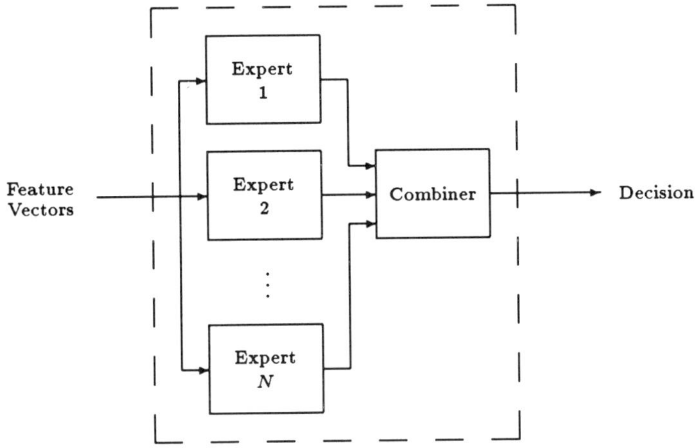

Figure 1 Data Fusion System

[8]. This chapter evaluates the linear and log opinion pool methods for speaker recognition. These methods are described in more detail as follows.

2.1 Linear Opinion Pool

The linear opinion pool is a commonly used data fusion technique that is convenient due to its simplicity. The linear opinion pool is evaluated as a weighted sum of the outputs for each expert:

$$P_{linear}(x) = \sum_{i=1}^{n} \alpha_i p_i(x), \tag{12.1}$$

where $P_{linear}(x)$ is the probability of the combined system, α_i are weights, $p_i(x)$ is the probability output by the ith expert, and n is the number of experts. For all experiments in this chapter, α is between zero and one and the sum of the α's is equal to one.

The linear opinion pool is appealing in that the output is a probability distribution and the weights α_i provide a rough measure of the expertise of the ith expert. However, it is noted that the probability distribution of the combiner output may be multimodal, which may impose a more complicated decision strategy. The linear opinion pool has been considered in speaker recognition for the combination of features [12], namely cepstrum and delta cepstrum features.

2.2 Log Opinion Pool

An alternative to the linear opinion pool is the log opinion pool. If the α weights are constrained to lie between zero and one and sum up to one, then the log opinion pool also outputs a probability distribution. However, as opposed to the linear opinion pool, the output distribution of the log opinion pool is unimodal [10].

The log opinion pool consists of a weighted product of the expert outputs:

$$P_{log}(x) = \prod_{i=1}^{n} p_i^{\alpha_i}(x). \tag{12.2}$$

Note that with this formulation, if any expert assigns a probability of zero, then the combined probability will also be zero. Hence, an individual expert has the capability of a "veto", whereas in the linear opinion pool the zero probability would be averaged in with the other probabilities.

One problem that both the linear and log opinion pools are subject to is the selection of the weights α_i. Several heuristic solutions [10] to this are to 1) use equal weights, i.e., $\alpha_i = 1/n$, 2) use weights proportional to a ranking, i.e., $\alpha_i = r_i / \sum_{r=1}^{n} r$, or 3) evaluate the weights over the range of zero to one for cross-validation data and select the best α_i. For the experiments in this chapter, plots for the third method are provided.

3 TEXT-INDEPENDENT SPEAKER IDENTIFICATION

Speaker identification systems can be either *closed set* or *open set*. Closed set speaker identification refers to the case where the speaker is known *a priori* to be a member of a set of N speakers. Open set speaker identification includes the additional possibility where the speaker may be from outside the set of N speakers. Open set speaker identification requires an additional thresholding step to determine if the speaker is "out of set".

An additional characteristic of speaker identification systems regards the form of spoken input, which can be text-dependent or text-independent. Text-dependent speaker recognition systems require that the speaker utter a specific phrase or a given password. Text-independent speaker identification systems identify the speaker regardless of the utterance. This section focuses on text-independent, closed-set speaker identification.

The classification stage for text-independent speaker recognition is typically implemented by building a parametric model for each speaker. Some parametric models express the speaker's acoustic space as a multimodal probability density function. Other models are based on discriminant boundaries that separate the acoustic space occupied by the speaker from that occupied by other speakers. The first approach is generally implemented with unsupervised training algorithms, whereas the latter method would use a supervised training algorithm.

Given a specific feature vector, a speaker model associates a number corresponding to the degree of match with that speaker. The stream of numbers obtained for a set of feature vectors can be used to obtain a likelihood score for that speaker model. For speaker identification, the feature vectors for the test utterance are applied to all speaker models and the corresponding likelihood scores are computed. The speaker corresponding to the model with the largest score is then selected.

The more common methods of constructing speaker models for text-independent speaker recognition use unsupervised training methods. These approaches include vector quantization [1], hidden Markov models [3], and Gaussian mixture models [4]. Here, only the data for that speaker is used to train that speaker's model. Alternative methods for building speaker models use supervised training, such as that in multilayer perceptrons [5], radial basis functions [6], and neural tree networks [7]. Here, a speaker's model is trained with the data from possibly all speakers in the population.

The fact that supervised and unsupervised classifiers base their discrimination on different criteria leads one to believe that they can be combined for improved performance. One such combination, based on the linear opinion pool, is evaluated in this section. The following sections review the NTN and VQ along with their application for speaker modeling in text-independent speaker identification. The hybrid NTN/VQ system is then presented and evaluated using the TIMIT database.

3.1 Vector Quantization

The vector quantization (VQ) method is based on an unsupervised training algorithm, i.e., the class label is not used. In this case, clustering is used to group the training data into its individual modes or classes. Numerous VQ algorithms exist, including the Linde-Buzo-Gray (LBG) [13] method and k-means algorithm [14]. The LBG method is used here.

The VQ algorithm can be used to construct a classification system for speaker recognition as follows [1]. Given the extracted feature vectors from a speaker, a codebook is constructed for that speaker. This process is repeated for all speakers in the population. For speaker identification, the feature vectors from a test utterance are applied to each of the codebooks. For a given codebook, the centroid that is closest to the test vector is found and the distance to this centroid is accumulated for that codebook. This process is repeated for all test vectors and the speaker corresponding to the codebook with the minimum accumulated distance is selected.

3.2 Neural Tree Networks

The modified neural tree network (MNTN) [7] is constructed with a supervised training algorithm. The NTN [15] is a hierarchical classifier that combines the properties of decision trees [16] and feed-forward neural networks [17]. Whereas the NTN is strictly a classification tree, i.e., only the leaf labels are used, the MNTN additionally uses probability measures at the leaf nodes. All subsequent references to the NTN in this chapter assume this modification is included.

The assignment of probability measures occurs within a technique called *forward* pruning. The forward pruning algorithm consists of simply truncating the growth of the tree beyond a certain level. For the leaves at the truncated level, a vote is taken and the leaf is assigned the label of the majority. In addition to a label, the leaf is also assigned a confidence. The confidence is computed as the ratio of the number of elements for the vote winner to the total number of elements. The confidence provides a measure of confusion for the different regions of feature space. The concept of forward pruning is illustrated in Figure 2.

A MNTN can be trained for each speaker in the population as follows. First, the MNTN for each speaker is presented with a training set that is comprised of the data for all speakers. Here, the extracted feature vectors for that speaker are labeled as "one" and the extracted feature vectors for everyone else are labeled as "zero". A binary MNTN for speaker i is then trained with this data. This procedure is repeated for all speakers in the population.

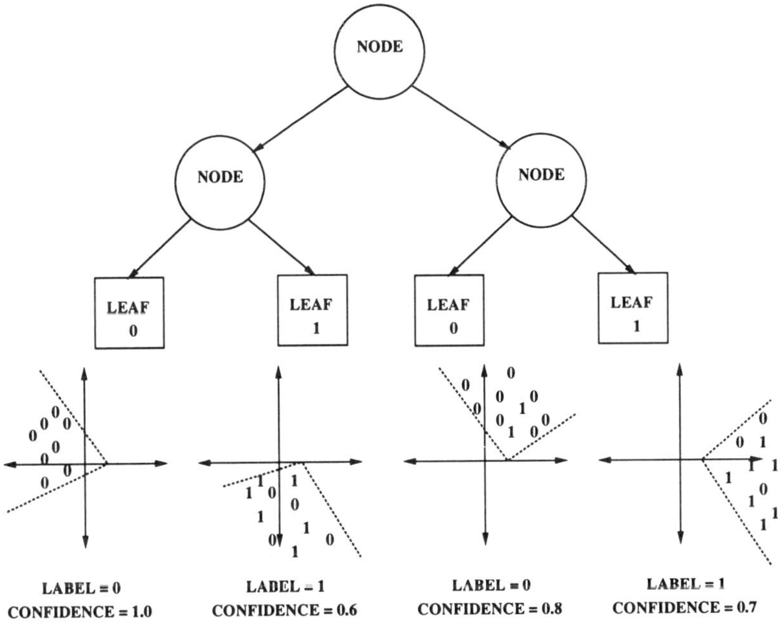

Figure 2 Forward Pruning and Confidence Measures

Specifically, a trained MNTN can be applied to speaker recognition as follows. Given a sequence of feature vectors x from the test utterance and a trained MNTN for speaker S_i, the corresponding speaker score is found as:

$$P_{MNTN}(x|S_i) = \frac{\sum_{j=1}^{M} c_j^1}{\sum_{j=1}^{N} c_j^0 + \sum_{j=1}^{M} c_j^1}, \tag{12.3}$$

where c^1 and c^0 are the confidence scores for the speaker and antispeaker, respectively. Here, the M and N correspond to the number of vectors classified as "one" and "zero", respectively.

3.3 Hybrid System

Data fusion principles are used to combine the outputs of the NTN and VQ speaker models. The method used here is the linear opinion pool. This consists of evaluating a weighted sum of the model outputs as illustrated in Figure 3. Here, the outputs of the NTN and VQ speaker models are multiplied by α and

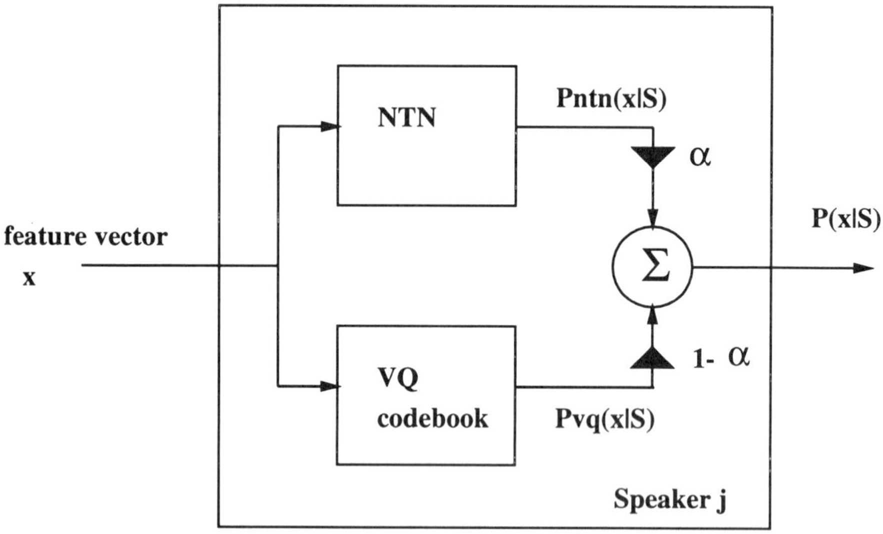

Figure 3 Data Fusion Approach

$1-\alpha$, respectively, where α lies between zero and one. Hence, when $\alpha = 0$ the system consists of solely the VQ-based classification system and likewise when $\alpha = 1$, only the NTN-based system is used.

To enable the VQ and NTN speaker models to be used simultaneously in such a system, several normalization steps must be used. First, the VQ distortion and NTN confidence must be converted to the same scale. Here, the VQ distortion and NTN confidence are normalized to lie on a scale of 0 to 1, where 1 denotes a perfect match. These scaled scores are analogous, though not necessarily equivalent, to probabilities.

The VQ distortion is normalized as follows:

$$P_{vq}(x|S_j) = e^{-(x-c_i)^2}, \tag{12.4}$$

where c_i is the centroid closest to x. The NTN label and confidence, which lies between 0.5 and 1.0, can be normalized to a single score as follows:

$$P_{ntn}(x|S_j) = \begin{cases} 0.5 * (1.0 + confidence), & if\ label = 1 \\ 0.5 * (1.0 - confidence), & if\ label = 0 \end{cases}. \tag{12.5}$$

These scores can now be combined and evaluated for speaker recognition tasks.

3.4 Experimental Results

Database

The database considered for the speaker recognition experiments is a subset of the DARPA TIMIT database. This set consists of 50 male speakers taken from the third dialect region. A VQ codebook and NTN are trained for each of 50 speakers.

The preprocessing of the TIMIT speech data consists of several steps. First, the speech is downsampled from a 16KHz to an 8 KHz sampling frequency. The downsampling is performed to obtain a toll quality signal. The speech data is processed by a silence removing algorithm followed by the application of a pre-emphasis filter $H(z) = 1 - 0.95z^{-1}$. A 30 ms Hamming window is applied to the speech every 10 ms. A twelfth order linear predictive (LP) analysis is performed for each speech frame from which twelve cepstral coefficients are derived.

There are 10 utterances for each speaker in the selected set. Five of the utterances are concatenated and used for training. The remaining five sentences are used individually for testing. The duration of the training data ranges from 7 to 13 seconds per speaker and the duration of each test utterance ranges from 0.7 to 3.2 seconds. It is noted that the duration of the test utterances is relatively short and that performance can be improved by increasing the duration.

Closed-Set Speaker Identification

The data fusion technique as applied to combining NTN and VQ speaker models is evaluated for closed-set speaker identification. Each VQ codebook is trained using the Linde-Buzo-Gray (LBG) [13] algorithm. Codebook sizes are evaluated for 64 and 128 codebook entries. Each NTN is trained with the data for all 50 speakers. The maximum NTN level is varied from six to eight levels. The performance is evaluated as a function of α and is shown in Figures 4 and 5. The identification performance of the six and seven bit codebooks (without the NTN) is 85% and 89%, respectively.

The individual performance of the VQ and NTN-based systems can be seen as those points corresponding to $\alpha = 0$ and $\alpha = 1$, respectively. For the six bit codebook experiment, it can be seen that the optimal choice of α yields a classification system that outperforms either individual system by at least 7%.

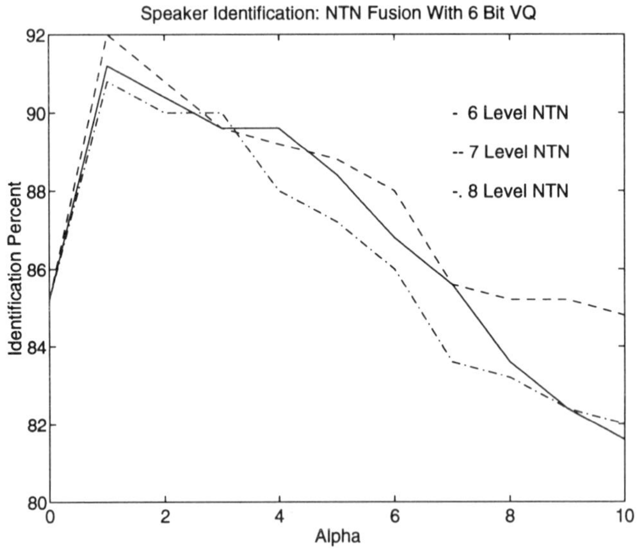

Figure 4 Closed Set Speaker Identification - 6 bit VQ

Similarly, for the seven bit codebook experiment the combined classification system outperformed either individual system by at least 4%.

4 TEXT-DEPENDENT SPEAKER VERIFICATION

The objective of speaker verification is to verify a person's claimed identity based on an utterance from that person. As with speaker identification systems, speaker verification systems can also be text-dependent or text-independent. Text-independent systems are more convenient from a user standpoint in that there is no need for a password. However, in order to completely model and evaluate the acoustic feature space of a speaker, substantially more data is necessary for training and testing than for text-dependent systems. This section focuses on text-dependent speaker verification.

Early work on text-dependent speaker verification utilized dynamic time warping (DTW) [18]. Later, it was found that hidden Markov models (HMMs)

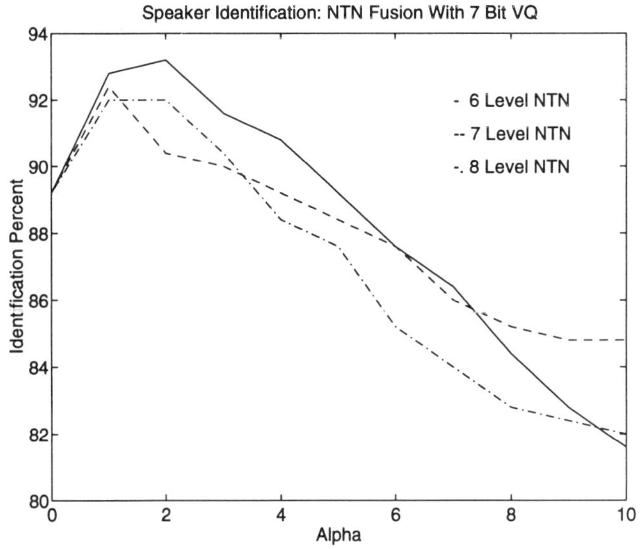

Figure 5 Closed Set Speaker Identification - 7 bit VQ

yielded improved performance over DTW techniques [2]. Several forms of HMMs including subword models [3] and whole word models [19], have since been considered. In general, HMMs have been considered the predominant technology for text-dependent speaker verification. However, one limitation of classification systems based on HMMs is that they generally require more data than other systems to sufficiently learn the model parameters. Thus, HMMs may not be suitable for applications that involve limited training and testing data.

A new system is proposed for text-dependent speaker verification. The new system uses data fusion principles to combine the results of two sets of text-dependent speaker models. The first set of speaker models utilizes neural tree networks (NTNs) [15]. NTNs are discriminant-based and provide an interspeaker measure. The NTNs have been successfully applied to text-independent speaker recognition [7] where, specifically, advantages were found for speaker verification along with tasks that involved limited training data. The second set of text-dependent speaker models uses a DTW-based technique. The DTW method uses a distortion measurement between the extracted features of a test utterance and those stored for that speaker during training. Hence it provides an intraspeaker measure. Additionally, it uses temporal information, which currently is not used in the NTN. Since the information provided by the NTN and DTW models is somewhat independent, they can be combined to yield a powerful classification system for text-dependent speaker verification. The NTN and DTW methods of speaker verification are briefly described as follows.

4.1 Neural Tree Network

There are several approaches for applying the NTN to *text-dependent* speaker recognition. One method is to train the NTN to discriminate between a password spoken by the target speaker and the same password spoken by other speakers. This can be interpreted as a "whole-word" NTN. Another method is to build "sub-word" NTN models [20]. In this case, the password of the target speaker will be segmented into sub-words, i.e., phonemes. A NTN will then be trained for each sub-word unit, where the anti-class data will consist of the data from other speakers' utterances of that sub-word unit. These models can then be concatenated to form passwords. The "whole-word" NTN will be considered here. Note that this method does not utilize the temporal information of the password. This information can be obtained, however, by combining the results of the NTN with a temporal-based model, such as the DTW method.

4.2 Dynamic Time Warping

The DTW algorithm is a distortion-based approach for time aligning the dynamics of two waveforms. For speaker verification, a reference template can be generated from several utterances of the password [18]. These utterances are combined to form a single reference template using a modified k-means algorithm [21]. Then during testing, a decision can be made to accept or reject the claimed identity based on whether or not the distortion falls below a predetermined threshold. To allow for subsequent fusion with other speaker models, such as NTNs, the DTW distortions must be converted to a compatible scale, i.e., a probability. We accomplish this by simply raising the negative distortion to an exponential.

4.3 Hybrid System

The task considered in this research focuses on limited training and testing data collected over a telephone channel. Specifically, four utterances of a password are used for training and the verification decision is based on one utterance of that password. This constraint confines the classification approaches that can be used in the system. The approach considered here is described as follows.

For the four training utterances, not only must a model be trained, but additionally an appropriate threshold must be determined. A resampling technique, namely the "leave-one-out" method, is used here to accommodate the small number of training exemplars. Three utterances are used to train a speaker model and the remaining utterance is used as an independent test case. This procedure is repeated four times, each time leaving out a different test utterance, to yield four models.

The combined classifier system uses the outputs of both sets of models for its final decision. Each individual classifier will provide its decision and a confidence to the combiner, which will output the final decision. This concept is illustrated in Figure 6.

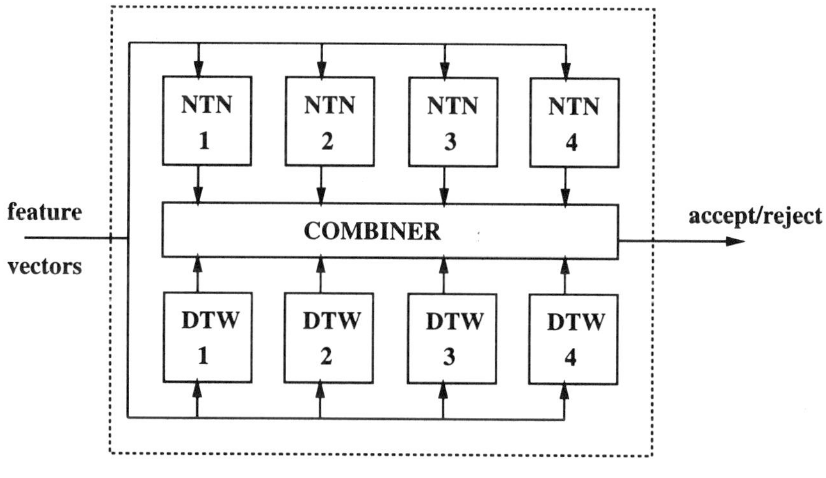

Figure 6 Verification system for claimed identity

The decision of each individual classifier is determined from the score as compared to a threshold. The threshold here is computed as a function of the *intraspeaker* and *interspeaker* scores. The intraspeaker score for each classifier is found as the performance on the left out utterance. Similarly, the interspeaker score is found by applying imposter utterances to the model and computing the average of the top scores (nominally five).

Each of the eight classifiers has its own threshold which is determined as follows:

$$T = x * interspeaker + y * intraspeaker, \tag{12.6}$$

where T is the threshold. Typical values for x and y are 0.8 and 0.2, respectively. During testing, each classifier will output a one or zero, corresponding to whether or not the score is greater than the threshold. Additionally, a *confidence* is output by each classifier. This confidence reflects the amount that a score exceeds or falls below the threshold. For the proposed system, the score of each classifier lies between zero and one with zero being the most confident reject and one being the most confident accept. Since all accepts lie between the threshold T and one, we define the accept confidence as:

$$C_{accept} = \frac{S-T}{1-T}, \tag{12.7}$$

where S represents the score. Similarly, for a reject, which lies between zero and the threshold T, we define the confidence as:

$$C_{reject} = \frac{T-S}{T}. \tag{12.8}$$

There are numerous methods for combining this information. One method is to simply take a vote on the different classifiers [8], i.e., if at least five classifiers verify the claimed identity, then accept it. Other methods utilize weighted sums or products of the model outputs, which are known as ***linear opinion pools*** and ***log opinion pools*** [10]. These two methods are evaluated here.

To simplify the combination of speaker models, the four NTN and DTW scores are combined into a single NTN and single DTW score. These accumulated scores are based on the confidences discussed previously. For the four NTNs (or DTWs), if the score is accepted, then a positive confidence is accumulated, else if the score is rejected, a negative confidence is accumulated. The final score is normalized to be between zero and one. Hence, the linear and log opinion pools will be used to combine two scores and not eight.

4.4 Experimental Results

Database

The database used to evaluate the system was collected over a telephone channel. All of the handsets utilized electret microphones. The speech acquired through the telephone channel is sampled at 8 kHz, and μ-law coded at 8 bits/sample. The speech signal is pre-emphasized with a pre-emphasis factor of 0.95. Features are extracted within 20 millisecond analysis windows having 5 millisecond shifts between consecutive analysis windows. The features extracted from the analysis windows consist of LP-derived cepstral coefficients.

The system was evaluated with data from 20 male speakers. Fourteen utterances of the state, "New Jersey" were collected from ten of these 20 speakers. Four utterances were used to train each model and the remaining ten were used for testing. For the remaining ten speakers, ten utterances of the state, "New Jersey" were collected and used as imposter utterances for each of the first ten speakers. This database allows for 100 true speaker trials and 1000 imposter trials. Note that a pre-existing database is used to provide the antispeaker data for the NTN training. This antispeaker database contained four utterances of "New Jersey" from roughly 40 male speakers. The 10 imposter speakers used to evaluate the false accept rate are not included in the antispeaker database.

Text-Dependent Speaker Verification

The operating curves for the NTN and DTW were evaluated and are shown in Figure 7. Here it is seen that the NTN performs significantly better than DTW, i.e., an equal error rate of 4.9% as compared to 9.8%. However, the equal error rate of the combined system using the linear opinion pool with $\alpha = 0.5$ is 3%, which is better than that of either method when used individually.

The error rates of the linear and log opinion pools are evaluated as a function of α. The performance of both methods is illustrated in Figure 8. For this experiment, the linear opinion pool performs slightly better than the log opinion pool for the optimum alpha values, though this result may not be significant for 1100 trials. The best equal error rates for each method, including the voting method described in the previous section, are provided in Table 1. Here it can be seen that all forms of data fusion yield a reduction in the equal error rate for either method used individually.

Table 1 Equal Error Rates

	NTN	DTW	linear	log	vote
EER	4.9%	9.8%	3.0%	3.6%	4.0%

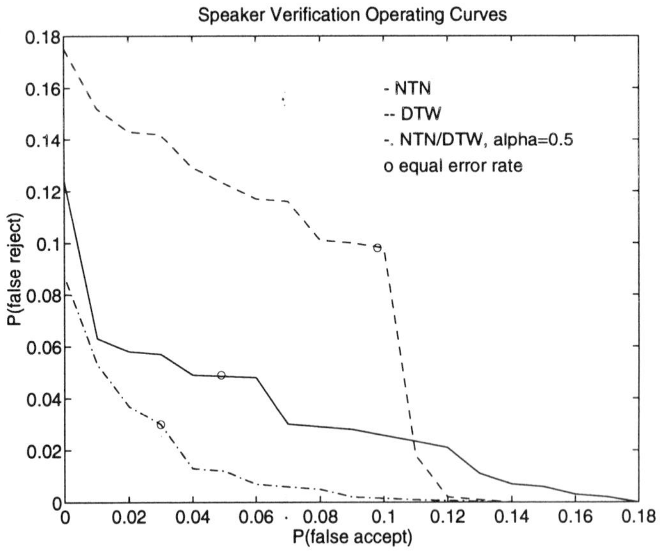

Figure 7 Speaker verification operating curves

5 CONCLUSION

Data fusion concepts have been evaluated for combining the results of classification systems designed with unsupervised and supervised training alogithms. This concept has been tested within the application of speaker recognition. Two separate tasks are evaluated, namely text-independent, closed-set speaker identification and text-dependent speaker verification. For the closed-set speaker identification task, the linear opinion pool method is evaluated for combining the outputs of vector quantization (VQ) and neural tree network (NTN) speaker models. The combined classification system is shown to outperform systems based on the VQ and NTN models when used individually. Particularly, the performance is increased by roughly 4-7% over that of either method. For the given experiments, this corresponds to a 50% reduction in error. For the speaker verification task, the linear and log opinion pool methods are evaluated for combining the outputs of dynamic time warping (DTW) and NTN models. The equal error rates for the optimal linear and log opinion pools are 3% and 3.6%, respectively. This surpasses the individual performance of either method, which is 4.9% and 9.8% for the NTN and DTW approaches, respectively.

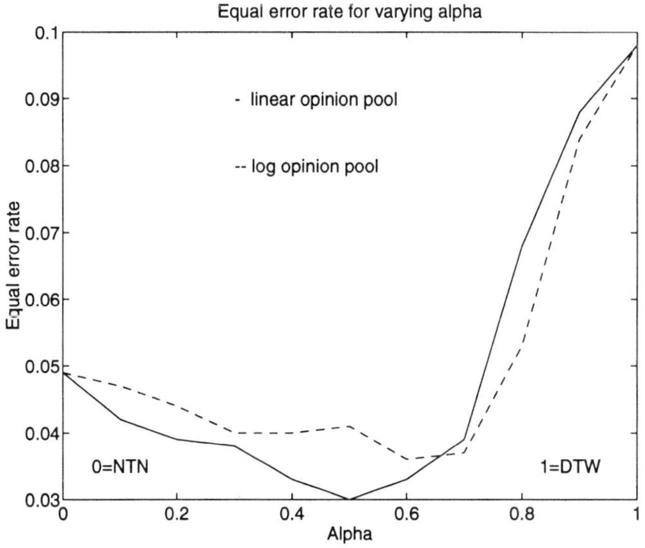

Figure 8 Equal error rate versus alpha

REFERENCES

[1] F. K. Soong, A. E. Rosenberg, L. R. Rabiner, and B. H. Juang. A vector quantization approach to speaker recognition. In *Proceedings ICASSP*, pages 387–390, 1985.

[2] J. M. Naik, L. P. Netsch, and G. R. Doddington. Speaker verification over long distance telephone lines. In *Proceedings ICASSP*, 1989.

[3] A. E. Rosenberg, C. H. Lee, and F. K. Soong. Sub-word unit talker verification using hidden markov models. In *Proceedings ICASSP*, pages 269–272, 1990.

[4] D. Reynolds. *A Gaussian Mixture Modeling Approach to Text-Independent Speaker Identification.* PhD thesis, Georgia Institute of Technology, February 1993.

[5] J. Oglesby and J. S. Mason. Optimization of neural models for speaker identification. In *Proceedings ICASSP*, pages 261–264, 1990.

[6] J. Oglesby and J. S. Mason. Radial basis function networks for speaker recognition. In *Proceedings ICASSP*, pages 393–396, 1991.

[7] K. R. Farrell, R. J. Mammone, and K. T. Assaleh. Speaker recognition using neural networks and conventional classifiers. *IEEE Trans. on Speech and Audio Processing*, 2(1), part 2, 1994.

[8] L. Xu, A. Krzyzak, and C. Y. Suen. Methods of combining multiple classifiers and their applications to hand-written character recognition. *IEEE Trans. on Systems, Man and Cybernetics*, 23(3):418–435, 1992.

[9] Henderson and Weitz. Multisensor knowledge systems. *International Journal of Robotics Research*, 7(6):114–137, June 1988.

[10] J. A. Benediktsson and P. H. Swain. Consensus theoretic classification methods. *IEEE Trans. on Systems, Man and Cybernetics*, 22(4):688–704, 1992.

[11] T. K. Ho, J. J. Hull, and S. N. Srihari. Decision combination in multiple classifier systems. *IEEE Trans. on Pattern Analysis and Machine Intelligence*, 16(1):66–75, 1994.

[12] F. K. Soong and A. E. Rosenberg. On the use of instantaneous and transitional spectral information in speaker recognition. *IEEE Trans. on Acoustics, Speech, and Signal Processing*, ASSP-36:871–879, June 1988.

[13] Y. Linde, A. Buzo, and R. M. Gray. An algorithm for vector quantizer design. *IEEE Trans. on Communications*, COM-28:84–95, 1981.

[14] R. O. Duda and P. E. Hart. *Pattern Classification and Scene Analysis.* Wiley, New York, NY, 1973.

[15] A. Sankar and R. J. Mammone. Growing and pruning neural tree networks. *IEEE Trans. on Computers*, C-42:221–229, March 1993.

[16] L. Breiman, J. H. Friedman, R. A. Olshen, and C. J. Stone. *Classification and Regression Trees.* Wadsworth International Group, Belmont, CA, 1984.

[17] R. Lippmann. An introduction to computing with neural nets. *IEEE ASSP Magazine*, pages 4–22, Apr. 1987.

[18] S. Furui. Cepstral analysis technique for automatic speaker verification. *IEEE Trans. on Acoustics, Speech, and Signal Processing*, ASSP-29:254–272, April 1981.

[19] A. E. Rosenberg, C. H. Lee, and S. Gokeen. Connected word talker recognition using whole word hidden markov models. In *Proceedings ICASSP*, pages 381–384, 1991.

[20] H. Liou and R. J. Mammone. Text-dependent speaker verification using sub-word neural tree networks. In *SPIE Proceedings, Conference on the Automatic Inspection and Identification of Humans*, July 1994.

[21] J. G. Wilpon and L. R. Rabiner. A modified k-means clustering algorithm for use in isolated word recognition. *IEEE Trans. on Acoustics, Speech, and Signal Processing*, 33:587–594, June 1985.

13

SPEAKER RECOGNITION OVER TELEPHONE CHANNELS

Yu-Hung Kao, Lorin Netsch
and P.K. Rajasekaran

Systems and Information Science Laboratory, Texas Instruments

1 INTRODUCTION

Being able to verify or determine the identity of a person by voice is very useful in many applications. For example, in telephone banking or calling card charging, user identity must be verified before the transaction can be authorized. Most systems use a PIN for authorization, but it can be forgotten or stolen. Other methods of authorization, such as finger prints or retinal scans, may be more secure than a PIN, they are not practical in many situations.

Speaker identification and verification over the telephone is especially important because it has very broad application potential. However, different handsets and transmission channels degrade the signal quality and make the identification more difficult. The amount of information needed for speaker templates (voice prints) of a large number of subscribers requires large hardware memory resources, data bandwidth and call setup time. We will address these problems in this chapter.

1.1 Definition

We first define the difference between speaker identification and verification:

- Speaker identification: For test speaker x, the task is to determine which one of the n registered speakers the test speaker x is.

- Speaker verification: For test speaker x who claims identity i (i is one of the n registered speakers), the task is a binary decision - accept or reject the claim.

Depending on whether speaker x is one of the n registered speakers or not, the problem is further classified into:

- Closed set: Speaker x is one of the n registered speakers. The system has seen all the speakers and a speaker set specific discriminant can be trained. The closed set task is easier, but not a realistic assumption for many applications.
- Open set: Speaker x is not necessarily one of the n registered speakers. This task is more realistic, especially in speaker verification.

In identification, the rate at which the system incorrectly identifies speakers is used to calibrate system performance. Verification performance is measured using both a plot of Receiver Operating Characteristics (ROC) and the equal error rate. The ROC plot displays the true speaker detection rate (rate at which the system correctly verifies a speaker) versus the impostor acceptance rate (rate at which the system incorrectly verifies an impostor) over a set of decision threshold settings. The equal error rate is defined as the rate at the decision threshold setting where the impostor acceptance rate equals the true speaker rejection rate.

2 CHALLENGES

Speaker identification and verification has been a topic of research for many years [1] [2] [3] [4] [5], and research activities continue to be devoted to the following areas:

2.1 Features

There are two sources of variation among speakers : 1) differences in vocal cords and vocal tract shapes; 2) differences in speaking styles. There are no acoustic cues specifically or exclusively tied to speaker identity. Most of the parameters and features used in speech analysis contain information useful for the identification of both speakers and spoken messages. The effectiveness of features is coupled with distance measures and statistical modeling methods. Good features for identification and verification should: 1) discriminate between speakers while being tolerant of intra-speaker variation; 2) be easy to measure from speech signals; 3) be stable over time; 4) occur frequently and naturally in speech; 5) be channel robust; 6) not be susceptible to mimicry.

In general, spectral features are more effective than prosodic features. Although prosodic features (pitch, stress, speed ...) are important speaker identity cues for human perception, they have not provided significant performance improvements when applied in automatic speaker identification. In addition, prosodic features are susceptible to mimicry.

Sambur [6] examined a total of 92 features, including formant frequencies and bandwidth for vowels, resonances during nasals and fricatives, F0 statistics and dynamics, and timing measurements (e.g. formant trajectory slopes in diphthongs, voice-onset time for stops). Using a *knock out* procedure, he found among the most useful features were F2 in nasals, F2-F3-F4 in vowels, and mean F0.

Spectral coefficients are the most commonly used features [1] [6]. A common measure for effectiveness of individual features is the *F-ratio*, which compares inter- and intra-speaker feature variances. A larger F-ratio usually indicates better discrimination capability. As long as the feature set covers all the spectral contents of the speech signal, the modeling can always take care of the discriminant issues. Typical spectral feature sets found useful are cepstral coefficients (coefficients of the inverse Fourier transform of the log spectrum), principal spectral components, and log area ratios, along with their time derivatives.

2.2 Channel and Handset Variation

Channel robustness is a critical issue for real world applications. Telephone signals are affected by handsets, transmission channels, and even speech compression algorithms. When training and test conditions are not matched, which is often the case, performance will degrade tremendously [7]. To deal with channel variation, we can either choose features invariant to it [8], design signal processing algorithms to compensate for the variation [9], or build noise models [10].

2.3 Free-Text vs. Fixed-Text

Speaker verification often employs a fixed-text mode, where users are required to utter a fixed *password*. This is useful in cooperative situations. In fixed-text mode, word-specific speaker discriminants can be tuned specifically for speaker variation [11]. Dimensionality reduction and speaker discriminating effects can be achieved because there is no phoneme variation.

In uncooperative situations, such as monitoring a conversation and identifying the speakers, no fixed utterance is available, so the problem becomes free-text. Free-text speaker identification and verification are more difficult than fixed-text because the variation due to different phonemes is much larger than that due to different speakers [1]. Free-text speaker identification must resolve the phoneme variation effect. Work has been done on classification of phonemes before speaker identification [12], which will convert a free-text problem into fixed-text.

2.4 Speaker Modeling

Vector quantization (VQ) [13] and continuous / discrete Hidden Markov Modeling (HMM) have been compared for speaker identification [14] and no significant difference was observed. The performance strongly depends on VQ codebook size or number of HMM mixtures. Larger codebook size or more mixtures provide more accurate identification; however, as the codebook size or number of mixtures increase performance will eventually saturate. The advantage of HMM is its ability to model temporal information. The result suggests that either temporal information is not very useful for speaker identification or the algorithm is not modeling temporal information correctly. However, if speech recognition is used in the speaker identification / verification algorithm [15], HMM can certainly provide a better time alignment than simply matching the acoustics with the closest VQ codeword. Recently, the idea of *cohort comparison* [16] has become popular and provides a method of speaker normalization.

3 TASKS AND CORPORA

In this chapter, we will discuss methods of implementing speaker identification and verification over telephone channels using VQ and continuous HMM speaker modeling and cohort comparison. We will present results of both free-text and fixed-text experiments applied to a variety of corpora.

Three telephone corpora were used in the experiments presented in this chapter. A fourth corpus was used in fixed-text experiments for additional statistical model training, but not testing.

3.1 King Corpus

The narrowband portion of the *King* corpus [17] is utilized in free-text identification experiments in this chapter. The corpus was collected in 10 sessions from 51 male speakers, 26 from San Diego and 25 from Nutley. The speakers were asked to talk about several topics, so that the speech is natural and spontaneous. The data were collected over long distance telephone lines. The data for the 25 Nutley speakers were much noisier than that of the 26 San Diego speakers. The speech material from each session is approximately 45 seconds long. The data were digitized at 8 kHz and 12-bit resolution. Sessions 1 to 5 and sessions 6 to 10 were collected under different environments. This division of data, known as *the great divide*, results in serious degradation of performance as observed in [8] when training on one set and testing on the other. We performed our experiments in three contexts: San Diego alone (26 speakers), Nutley alone (25 speakers), and all 51 speakers combined. Further, the experiments were carried out across *the great divide* producing the most challenging test condition. Because it is a spontaneous monologue corpus, only free-text experiments were performed.

3.2 SPIDRE Corpus

The SPIDRE (SPeaker ID Research and Evaluation) corpus is designed specifically for free-text speaker identification experiments. It consists of a set of spontaneous conversations selected from 45 speakers (both male and female) from the SWITCHBOARD corpus [18]. Only the one side of each conversation corresponding to the 45 speakers is used. For each speaker there are four calls and three different handsets were used for these four calls, so two of the four calls are from the same handset. There is no handset overlap between speakers, hence $3 \times 45 = 135$ handsets were used. The corpus was separated into 4 sets, each set containing one call from each speaker. Sets 1 and 2 are from the same handset, while sets 3 and 4 are from different handsets. In the experiments described in this chapter, set 1 was used for training, set 2 for the same-channel test, and sets 3 and 4 for cross-channel tests. The statistics from set 1 indicate that the average length of a call is 362 seconds, 35% of which is speech from the desired speaker. Other sets have similar statistics.

3.3 Digit Verification Corpus

The Digit Verification corpus consists of speech from 25 female and 25 male subjects. Each subject provided 25 calls over long-distance telephone lines, and was encouraged to use a wide variety of handsets and environments. During the call, each subject provided three tokens of a unique familiar ten-digit string (a leading digit plus a social security number), and three tokens of the unique digit string of another subject of the same sex.

3.4 Additional Training Corpus

The Additional Training corpus consists of 36 female and 70 male subjects. Each subject provided 25 to 40 calls over long-distance telephone lines, and was encouraged to use a wide variety of handsets and environments. During the call, each subject provided three tokens consisting of a four-digit string and two other non-digit words.

4 FREE-TEXT IDENTIFICATION AND VERIFICATION

This set of experiments used a free-text method for all three corpora. VQ codebooks were used to model the acoustic space of speakers after removal of silence portions of the speech signal.

4.1 Robust Methods

The free-text experiments focused on several methods of robust front-end processing and cohort comparison. These methods are designed to increase the separability of speakers and reduce channel variation effects.

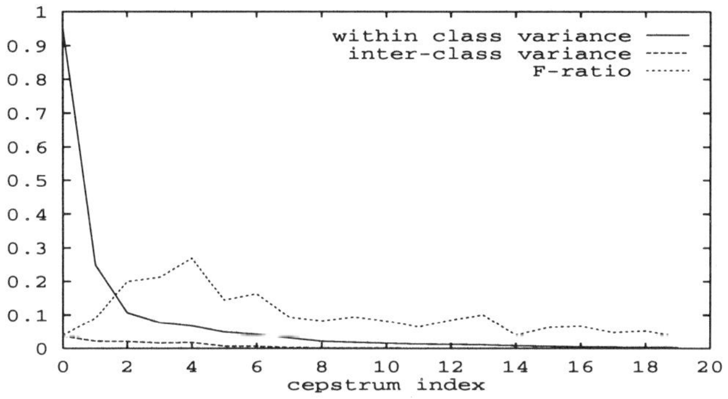

Figure 1 Variance analysis for cepstrum

Bandpass Liftering

The F-ratio is a common measure to calibrate the discrimination power of features. Figure 1 shows the variance analysis for 20 dimensions of cepstral coefficients. It is clear that both within-class and inter-class variances decrease as cepstrum index increases.

Studies of variance of cepstral coefficients of speech indicate *diminishing* discriminating power of the higher order coefficients. This shows that the variability of higher order coefficients is an inherent disadvantage of the linear prediction process [19]. On the other hand, channel variation often affects the lower order coefficients much more than higher order coefficients. For example, the effect of different channel frequency responses is usually most prominent in the first couple of cepstral coefficients.

Because the higher order coefficients have less discriminating power and the lower order coefficients are more susceptible to channel variation, as an engineering compromise, we use a window function to de-emphasize both higher and lower order coefficients. We call this approach bandpass liftering (BPL) [19]. The window used is:

$$w(k) = 1 + h \sin(\pi k / L) \tag{13.1}$$

where $h = L/2$, $k = 1, 2, \ldots, L$ and $w(k) = 0$ for other k.

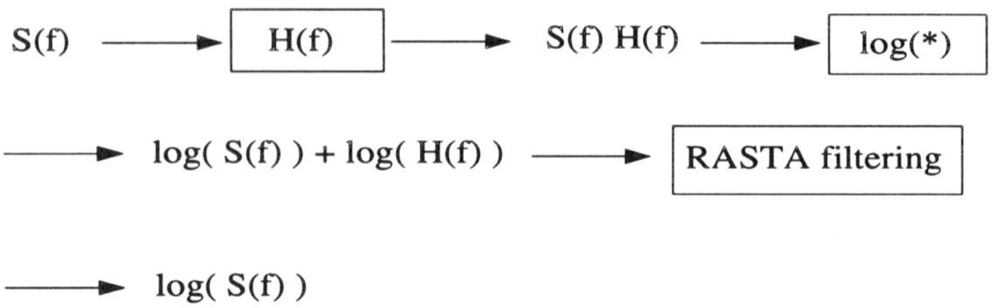

Channel effect H(f) is usually stationary or slowly varying compared to S(f), therefore it can be filtered out by the RASTA (RelAtive SpecTrA) filter. Log(*) operation is necessary before the RASTA filtering.

Figure 2 RASTA filtering

RASTA Filtering

Channel variations come from different frequency responses of the channels and telephone handsets. Compared to speech signals, these channel effects are *stationary* or *slowly varying.* We can use a bandpass filter in the log spectral domain to remove the lower frequency components of the spectrum and still preserve the speech signal [20]. Of course, we cannot remove too much of the lower frequency spectrum. It may contain important speech information. The proper design of the filter requires analysis and experimentation.

Figure 2 shows Relative Spectra (RASTA) filtering being applied to the log spectrum. This is a form of homomorphic deconvolution, in which an estimate of the speech signal $S(f)$ is made from the corrupted spectrum, $S(f)H(f)$.

In several experiments, we used the cepstrum to form feature vectors. In the cepstral domain, the channel effect is an *additive* term, so RASTA filtering can be applied directly to the cepstral coefficients. Several filters were evaluated, and it was found that the performance was not very sensitive to different designs. The following filter was used in experiments reported in this chapter:

$$H(z) = \frac{0.2 + 0.1z^{-1} - 0.1z^{-3} - 0.2z^{-4}}{(1 - 0.98z^{-1})z^{-4}} \tag{13.2}$$

One way to measure the effectiveness of channel normalization algorithms is by inspecting feature *scattergrams.* However, a stereo corpus, where speech data transmitted through two channels are recorded simultaneously, is required to compute the scattergrams. In our experiments, we used the CSR (Continuous Speech Recognition) corpus, where each utterance is recorded simultaneously by two different microphones (Sennheiser vs. Crown or SONY). Because of the simultaneous recording of two channels, the *sample by sample* match makes exact comparison of channel effects possible. Here we compare the cepstral coefficients, frame by frame, dimension by dimension.

For each frame, we used cepstral coefficient c_i from microphone 1 as the x coordinate, and c_i from microphone 2 as y coordinate to construct a point in the scattergram. As shown in Figures 3 and 4, it is clear that the scattergrams after RASTA filtering are more aligned with the line: $x = y$. This means the cepstral coefficients from the two different microphones are *normalized* to be equal to each other. The improvements can be quantified by computing the distance between each point and the line $x = y$, then normalizing by the variance of the distribution of all the points. We observed consistent improvements for all 20 cepstral dimensions. The most significant improvement can be observed in dimension 1, as shown in Figures 3 and 4.

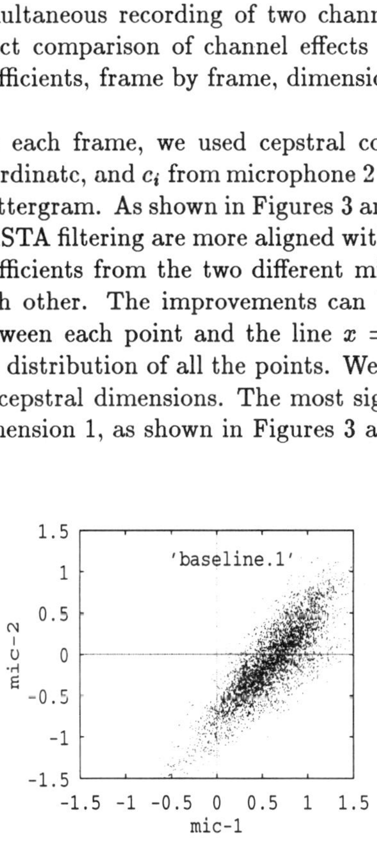

Figure 3 Baseline

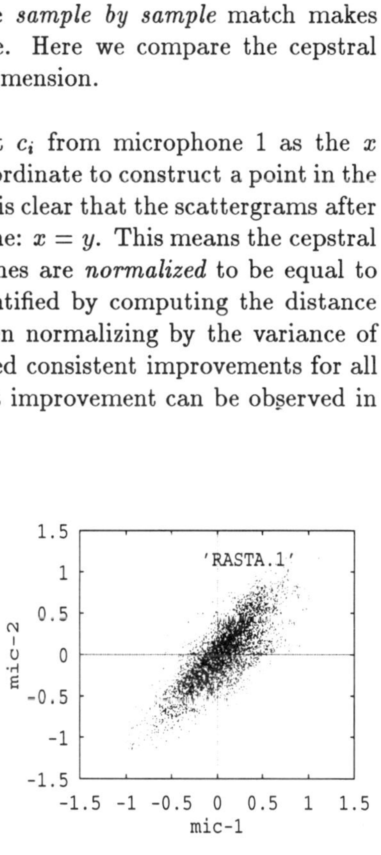

Figure 4 RASTA

Cohort Comparison vs. Thresholding

For speaker verification, when speaker x claims identity i, we compute the score of the input utterance with respect to speaker model i. If the score is better than a specific threshold, then we confirm the claimed identity, otherwise the claim is rejected. A low true speaker rejection rate and a low impostor acceptance rate are desired; but as we change the threshold value, trade-off is being made between these two error rates - one decreases, while the other increases.

When we used a threshold as the accept / reject criterion, we were surprised to see that RASTA, which performs so well in our speaker identification experiments, actually *degraded* speaker verification performance. We found that the explanation lies in the difference between *absolute threshold* vs. *cohort comparison.*

We inspected the scores of both true speaker attempts and impostor attempts before and after RASTA filtering, and found that the impostor scores improve *more* than the true speaker scores. In the absolute threshold approach, if true speaker scores and impostor scores improve equally, the performance stays the same. If true speaker scores improve more than impostor scores, the performance becomes better. However, when impostor scores improve more than true speaker scores, the performance becomes worse. In the speaker identification experiment, the scores are not compared with an absolute threshold, but compared with *peer scores*, that is, the score with respect to speaker model i is compared with scores with respect to other speaker models. In this way, the peer scores all depend on the input utterance, and they all adapt *together.* Uneven improvements are no longer a problem, since they are *self-normalized.* Following this idea, we changed the verification algorithm from comparing with an absolute threshold to comparing with peer scores. Instead of computing the score of the claimed identity model only, we compute scores with respect to all the registered models. We then determine the rank of the score using the claimed identity model, and subject the rank to a threshold to determine verification acceptance or rejection.

4.2 Free-Text System

The front-end of the system extracted 20 cepstral coefficients from a 14-th order LPC analysis (20 ms frame period, 30 ms window) on speech data sampled at 8KHz. A simple energy threshold was used to discard non-speech. A 30-element codebook was trained for each speaker as the speaker model. Experiments were performed using cepstral coefficients without (baseline) and with robustness processing. The results were tabulated and compared.

Closed set speaker identification and open set speaker verification were performed on all three corpora. For brevity, only selected results are presented here. However, a consistent, significant trend was observed in all experiments.

In closed set speaker identification, test utterances were compared frame by frame with each speaker model. The best codeword match was selected for each model, and distortions were accumulated to make the final decision.

For open set speaker verification, half of the speakers in the corpus were used as registered targets, and the other half as impostors. Test utterances were compared with all the registered target models, and the accumulated distortions were tallied to compute the rank of the claimed identity. If the rank was better than a certain threshold, then verification was declared successful; otherwise, the speaker was declared an impostor. The thresholds were adjusted over a range to generate the ROC plot (true speaker detection vs. impostor false alarm).

A closed-set speaker verification experiment was performed using the Digit Verification corpus. This experiment applied the cohort comparison ranking algorithm.

4.3 Results

King Corpus

Free-text, closed set speaker identification results using the King corpus are shown below. Robustness processing shows significant and consistent improvements.

The table below shows results within *the great divide*: we trained on sessions 1, 2, 3, and tested on sessions 4, 5; and trained on sessions 6, 7, 8, and tested on sessions 9, 10.

Identification error rate			
Front-end	San Diego-26	Nutley-25	All-51
Baseline	18.3%	65%	41.2%
BPL	14.4%	53%	33.3%
RASTA	8.6%	50%	28.9%
BPL + RASTA	5.8%	39%	22.1%

Below are results across *the great divide*: we trained on sessions 1, 2, 3, and tested on sessions 9, 10; and trained on sessions 6, 7, 8, and tested on sessions 4, 5.

Identification error rate			
Front-end	San Diego-26	Nutley-25	All-51
Baseline	92.3%	64%	80.4%
BPL	63.5%	54%	63.2%
RASTA	57.7%	47%	56.4%
BPL + RASTA	22.1%	35%	41.2%

SPIDRE Corpus

Free-text, closed set speaker identification results using the SPIDRE corpus are shown below. Set 1 was used for training. Set 2 was a test using the same handset. Sets 3 and 4 were tests using differing handsets.

Identification error rate			
Front-end	Set 2	Set 3	Set 4
Baseline	15.6%	77.8%	73.3%
BPL	15.6%	71.1%	64.4%
RASTA	11.1%	62.2%	51.1%
BPL + RASTA	6.7%	48.9%	40.0%

In the next table we computed the error rate reduction attributed to each robustness processing technique, both *with* and *without* the other technique present:

Error rate reduction			
Front-end	Set 2	Set 3	Set 4
Baseline ⇒ BPL	0%	9%	12%
RASTA ⇒ RASTA+BPL	40%	21%	22%
Baseline ⇒ RASTA	29%	20%	30%
BPL ⇒ RASTA+BPL	57%	31%	38%

We observed that the error rate reductions of bandpass liftering or RASTA techniques are larger when accompanied by the other technique. The same computation has been done in the King results and the same phenomenon can be observed. Usually when two different techniques are combined, the benefit from both taken together is less than the sum of the benefits of each taken separately. Yet in our experiments the reverse is true. If two techniques are working toward the same goal, then combining them may result in less than additive improvement, because the problem is *partly* solved by the other technique. In our case, since BPL is a static weighting along the cepstral dimension, it deemphasizes the highly variant and noisy cepstral coefficients. On the other hand, RASTA is a dynamic filtering method along the time dimension, and it aims to remove static channel effects. As for why one works better *with* the other present, we suspect that when the data are very bad, it is difficult for one technique to *demonstrate* the effect. For example, the lower components of the cepstrum have high variance and are very susceptible to channel variation; without deemphasizing them, no matter how we *fix* other components with RASTA, the error score contributed by these lower components is too large to overcome.

Digit Verification Corpus

Free-text, closed set speaker verification experiments were performed using the Digit Verification corpus. Only the male portion of the corpus was used. In training, the 25 first impostor utterances from each speaker were used to train the speaker model. Because these 25 utterances include all the 25 different ID's, the training material covers all the phonetic events. This is crucial in the free-text mode, because the phoneme effect is larger than the speaker effect. In the test, second and third true speaker utterances were used to compute the true speaker rejection rate, providing 1250 attempts. Second and third impostor utterances (also 1250 attempts) were used to compute the impostor acceptance rate.

Figures 5 and 6, show that the equal error rate was reduced to less than half in all four front-end methods when using cohort comparison instead of an absolute threshold. The following table shows the equal error rates in Figures 5 and 6.

	Equal error rate	
Front-end	absolute threshold	cohort comparison
Baseline	13%	5.5%
BPL	12%	2.5%
RASTA	21%	6%
BPL + RASTA	18%	2.5%

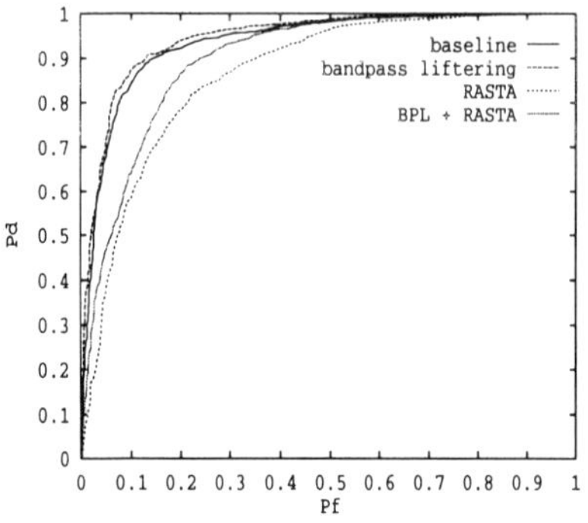

Figure 5 ROC plot using absolute threshold

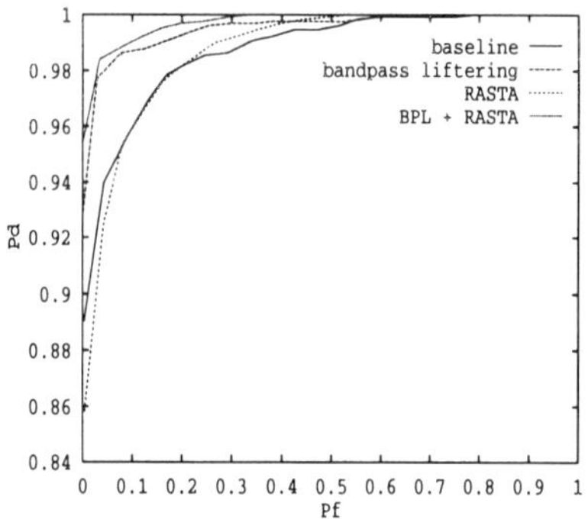

Figure 6 ROC plot using cohort comparison

5 FIXED-TEXT VERIFICATION

5.1 Objective

This section introduces Temporal decorrelation (TD) voice verification, a method that may be used to reduce the amount of speaker-specific data necessary to perform fixed-text verification. It can also provide a level of robustness to signal distortions, such as in the telecommunications network, where use of different telephone handsets and channels cause the distortion. The TD method implements a post-processing step occurring after a recognition task utilizing Hidden Markov Models (HMM) for words. Since there is significant correlation between observed speech feature vectors corresponding to the states from a known HMM word model, it should be possible to reduce the redundant speaker-dependent information. Thus, rather than calculating an utterance likelihood based on the individual statistics of each state of the HMM word models, TD constructs the likelihood based on statistics of entire words. It also seems plausible that the mechanisms causing distortion are consistent temporally, since only a single handset or channel is used during a call. Hence, it should be possible to find statistical features that span entire words and which are relatively independent of the distortions encountered.

5.2 Description of TD Method

Typical HMM-based verification processing, as shown in Figure 7a, implements HMM recognition to calculate the likelihood of the input speech given speaker-dependent verification models. If the likelihood is lower than a threshold, then the system declares that the speaker is an impostor. The TD method is shown in Figure 7b. In this method, the first step is again speech recognition using HMM models. However, the TD method uses speaker-independent HMMs. The speech is aligned according to a Viterbi maximum likelihood algorithm. Often, the verification phrase spoken is known by virtue of knowledge of the claimed identity. In this case, recognition may be grammatically constrained to the known phrase. This provides a significant advantage to the recognizer, resulting in excellent ability to locate the spoken phrase in noisy speech. No specific utterance extraction algorithm is necessary.

The speaker-independent HMM word models constructed for use in the TD method utilize single-mixture multivariate Gaussian observation distributions. The HMM model topology allows for variation in speech rate and also ensures that at least one input speech feature vector will be aligned with each observation distribution of the HMM. The number of states in the model varies proportionally to the nominal duration of the spoken word. Multiple states may share an observation distribution. A typical example HMM topology is shown in Figure 8. The vertical states in the example model all share the same observation distribution. This model was designed such that the number of observation distributions was half of the nominal duration in input frames. Beginning and ending observation topology was restricted to explain a finite duration of speech. This was done to limit the effect of leading and trailing non-speech on observation distribution parameters.

The output of the recognizer provides a mapping of input frames of speech feature vectors to HMM states, based on the HMM observation distributions and transition probabilities. The mapping is of prime importance in the TD method. It enables construction of word-level feature vectors after recognition alignment. Post-recognition word-level processing constructs a single word-level feature vector for each word of the input utterance. As illustrated in Figure 9, each word-level feature vector is formed by first averaging all input frame feature vectors mapped to each HMM observation distribution to form an averaged input utterance feature vector for each distribution. The averaged feature vectors corresponding to each word are concatenated to form a word-level concatenated vector. Note that each word-level concatenated vector has

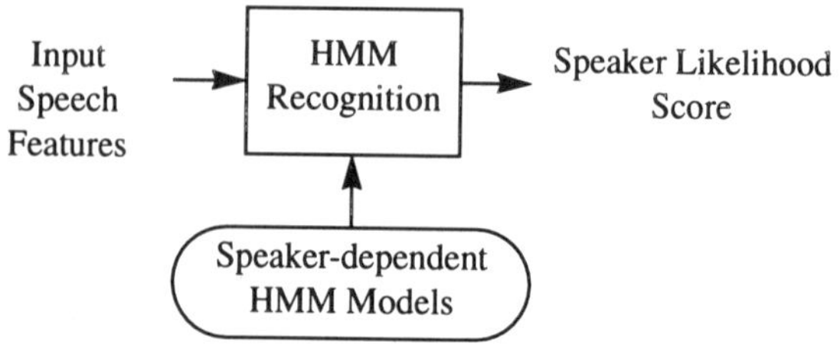

(a) Typical HMM-based speaker verification

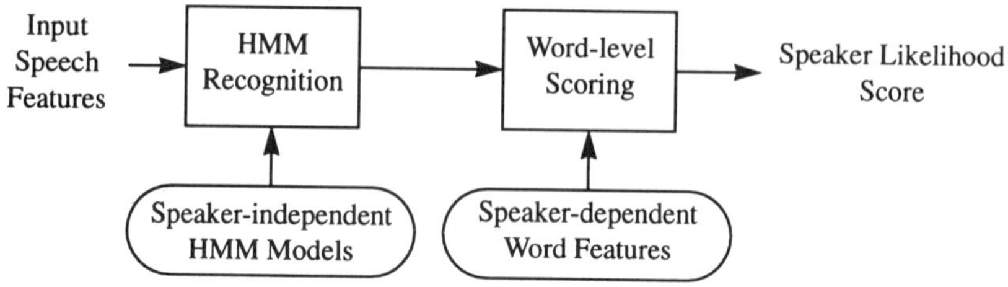

(b) Temporal Decorrelation speaker verification

Figure 7 Verification processing comparison

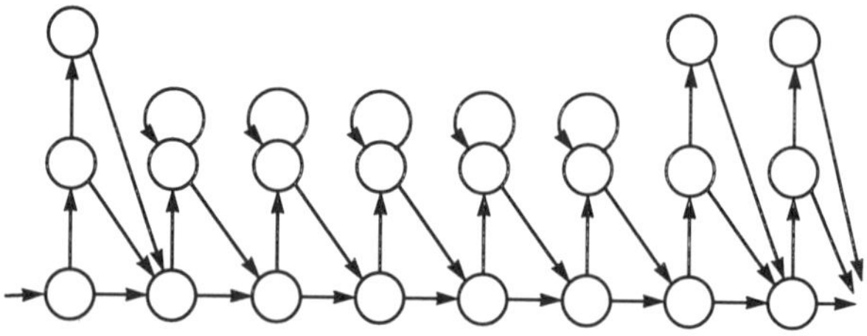

Figure 8 Example TD HMM model topology

a dimension which is the product of the number of frame-level features and the number of observation distributions in the HMM for the word.

The word-level concatenated vectors corresponding to each of the words recognized by the HMM alignment each undergo a temporal decorrelating linear transformation (TDT) unique to each word to form word-level feature vectors for each word. The TDTs reduce dimensionality of speaker-dependent information by producing decorrelated feature vectors for each word having a dimension much smaller than the concatenated vector.

The TDTs are designed to normalize the covariance of the within-speaker statistics and also provide a linear discriminant measure between speakers, which are used to calculate the likelihood. The TDT for each word is calculated prior to performing any verification using speech data in which many speakers provide many tokens of the word. First a sample mean and covariance matrix is calculated corresponding to the concatenated vector for each word and

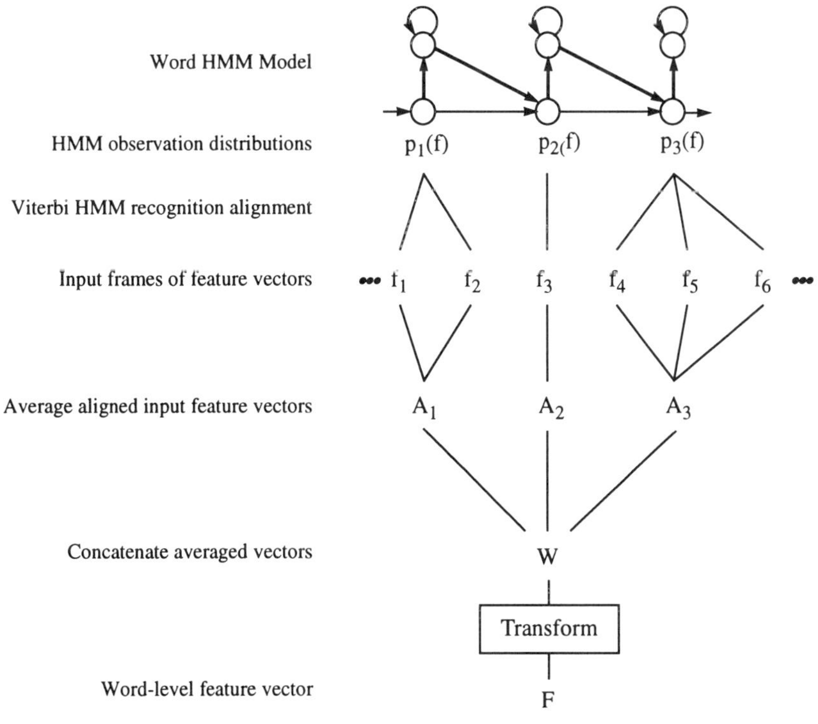

Figure 9 Forming word-level features

speaker. Pooling the covariance matrices for each speaker results in a single *pooled within-speaker* covariance matrix for each word. The covariance matrices and means corresponding to the word for all speakers are then used to calculate a *total* covariance matrix which represents the covariance of all concatenated vectors from all speakers. Simultaneous diagonalization by eigenvector analysis [21] is performed using the *pooled within-speaker* and *total* covariances. The resulting word-level features which best discriminate between speakers are retained to form the TDT for each word.

Each speaker enrolls by speaking a uniquely assigned phrase several times. The average of each word-level feature vector is calculated over all enrollment utterances, resulting in a set of speaker-dependent word-level reference vectors.

During verification, a subject wishing to be verified first provides some claim to identity. This is used to obtain the set of word-level reference feature vectors for that identity. The subject then speaks the uniquely assigned phrase, and word-level feature vectors are formed from the input speech. The input and reference word-level feature vectors are compared to generate a likelihood measure, typically using a Euclidean distance. The likelihood measure for each word is weighted by the relative duration of each word to obtain a final single likelihood measure which is compared to a threshold to make the true speaker/impostor decision.

5.3 Experimental Results

The TD method has been tested using the digits *zero* through *nine* and *oh* as words. The TD method was tested using the Digit Verification corpus for true-speaker and impostor data. The Additional Training corpus was only used as a source for additional statistical training of the TDTs.

For all experiments the input frame-level feature vectors and HMM observation distributions had a dimension of 16. Frame level features were derived from initial frame-level energy, filter-bank, and filter-bank difference data [22]. The TDTs reduced the word-level concatenated vectors for each word to a 20 element feature vector.

Enrollment was accomplished using the first call for each of the test speakers. Hence, there were three tokens of each test speaker's ten-digit phrase used for enrollment. Enrollment consisted of aligning the test speaker's input speech vectors with the speaker-independent HMM models, and calculating the aver-

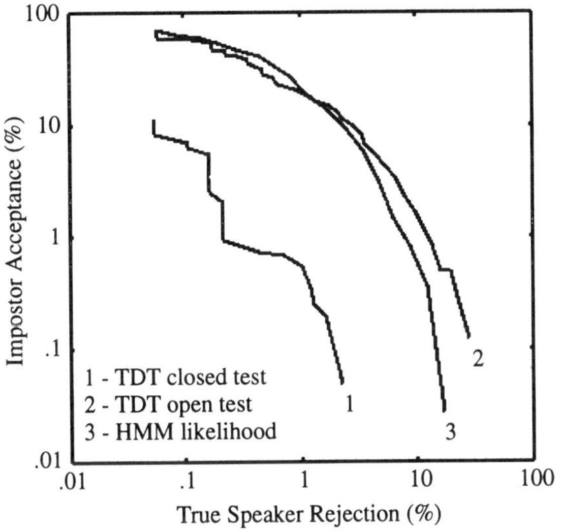

Figure 10 Digit Verification ROC

age word-level feature vector for each word. The averaged word-level feature vectors were stored as the reference feature vectors for the speaker.

After enrollment, true speaker verification scoring utilized the remaining 24 calls for each test subject. For each verification token, word-level feature vectors were calculated for each word, and a word score was formed as the Euclidean distance between the input word-level feature vector and the stored reference word-level feature vector. The final verification token score is the sum of each word score weighted by the relative duration of the input word.

To simulate dynamic reference vector updating, the reference word-level feature vectors were updated using the last token of each true speaker call after verification scoring was done for the call. Updating consisted simply of averaging the input word-level feature vectors with the reference word-level feature vectors.

Impostor verification scores were calculated after all true speaker verification scoring and reference updating. Scoring for the impostors was performed in a manner similar to true speaker scoring. Impostors were always the same sex as the true speaker and spoke the same digit string. Word-level reference vectors were not updated during impostor testing.

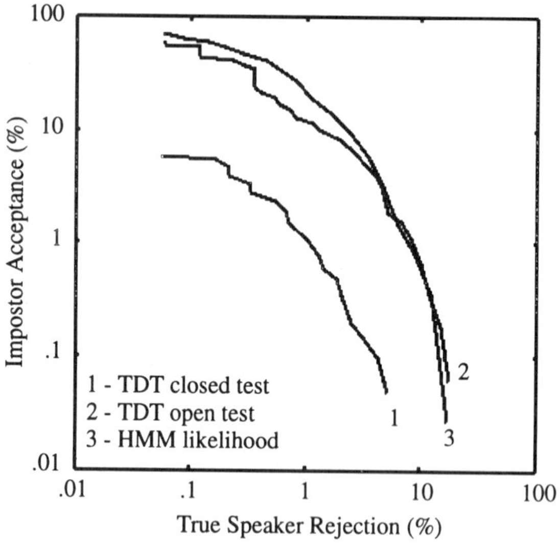

Figure 11 Augmented Covariance ROC

This first experiment used the Digit Verification corpus only to form the TDTs. The TDTs were formed using data from the first 13 male and 13 female subjects. A single TDT was calculated for each digit using data from both males and females. After forming the TDTs, two experiments were performed. A closed set experiment used the first 13 male and 13 female subjects to calculate true speaker and impostor scores. An open set experiment used the last 12 male and 12 female subjects. The resulting ROC for each experiment is shown in Figure 10. Also shown for reference is a closed set ROC based on calculation of likelihood using each typical HMM likelihoods.

A second set of experiments used the same test subjects as the first experiment, but utilized the Additional Training corpus digit data to augment the word-level concatenated vector *within-speaker* covariance and *total* covariance estimate in calculation of the TDTs. The resulting ROC for each experiment is shown in Figure 11. Also shown again for reference is the same ROC based on individual HMM models.

From the initial ROC results it can be seen that the word-level feature verification algorithm for the closed set test provided improved performance over the closed set HMM likelihood approach. In addition, the speaker-specific information needed per word has been reduced significantly. For an HMM observation distribution metric algorithm, one must store a 16-element feature vector corresponding to each distribution in the HMM (typically this results in 128 values for the digit HMM structure used in these experiments). This should be compared with only 20 word-level features. The word-level feature performance for the open set test in Figure 10, however, did not provide improved performance over the HMM observation model approach.

Of importance in the results is the discrepancy between the closed-set and open-set results. The superior performance of the closed-set results in Figure 10 implies that the TDTs are matched to the closed-set subjects used to generate the TDTs. This has implications for speaker identification algorithms, and methods of enrollment. When data from the Additional Training corpus were used to augment word-level covariance estimates in Figure 11, as expected, closed-set performance degraded while open-set performance improved. This indicates the challenge in estimating the TDTs, which is that estimation of the large dimension of word-level covariance matrices requires large amounts of data. Presently, additional collection of digit data over long-distance telephone lines is under way that will improve covariance estimates. One can at least be assured that the best open-set performance lies somewhere between the closed-set and open-set results in Figure 11.

6 CONCLUSIONS

In this chapter we have presented methods for free-text and fixed-text speaker identification and verification over telephone channels. We have discussed challenges that are faced due to the variability of the speech signal in a telephone environment. VQ free-text speaker identification has been shown to be more robust when it incorporates bandpass liftering and RASTA filtering. The concept of cohort comparison used in speaker identification was adapted to form a ranking algorithm which was shown to significantly improve performance in a closed-set VQ-based speaker verification test.

We introduced Temporal Decorrelation, a method of using HMMs in fixed-text speaker verification to greatly reduce the amount of speaker-dependent information that must be retained. This method also shows promise for improved verification performance over the simple HMM likelihood approach, and avoids the necessity of making an explicit speech or silence decision required by VQ methods. Comparison of closed-set testing of TD with free-text VQ cohort ranking indicates improved performance due to the advantage of fixed-text alignment, even though TD testing did not use cohort ranking.

The drawback of the TD method is the large amount of training data necessary to generate decorrelating transformations for each word used in verification. Work is under way to gather additional training data over telephone channels to improve the transformations.

Even with the above improvements, the telephone channel still remains a challenge. Methods of compensating for channel effects, and development of robust features and processing methods will continue to be important areas of research.

REFERENCES

[1] S. Furui, "Research on Individuality Features in Speech Waves and Automatic Speaker Recognition Techniques," Speech Communication, Vol. 5, 1986, pp. 183 - 197.

[2] G. Doddington, "Speaker Recognition - Identifying People by Their Voices," Proceedings of the IEEE, Vol. 73, No 11, Nov. 1985.

[3] D. O'Shaughnessy "Speaker Recognition," IEEE ASSP Magazine, Oct. 1986.

[4] A. Rosenberg "Automatic Speaker Verification: A Review," Proceedings of the IEEE, Vol. 64, No 4, Apr. 1976.

[5] B. Atal, "Automatic Recognition of Speakers from Their Voices," Proceeding of IEEE, Vol. 64, No 4, Apr. 1976.

[6] M. Sambur, "Selection of Acoustic Features for Speaker Identification," IEEE Transactions on ASSP-23, No. 2, Apr. 1975, pp. 176 - 182.

[7] Y.H. Kao, J. Baras, and P.K. Rajasekaran, "Robustness Study of Free-Text Speaker Identification and Verification," ICASSP 1993, pp. 379 - 382.

[8] H. Gish, "Robust Discrimination in Automatic Speaker Identification," ICASSP 1990, pp. 289 - 292.

[9] A. Acero, "Acoustical and Environmental Robustness in Automatic Speech Recognition," Ph.D. Dissertation, Carnegie Mellon University, 1990.

[10] R.C. Rose, J. Fitzmaurice, E.M. Hofstetter, and D.A. Reynolds, "Robust Speaker Identification in Noisy Environments Using Noise Adaptive Speaker Models," ICASSP 1991, pp. 401- 404.

[11] J.M. Naik, L.P. Netsch, and G.R. Doddington, "Speaker Verification Over Long Distance Telephone Lines," ICASSP 1989, pp. 524 - 527.

[12] T. Matsui and S. Furui, "A Text-Independent Speaker Recognition Method Robust Against Utterance Variations," ICASSP 1991, pp. 377 - 380.

[13] F. Soong, A. Rosenberg, B.H. Juang, and L. R. Rabiner "A Vector Quantization Approach to Speaker Recognition," AT&T Technical Report, 1976.

[14] T. Matsui and S. Furui, "Comparison of Text-Independent Speaker Recognition Methods Using VQ-Distortion and Discrete / Continuous HMM," ICASSP 1992, Vol. 2, pp. 157 - 160.

[15] L. Netsch and G. Doddington, "Speaker Verification Using Temporal Decorrelation Post Processing," ICASSP 1992, Vol. 2, pp. 181 - 184.

[16] A.L. Higgins and L.G. Bahler, "Text-Independent Speaker Verification by Discriminant Counting," ICASSP 1991, pp. 405 - 408.

[17] "Speaker Recognition Feasibility Study", Final Report, Contract No. 85F101500, January, 1987.

[18] J.J. Godfrey, E.C. Holliman, and J. McDaniel, "SWITCHBOARD: Telephone Speech Corpus for Research and Development," ICASSP 92, Vol. 1, pp. 517 - 520.

[19] B.H. Juang, L.R. Rabiner, and J.G. Wilpon, "On the Use of Bandpass Liftering in Speech Recognition," IEEE Transactions on ASSP-35, No. 7, July, 1987.

[20] H. Hermansky, N. Morgan, A. Bayya, and P. Kohn, "RASTA-PLP Speech Analysis Technique," ICASSP 1992, Vol. 1, pp. 121 - 124.

[21] K. Fukunaga, "Introduction to Statistical Pattern Recognition," Academic, 1972.

[22] G. Doddington, "Phonetically Sensitive Discriminants for Improved Speech Recognition," ICASSP 1989, Vol. 1, pp. 556 - 559.

PART 4

TEXT TO SPEECH SYNTHESIS

14

APPROACHES TO IMPROVE AUTOMATIC SPEECH SYNTHESIS

Douglas O'Shaughnessy

INRS-Telecommunications,
University of Quebec

ABSTRACT

For several years now, there have been automatic text-to-speech systems for several languages which yield intelligible but unnatural synthetic speech. Quality inferior to that of human speech is usually due to inadequate modeling of human speech production in coarticulation, intonation, and vocal-tract excitation. We will examine the current approaches in these areas, discuss the compromises that are often made, and suggest ways for improvement.

1 INTRODUCTION

Text-to-speech synthesis is the automatic generation of a speech signal, starting from a normal textual input [26],[28],[1],[11],[18], [32],[33]. In the last few years, such synthesizers have become widely available for several languages, in forms ranging from inexpensive packages for use with personal computers to stand-alone systems. The former produce mostly-intelligible (but highly unnatural) speech, while the latter yield very intelligible (but still somewhat unnatural) speech. We are still far from having transparent synthetic speech, where listeners cannot distinguish it from human speech. Of course, very natural speech is possible from computers if the desired vocabulary is very limited, and the system merely concatenates stored speech units, having been previously spoken and directly stored in memory as coded speech. We limit the discussion here to true text-to-speech systems which accept any input text in the chosen language (including new words and typographical errors); the alternative 'voice response systems' of very limited vocabulary are essentially voice coders of much simpler complexity, but also inflexible and quite limited in applications. The explosion of commercial synthesizers in the last few years is due to both advances in computer technology and improvements in the methodology of speech synthesis. This chapter discusses technical aspects of synthetic speech devices, as well as their applications and limitations, and we note where improvements are needed.

2 BACKGROUND

Speech synthesis is the automatic generation of speech waveforms based on an input of text to synthesize and on previously-analyzed digital speech data. The critical issues for current speech synthesizers concern trade-offs among the conflicting demands of maximizing speech quality, while minimizing memory space, algorithmic complexity, and computation speed. While simple text-to-speech (TTS) is possible in real time with low-cost hardware, there is a trend toward using more complex programs (tens of thousands of lines of code; megabytes of storage). Special digital signal processing chips are necessary for use with slower computers (under 10 million instructions/second) or for applications that require several output channels. TTS systems constructively synthesize speech from text using linguistic processing and concatenating small speech units (e.g., phonemes). Real-time systems which generate synthetic speech from normal text have existed for several languages for some years now. Such speech is generally intelligible, but lacks naturalness. Quality inferior to that of human speech is usually due to inadequate modeling of three aspects of human speech production: coarticulation, intonation, and vocal-tract excitation. We will examine current approaches to synthesis in these areas, discuss the compromises that are often made to simplify speech modeling, and note ways in which synthetic speech quality is being improved.

Speech is often modeled as the response of a time-varying linear filter (corresponding to the vocal tract from the glottis to the lips) to an excitation waveform consisting of broadband noise, a periodic waveform of pulses, or a combination. A periodic pulse excitation gives a mechanical tone to the synthetic speech. Excitation which is less periodic, e.g., multipulse patterns in linear predictive coding (LPC) synthesis, or which uses pulses of variable shape may yield more natural quality, but is difficult to control in different phonetic environments. The pulses represent puffs of air from the lungs between closures of the vocal cords, which occur regularly with "pitch periods" ranging from 3–13 ms. Most sounds (e.g., vowels) employ vibrating vocal cords and are called "voiced." "Unvoiced" consonants keep the vocal cords open, permitting air to pass freely into the vocal tract until it meets a narrow constriction where a broadband noise source is generated. The phonemes of a language (abstract linguistic units – about 30–40 in number) are realized acoustically by "phones" (there is an infinite number of phones – each spectrally different from the others). Different phones are characterized by the type of excitation used and by the dynamic configuration of the articulators in the vocal tract (i.e., the positions of the tongue, lips, and velum). The speech signal also conveys information about the identity and emotions of the speaker, as well as syntactic and semantic information, to help a listener decode the sequence of phones into a meaningful message. These latter factors are harder to quantify, and involve acoustic features that are often poorly reproduced in synthetic speech.

During speech the articulators are in constant motion, moving from positions appropriate for one phoneme to those for the ensuing phoneme. Due to physical constraints of mass and speed, the motions of the articulators during a phone are affected by the positions taken in the immediately preceding sounds. Each phoneme is characterized by articulatory target positions, which are often not reached due to undershoot in rapid or even moderate-rate speech. This "coarticulation" is a complex process which smooths the transitions between sounds and assists the hearer to overcome the effects of interfering sounds. Anticipatory modification of articulator positions also occurs in which articulators not needed in the production of a current phone move toward positions corresponding to an ensuing phone. For example, in the word 'stew' (phonemically /stu/), the lips start rounding in anticipation of /u/ during /st/, since /st/ only requires positioning of the tongue tip; the initial portion of /u/ is affected by the preceding /t/ (the pronunciation of /u/ is quite different in the words 'boo' and 'coo'). Coarticulation has been modeled by storing spectral transitions in "diphone" units, or in the form of phonetic rules. The former requires more memory and is less flexible in varying speaking rate, but tends to yield more natural spectral patterns than rule-based synthesis. An alternative to spectral-based synthesis is the concatenation of small waveform units with appropriate adjustment and smoothing at the unit boundaries.

The speech production process is often described in terms of time-varying parameters. A parametric synthesizer describes in a few numbers (parameters) the control functions for a device that produces the actual waveform, as opposed to a waveform synthesizer, which works on the temporal speech signal directly. Parametric synthesizers are popular because they are storage efficient and usually deal with perceptually relevant aspects of the speech signal. The ear is relatively insensitive to phase distortion, performing a rough amplitude spectrum analysis, with logarithmic resolution both in amplitude (decibels) and in frequency (the mel or Bark scale). Most parametric synthesizers concentrate on modeling well the high-amplitude parts of the spectrum (e.g., the resonance "formants"), to which the ear is most sensitive. These synthesizers separate excitation and vocal-tract information in speech modeling, permitting easy manipulation of the underlying parameters of speech production (e.g., F0). One pays a price for such flexibility and reduced memory costs, however. Parametric synthesis normally produces lower quality speech than most waveform synthesizers because of simplistic modeling of natural speech production. Small speech units (e.g., phones) are typically used since speech quality does not increase sufficiently with larger units to justify the increased memory.

Besides coarticulation and excitation, the third problem area for synthesis is intonation. Natural intonation depends in complex ways on linguistic aspects of the text to be spoken; most synthesizers use simplistic natural language analysis, which does not furnish enough information to specify intonation adequately. In a person's normal speaking manner, each phone has an average duration and intensity (e.g., vowels are typically longer and louder than consonants), which are strongly affected by its phonetic and linguistic context. Variations in the duration and intensity of a phone and, if voiced, in its fundamental frequency (F0 = inverse of the pitch period) lead to the perception of length, loudness, and pitch, respectively, which comprise "intonation." Intonation carries information about the syntactic structure of the utterance, about which words the speaker emphasizes, about the emotional state of the speaker, and about the phonemic identity of the individual sounds.

Most synthesizers reproduce speech of bandwidth in the range of 0–3 kHz (e.g., for telephone applications) or 0–5 kHz (for higher quality). Frequencies up to 3 kHz are sufficient for vowel perception since vowels are adequately specified by the lowest three vocal tract resonant frequencies (formants F1–F3). The perception of some consonants, however, is slightly impaired if energy in the 3–5-kHz range is omitted. Frequencies above 5 kHz are useful to improve speech clarity and naturalness but do little to aid speech intelligibility. If we assume that the synthesizer reproduces speech up to 5 kHz, a rate of 10 000 samples/s is needed. Since linear pulse code modulation (PCM) requires about 12 bits/sample for toll-quality speech, storage rates for simple voice-response systems exceeding 100 kbits/s result, which are prohibitive except for applications with very small vocabularies. Thus we examine parametric synthesis in the following.

3 SYNTHESIS METHODS

Synthesizers can be classified by how they parametrize the speech for storage and synthesis. High-quality speech (voice response) requires synthesis using waveform coders and large memories. More efficient but lower-quality TTS systems use vocoders; such systems are considered *terminal analog synthesizers* because they model the speech output of the vocal tract without explicitly taking account of articulator movements. A third way to generate speech is *articulatory synthesis*, which represents vocal tract shapes directly, often using data derived from X-ray analysis of speech production [8]. Due to the difficulty of obtaining accurate three-dimensional vocal tract representations and of modeling the system with a limited set of parameters, this last method has yielded lower-quality speech and has yet to see commercial application.

A synthesizer algorithm consists of (a) storage of speech parameters, obtained from natural speech and organized in terms of speech units and (b) a program of rules to concatenate these units, smoothing the parameters to create time trajectories. English has 22 consonants and about 15 vowels/diphthongs; storing phone units would thus require at least 37 parameter vectors. These vectors would typically include a spectral representation of the vocal tract (e.g., LPC coefficients, or formants and their bandwidths) and excitation information (voiced-unvoiced, amplitude(s) for voicing, aspiration, etc.). Assuming each vector has 10–20 elements, several hundred stored parameters would suffice to generate all speech. However, the stored vectors represent only target parameter values, derived from pronunciation in isolation or in a neutral context. Phoneme synthesis trades off minimal storage for a very complex program to interpolate the target vectors to simulate natural coarticulation.

Theoretically, English needs $38 \times 38 = 1444$ diphones (all pairwise combinations of 37 phonemes and silence), but some combinations are not used in current English; so only about 1200 are necessary. Each diphone needs at least two vectors, representing two terminal phonemic states at the start and end of the diphone, but many require 3–5 vectors to accurately model the parameter trajectories. Whenever a diphone contains a spectral discontinuity (e.g., at voiced-unvoiced boundaries), simple interpolation of two terminal vectors cannot adequately describe the parameter behavior during the diphone. At least 3000 diphone vectors are needed to generate high-quality synthetic speech [22].

The choice of synthesis method is influenced by the sizes of the vocabulary and of the stored unit; e.g., a word-concatenating synthesizer for general TTS can require more than 700 Mbits of memory, even when using an efficient waveform coder (e.g., adaptive differential PCM – ADPCM). Because they must model all possible utterances, unlimited text systems are more complex and yield lower-quality speech than voice response systems. Since there about 20,000 English syllables, each about 250 ms in duration, storing all syllables would be tedious and would require about 50 Mbits using a complex waveform technique (e.g., sub-band coding) or about 10 Mbits using parametric methods (e.g., LPC). Due to these large memory requirements, the waveform method has been used only with small vocabularies (a few minutes of speech). Advances in memory technology are likely to increase the vocabulary of low-cost waveform synthesizers; however, for greater flexibility, methods using small units such as phonemes are necessary. As the unit size shrinks, however, special techniques are needed to alter intonation or smooth speech units at boundaries.

While advances in memory technology will increase the vocabulary of low-cost waveform synthesizers, parametric methods are necessary for TTS due to its large vocabulary. Parametric synthesis normally produces only lower 'synthetic-quality' speech because of inadequate modeling of speech production. TTS uses small speech units (e.g., phones or diphones) since speech quality does not increase sufficiently with larger units to justify the much larger memory needed. Coding each of about 1200 diphones in English with 2–3 frames of parameters requires storage of about 200 kbits [12]. More popular are phoneme synthesizers, which normally store 1–2 sets of amplitude and spectral parameters for each of the 30–40 phonemes in as few as 2 kbits.

4 SIZE OF THE STORED SPEECH UNIT

The simplest synthesis method concatenates stored words or phrases. This approach can yield high-quality speech but is limited by the need to store in memory all the phrases to be synthesized after they have been spoken either in isolation or in context sentences. For maximum naturalness in the synthetic speech, each word or phrase must be pronounced originally with timing and intonation appropriate for all sentences in which it could be used. Thus if a word could occur in several syntactic structures, its pronunciation should be recorded using sentences simulating the various contexts. Also, merely concatenating words originally spoken in isolation usually leads to degraded intelligibility and naturalness. Among other factors (e.g., spectral amplitude), the duration of stored units must be adjusted in concatenation since their length varies in sentential context (especially for units smaller than words).

For synthesis of unrestricted text, advanced systems generate speech from sequences of basic sounds, which substantially reduces memory requirements since most languages have only 30–40 phonemes. However, the spectral features of these short concatenated sounds (50–200 ms) must be smoothed at their boundaries to avoid jumpy, discontinuous speech. The problem is that the pronunciation of a phoneme in a word or phrase is heavily dependent on its phonetic context (e.g., on neighboring phonemes, intonation, and speaking rate). The smoothing and adjustment process as well as the need to calculate an appropriate intonation for each context results in complex synthesizers with less natural output speech. While most synthesizers have been based on word or phone concatenation, other possibilities link intermediate-sized units of stored speech such as diphones [2]. Diphones are obtained by dividing a speech waveform into phone-sized units, with the cuts in the middle of each phone (to preserve the transition between adjacent phones). When diphones are concatenated in a proper sequence (so that phone spectra on either side of a boundary match), smooth speech usually results because the merged sounds at the boundaries are spectrally similar; e.g., to synthesize the word *straight*, the diphone sequence /#s-st-tr-re-et-t#/ would be used (# denotes silence).

Proper interpolation of spectral parameters at the boundaries between units is very important for the short units (phones or diphones) that are usually employed, and decreases in importance as concatenation units increase in size, because of the decrease in number of boundaries per second in the synthesized speech. Smoothing is much simpler when the joined units approximately match spectra at the boundaries. Since diphone boundaries interface spectra from similar sections of two realizations of the same phoneme, their smoothing rules are fairly simple. Systems that link phones, however, must use complex smoothing rules to represent coarticulation (the overlapping and nonlinear movements of tongue and lips) in the vocal tract. Not enough is understood about coarticulation to establish a complete set of rules to describe how the spectral parameters for each phone are modified by its neighbors. Diphone synthesizers try to circumvent this problem by storing the parameter transitions from one phone to the next since coarticulation primarily influences only the immediately adjacent phones. Unfortunately, there are many cases where coarticulation extends over several phones. Improved quality is possible by using multiple diphones depending on context, effectively storing "triphones" or "quadphones" (portions of speech waveforms comprising up to parts of four successive phones), which may substantially increase memory requirements. Many coarticulation effects can be approximated by simple rules (e.g., lowering all resonant frequencies during lip rounding), but others (e.g., the undershoot of phoneme target positions, which occurs in virtually all speech) are much harder to model accurately.

Diphone synthesis requires more memory and is less flexible in varying speaking rate or speaker style than rule-based synthesis, but tends to yield more natural spectral patterns. While most synthesizers use concatenative speech units of uniform size, a recent trend has been to use units of longer length where appropriate; a recent Japanese synthesis system has a non-uniform unit selection scheme, to automatically choose speech units using objective measures [27]. As measures, they employed a contextual spectral difference, a prototypicality of each segment for its context, the spectral gap between adjacent phonemes, and local spectral discontinuity.

5 WAVEFORM CONCATENATION

As an alternative to spectral-based synthesis, waveform synthesis can yield very good quality synthetic speech, at the cost of increased memory and speaker inflexibility. Most synthesizers have adopted the spectral approach, where the units are described in terms of Fourier amplitude. The reasons for this lie in the more efficient coding of speech spectrally, and the flexibility to modify the stored units under changes in prosody or speaker. While concatenating speech units using stored waveforms has been difficult when trying to smooth transitions or otherwise modify the units depending on context, it can yield good results with appropriate adjustment and smoothing at the unit boundaries. Recently, quite good synthesis quality has been obtained with such a concatentation method called PSOLA (pitch-synchronous overlap-and-add): it overlaps and sums small waveform units, using typically diphones as the units. Higher quality has recently been obtained through the use of longer units such as triphones and quadphones. These units are often obtained automatically from training speech samples from desired talkers, although a careful initial segmentation of training speech is often required by a trained phonetician. Like other waveform or diphone synthesis methods, PSOLA cannot be modified to simulate voices other than that of the training speaker, and therefore is not a speaker-flexible synthesis method. An an example of PSOLA, a recent German system employed 2310 diphones and 440 additional special units containing glottal stops [2]. A current French system has 1290 diphones, plus 751 triphones and 288 quadphones (the latter involving sonorant consonants). An automatic segmentation using hidden Markov model recognition techniques provided unit boundaries which differed from the phonetician's markings by less than 30 ms in 90% of the cases.

Since one voice suffices for many TTS applications, PSOLA provides a useful alternative to spectral-based TTS. The original time-domain PSOLA suffered from: 1) the need to manually mark pitch periods (for optimal performance) in the original speech analysis, 2) problems in spectral and prosodic smoothing at unit boundaries, and 3) requiring a large memory for the stored units (e.g., 80 kilobits/s) [5]. Recent modifications have lessened some drawbacks, while retaining low computation at synthesis time (7 operations/sample).

6 SPECIFYING THE PARAMETER TRAJECTORIES

Many synthesizers can generate very good speech if appropriate parameters are supplied, based on an automatic analysis of the input text. For example, signals indistinguishable from natural human speech can be produced with parallel formant synthesizers that allow time variation of F1-F3 and their bandwidths and that permit mixed excitation for unvoiced fricatives. Synthesizers with both parallel and cascade paths can do the same if a nasal antiresonator is included with the formant resonators. With the possible exception of nasals and voiced fricatives, conventional LPC synthesizers can likely also generate such natural speech. The major difficulty in all cases, however, is that such high-quality speech has not been possible automatically by rule, but only semi-automatically if algorithmically selected parameters are modified frame by frame to match human speech patterns (manual "synthesis by art"). Actual TTS synthesis models are too simple in general (compared to the speech production processes they simulate), and smoothed parameter tracks obtained from speech analysis (or by rule) must usually be adjusted to closely match natural speech. The frame-by-frame parameter values that lead to the best speech often exhibit apparently random discontinuities that are difficult to predict by rule.

The challenge for synthesis researchers is to accurately model the parameter variations found in natural speech so that they can be generated by rule from an input text. The basic approach in phonemic formant synthesis has been to store target parameters for each phoneme and to interpolate parameter trajectories between target sets, simulating coarticulation effects. To avoid context effects, the targets for continuant phonemes (e.g., vowels and fricatives) are often based on data from isolated utterances of individual phonemes [17]. Transient phonemes (e.g., plosives) require adjacent phonemes, and their targets can be derived from utterances with neutral adjacent phonemes (e.g., schwa vowels) or from averaged data using different contexts. *Allophonic synthesis* involves an extension of this basic approach in which targets are stored for each allophone, directly incorporating contextual effects into the stored parameters.

In transitions between acoustic segments (e.g., phones or parts of phones), each segment is specified by context-independent targets for spectral parameters that are modified by the context effects of coarticulation. Vowel-to-vowel transitions are good examples where spectral parameters smoothly move from values dominated by the first vowel to values dominated by the second. Whereas the primary cues for vowels appear in the spectra toward the middle of their durations, plosives rely on parameter transitions near closure onset and release (i.e., at the ends of neighboring phones) to cue place of articulation and voicing. Linear interpolation of targets between segments is generally inappropriate to model sudden acoustic changes such as voiced-unvoiced switches, amplitude changes between obstruents and sonorants, or spectral jumps at nasal or plosive onsets and releases. If, however, parameter values are specified for segment boundaries as a function of the manners of articulation of the phonetic segments involved, piecewise-linear parameter contours (one from the middle of the first segment to the boundary and another from the boundary to the middle of the next segment) may suffice. Phones can be ranked in increasing order of abruptness, e.g., vowels, semivowels, liquids, fricatives, nasals, and plosives, in terms of how important transitions are to their identification. Such a ranking can then specify parameter values at segment boundaries involving different phones, where phones with higher abruptness would dominate the transition.

If the synthesizer concatenates phonetic units whose spectra are very different at the joining boundary, substantial smoothing is necessary to avoid discontinuous speech. Acceptable speech is possible by concatenating large stored units (e.g., words) with a limited amount of smoothing because most of the phone boundaries are well modeled within each word. One approach to word concatenation linearly interpolates stored parameter tracks within a time window of 40–100 ms straddling the boundary between the words. It is inadequate to simply discard the stored parameter trajectories within this smoothing interval and connect the interval endpoints because substantial parameter variation often occurs at the edges of phonetic units.

7 SYNTHESIZER OPERATION

Most TTS synthesizers use a digital model with second-order filters, each corresponding to a formant resonance of the vocal tract, which employs a combination of cascade and parallel structures [17]. In the simplest version, the excitation is either a periodic train of impulses (for most voiced speech), pseudo-random noise (for unvoiced speech), or periodically shaped noise (for voiced fricatives). The vocal tract is usually modeled as a cascade of second-order digital resonators, each representing either a formant or the spectral shape of the excitation source. The cascade filter structure approximates the speech spectrum well for vowels and can be controlled with one amplitude parameter. Advanced synthesizers allow the first four formant frequencies and first three bandwidths to vary as a function of time (variation in higher formants has little perceptual effect); simpler systems compromise on speech quality, varying only F1-F3, with all bandwidths fixed. Lack of time variation in the bandwidths yields a degradation that is most noticeable in nasal sounds.

7.1 Linear Predictive Coding Synthesis

Commonly used for synthesis, linear predictive coding (LPC) has a simpler structure than synthesis based on the formant resonances of the speech spectrum, because all the spectral properties of speech (except periodicity) are included in a set of LPC coefficients, which are calculated automatically in the analysis of natural speech. (Programs to track formants often make mistakes, which can lead to serious speech degradation in synthesis.) The number p of coefficients is typically 10–12, i.e., two for each formant resonance in the 4–5 kHz bandwidth of the speech signal, plus a few for modeling spectral zeros and the source of vocal tract excitation. The LPC synthesis filter is usually excited by either a periodic (typically impulsive) source or a noise source, depending on whether the analyzed speech is estimated to be voiced or not. The excitation is specified by a gain factor, a voiced-unvoiced (periodicity) bit, and (if voiced) a fundamental frequency value. This commonly-used excitation is a gross simplification of the actual LPC 'residual' signal, which can be obtained by passing the actual speech through a filter which is the spectral inverse of the LPC synthesis filter. ADPCM, of course, employs a coded version of the residual in its excitation of the synthesis filter, and obtains excellent quality speech. For TTS, however, required modifications of intonation render simple direct use of the residual as excitation unfeasible.

In parametric synthesis (e.g., LPC), speech units are divided into short *frames* (10–30 ms) of samples; the speech signal is analyzed through a 'time window' to obtain a spectral representation of the unit. Parametrization of successive frames adequately models the speech if the frames are short compared to vocal tract motion. For each frame, LPC analysis produces a set of p real-valued predictor coefficients a_k, which represent an optimal estimate to the spectrum of a frame of speech using p poles (and no spectral zeros). Such information can be represented in different formats, some more useful for quantization or more physically interpretable than others. *Reflection coefficients* k_m provide a popular way to represent LPC information, which can also be related to acoustic tube models of the vocal tract. By verifying that all k_m for each frame have magnitudes less than one, a stable LPC synthesis filter is guaranteed. Filters with feedback (e.g., as in LPC or formant synthesizers) usually do not have stability problems except when parameters change too abruptly.

Spectral and excitation parameters are fetched from the stored speech units, typically periodically (every 10-30 ms frame). Often the parameters from successive frames are linearly interpolated during a frame, to allow more frequent updates to the synthesizer. In natural speech, the parameters do not change linearly over the course of a frame, but given storage limitations which often specify a minimum frame duration, linear interpolation usually improves synthesis quality, as compared to abrupt transitions between frames. Since the synthesis filter is time-varying, updating the parameters just before a pitch impulse excitation minimizes difficulties; at this moment the speech signal and the filter registers have lowest amplitude, and little filter instability will occur. Otherwise clicks may appear in the output speech when the spectrum changes rapidly. The filter memory cannot simply be set to zero when the coefficients are changed, without leaving unnatural gaps in the output.

7.2 Formant synthesizers

The other popular type of parametric synthesizer employs formant resonances and bandwidths to represent the stored spectrum. The vocal tract filter is usually represented by 10-12 spectral parameters (the same number as for LPC, but quite different in interpretation). Formant synthesizers have an advantage over LPC systems in that bandwidths can be more easily manipulated and that zeros can be directly introduced into the filter (using a second-order inverse resonator). Locating the poles and zeros automatically in natural speech is, however, a more difficult task than automatic LPC analysis.

Unvoiced sounds have an excitation source within the mouth cavity, primarily exciting the vocal tract between the source and the lips. With this shortened acoustic tube, most energy is at higher frequencies. A parallel bank of resonators is often used for these sounds, so that each filter can be individually controlled for amplitude [12]. Indeed, there has been considerable controversy over whether a cascade of resonators (often used for vowels) is even necessary, given that a parallel bank of filters can synthesize both vowels and consonants. Adding the outputs of parallel branches can introduce unwanted spectral zeros and phase effects into the synthetic speech, but recent research has alleviated some of these problems [7].

Nasals usually require an extra resonator, because the acoustic path including the nasal cavity is longer than for vowels, which increases the number of resonances in the speech bandwidth (because resonance frequency varies inversely with acoustic tube length). Spectral zeros are found in nasals at those frequencies where the mouth cavity presents zero impedance to the pressure waves coming out of the pharynx. Since the mouth cavity is closed, it acts as a side-branch to the main air flow path out the nostrils. A second-order antiresonator is often used to model the lowest frequency zero. The extra resonator and antiresonator can be considered as a pole/zero pair, with cancelling effects except when nasals are to be simulated.

7.3 LPC vs. formant synthesis

LPC synthesis has a simpler structure than formant synthesis because all spectral properties of speech (except periodicity) are included in the LPC coefficients, calculated automatically in the analysis of natural speech. A lattice filter is often used for synthesis because the filter coefficients can be linearly interpolated between frames without yielding an unstable filter. The excitation is specified by a gain parameter, a voiced-unvoiced bit, and (if voiced) an F0 value. For simplicity, most LPC systems do not allow mixed excitation for voiced fricatives.

The choice between formant and LPC synthesis is governed by tradeoffs. LPC has a simpler and fully automatic analysis procedure and a simpler synthesis structure than the formant approach. On the other hand, our understanding of formant behavior facilitates parameter interpolation at segment boundaries in formant synthesis, whereas interpolation of LPC parameters is usually very limited because each coefficient affects a wide range of speech frequencies in a complex fashion. In addition, formant synthesis is more flexible in allowing simple transformations to simulate several voices. Shifting formant frequencies in formant synthesis can easily alter speaker-related aspects of voice, whereas the LPC reflection coefficients would have to be converted into pole locations to allow such modification. Because spectral valleys are modeled more poorly than peaks in LPC analysis, formant bandwidths are often improperly estimated, whereas direct (albeit difficult) bandwidth determination from spectra in the formant approach can yield better estimates. The all-pole assumption of standard LPC leads to less accurate spectral estimation for nasal sounds than for other phones, whereas antiresonators in formant synthesis permit direct modeling of the most important spectral zeros in nasals. Mostly, the LPC vs. formant choice trades off the simple, automatic analysis of LPC against the higher speech quality possible with more complex formant synthesis.

7.4 Excitation Modeling

Since voicing and aspiration originate at the glottis, articulatory synthesizers (employing a vocal tract model of short tube sections, each of uniform area) introduce such excitation at the location in the model representing the vocal cords. Natural frication, on the other hand, is generated just forward of a narrow constriction in the vocal tract, and thus a random noise source can be automatically inserted into the model anterior to the cross section of minimum area. For typical frication, intensity is inversely proportional to the square of the constriction diameter and can be directly calculated as a function of vocal tract cross-sectional areas; however, the areas are rarely known with sufficient precision to yield accurate models. Simple periodic waveforms can be used for voiced excitation, or laryngeal articulation can be simulated more precisely with a coupled two-mass model. Instead of specifying amplitudes for voicing, aspiration, and frication separately, such a comprehensive model would specify articulatory parameters such as subglottal pressure, vocal cord tension, and a neutral or rest area for the glottis.

With either LPC or formant synthesis, virtually all commercial systems use a simple model for most voiced excitation, i.e., waveforms repeated periodically at the F0 rate, often just a unit-sample train. This assumes that the vocal tract is excited once per pitch period and that the filtering action of the vocal cords (to create the puffs of air just above the glottis that resemble a rectified sine wave in time) is stationary throughout the analysis frame and can be well modeled by a few poles or zeros. The effects of glottal zeros in all-pole LPC appear in modifications to the positions of some poles, whereas formant synthesis may directly combine poles and zeros in a specific glottal filter.

Since using a unit-sample train as excitation assigns all harmonics the same zero phase, which leads to a more "peaky" waveform than natural speech, some synthesizers employ excitation signals $e(n)$ more complex than impulses (e.g., approaching the actual LPC residual). Using some form of the residual as excitation (e.g., a multipulse version [3] or a lowpass version – RELP [21]) is also increasingly common in TTS. For example, one recent study found that, at least for the intelligibilty of voiced consonants, RELP was better than PSOLA [21]. F0 variation causes some difficulty with these approaches, since the $e(n)$ signal must be truncated, overlapped, or otherwise shortened when the pitch period becomes less than the duration of the stored $e(n)$. To avoid spectral distortion due to truncation, most of the energy in $e(n)$ should be concentrated in the first 2–3 ms (since virtually all pitch periods are longer), or $e(n)$ could be an "embedded" all-pass signal where truncation does not seriously affect its magnitude spectrum. Unfortunately, exciting the synthesis filter with a fixed waveform $e(n)$ improves speech quality only marginally, compared with speech using a zero-phase impulsive excitation. There is sufficient variability in the natural glottal waveform to make any completely periodic excitation a significant simplification, which is a major cause of the lack of naturalness in speech from parametric synthesis. Fundamentally, we do not understand how to precisely model the excitation to fit different vocal tract responses and pitch periods, and the resulting synthetic speech suffers in naturalness. Another difficulty is that the basic speech production model most commonly used in TTS assumes strict separability and independence of the excitation and vocal tract filter effects; another simplifying assumption is that there is only plane wave acoustic propagation in the vocal tract; neither of these assumptions are completely valid, but dispensing with them renders speech modeling much more complex.

7.5 Waveform Excitation for Voicing

An alternative excitation is a simple stored $e(n)$ that matches the general time-domain characteristics of a glottal pulse. Here, $e(n)$ may be represented by a small set of parameters determining the relative timing of the rise and fall of the glottal pulse, which can be adjusted for different speech intensities and pitches. Waveforms with discontinuities, such as sawtooth or rectangular pulses, are inappropriate because they have −6-dB/oct decay rates. The desired −12-dB/oct falloff is accomplished through a discontinuity in slope at the time of vocal cord closure, which is abrupt (the gradual opening of the vocal cords, on the other hand, causes no slope discontinuity).

One problem with using a filtered impulse to simulate a glottal pulse is that the filter output achieves a maximum at a fixed time after the impulse input, whereas natural glottal pulses tend to peak at a fixed percentage of the pitch period, reflecting roughly proportional open and closed glottal phases. In practice, the filter characteristics are chosen so that the synthetic and natural glottal signals match for an average value of F0. If F0 deviates significantly from this average, the speech may change quality, corresponding to the differences between soft speech and shouting. Shouting, besides raising speech amplitude, causes the glottal closed phase to be a relatively longer part of the pitch period. The glottal excitation is often described in terms of an open quotient (OQ), i.e., the ratio of the open glottal interval to the closed glottal interval, as a percentage of the fundamental period ($T_0 = 1/F_0$). By varying OQ and a skew measure (a measure of the symmetry of the glottal pulse, which tends to rise more slowly than it falls – due to the more rapid closure of the vocal cords), researchers have recently produced more natural-sounding synthetic speech [13],[14],[20],[24].

We can also use versions of actual LPC residual waveforms for vocal tract excitation. Exciting the LPC synthesis filter with its corresponding residual signal (as in ADPCM) reconstructs exactly the original speech (except for quantization noise). RELP exploits this fact by storing a lowpass version of the residual and then using nonlinear techniques to approximate the residual for synthesis. Such bit-saving economy is probably unnecessary for many synthesis applications. Instead, the problems in synthesis concern how many residual patterns to store and how to modify them under changes in duration and F0; e.g., excellent speech can result if, for each combination of LPC coefficients and desired F0, the corresponding residual excites the appropriate filter. Given thousands of possible LPC vectors (corresponding to perceptually different speech sounds) and perhaps 100 perceivable pitches, the number of residuals to store and then access in real time could lead to memory exceeding 8 Mbits, even with an efficient multipulse representation. This memory estimate ignores the fact that residuals also vary in a complex fashion with amplitude.

8 SYNTHESIS OF FEMALE SPEECH

Significant research has been pursued in the last few years to obtain synthetic female voices of similar quality to that of synthetic male speech. While there is little evidence that human female speech is less intelligible than male speech, coders and synthesizers have not modeled female speech well. The most obvious differences are related to the shorter vocal tract and smaller vocal cords of women (with their yet smaller vocal apparatus, the speech of children is more problematic). This leads to fewer resonances in a given transmission bandwidth (e.g., over the telephone) and to more widely spaced harmonics, with correspondingly less clear definition of formants when using traditional spectral analysis methods. It seems that the human ear must use spectral analysis techniques different from those of speech researchers, because human listeners understand female speech very well. Also, the glottal excitation that women use appears to be more breathy in general than for men, which leads to more randomness in their speech waveforms.

Addressing these differences recently, researchers have obtained much more natural synthesis of female speech [19],[13]. In particular, they have adjusted the glottal puff excitation to have a shorter closed phase for female speech and have adjusted the positions of vocal tract poles and zeros to better correspond to differences in shorter vocal tracts. Traditionally, researchers have simply modeled female vocal tracts as 17-cm male vocal tracts scaled down by 10–15%, which is a reasonable first-order approximation, but ignores the fact that male-female differences do not occur uniformly along the vocal tract; e.g., it is mostly in the lower pharyngeal area that the major difference occurs in length.

9 SYNTHESIS OF INTONATION

One major difficulty in synthesizing speech is that of determining a natural intonation corresponding to an input text. When the stored synthesis units are large (e.g., phrases), F0 and intensity are usually stored explicitly with the spectral parameters in LPC and formant synthesizers or implicitly in the residual signals of most waveform synthesizers. When concatenating units the size of words or smaller, unnatural speech occurs unless the intonation stored for each unit is adjusted for context (except for word concatenation, when generating simple lists of words). The need for rules to determine intonation directly from the input text is a major difficulty in generating natural speech with small stored speech units. Each language has its own intonation rules, which can be discovered only by comprehensive analysis of much natural speech.

The domain and method of variation are different for F0, duration, and intensity. It is of little use to store F0 for units smaller than a word, since natural F0 patterns are determined primarily at the level of words and phrases: F0 undergoes major changes on the lexically stressed syllables of emphasized words, with F0 rises and falls marking the starts and ends of syntactic phrases, respectively. Even at the word level, substantially different F0 contours can occur depending on the word's position in a sentence. Thus small-vocabulary synthesizers should record each word or phrase in sentence positions appropriate for all possible output contexts. Most F0-by-rule algorithms [23] operate on a sentence-by-sentence basis, assigning a basic falling pattern following the declination effect, with superimposed F0 excursions above (and sometimes below) the declination line. F0 may be specified either in terms of explicit contours or simply via targets, whose values then must be interpolated to yield contours. In one approach, two falling declination lines, a high *topline* and a low *bottomline* (the latter with lesser negative slope), gradually converge toward the end of each sentential utterance [4]. If stressed syllables and major phrasal boundaries can be identified in the input text, F0 can be assigned to follow one of the declination lines except during stressed syllables, where it switches levels, rising at the start of a phrase and falling at its end [9]. This is a reasonable first-order F0 model in many cases, especially for Germanic languages, including English. While the reality of declination lines in conversational speech is still controversial, at least the bottomline is found in most F0 algorithms.

Intensity is usually stored as a parameter, when using units as small as phones, since the primary variations in intensity are phoneme-dependent (there are however tendencies for stressed syllables to be more intense and for sentence-final syllables to be less intense). Stressed syllables, however, tend to be more intense, and sentence-final syllables less intense, than average. Thus intensity should be modified at the syllable and sentence level when employing smaller synthesis units; e.g., in word concatenation, the stored intensity parameters for each word reflect lexical stress, but each word's intensity should be adjusted for sentential stress. In diphone concatenation, intensity should be modified for both lexical and sentential stress.

In speech production, the duration of any acoustic event is a function of many factors. The syntactic structure of each sentence has strong effects on the durations of its pauses and words. Words considered important by the speaker are usually lengthened, primarily in the durations of their lexically-stressed syllables. One's speaking rate obviously affects phone durations, but in a nonlinear way, with vowels more subject to variation than consonants. The number of syllables in a word affects its component durations, with average phone durations decreasing with an increasing number of syllables.]Duration is usually encoded implicitly in stored speech data. In parametric synthesizers, the parameters are evaluated once every frame. Variable-frame-rate systems take advantage of the tendency for many speech sounds to settle into a steady state for several frames; the frame size is adjusted using some measure of spectral continuity between consecutive frames. For diphone and phone concatenation synthesis, however, durations must be determined explicitly (from analyses of the input text) since diphones and phones are stored with arbitrary durations.

Although both accurate duration and F0 modeling are important in synthetic speech quality, more F0 models have been proposed than have comprehensive duration models. Compared to F0, duration is more tightly coupled to phonemics and can be roughly specified by a few simple rules. Vowels are generally longer than consonants, fricatives longer than stops, unvoiced consonants longer than voiced ones, stressed syllables longer than unstressed ones, and consonants shorter in clusters. Context affects duration too: vowels are longer before voiced consonants than before unvoiced ones, syllables lengthen at the end of major phrases, etc. When speaking rate varies, vowels tend to expand or compress more than consonants do, leading to the theory that sounds may have a certain minimum duration beyond which they may not be compressed without loss of speech quality [16]. Given the multitude of factors (syntactic, semantic, phonetic, phonological) that affect duration, there has been controversy over the best way to combine effects in calculating durations for speech synthesis [30]; e.g., the different incremental effects might add to or multiply a base duration, or they could be modeled as fixed durational increments or percentages of a phoneme's duration.

10 EVALUATION OF TTS SYSTEMS

It has always been problematic to formally evaluate synthetic speech. Automatic objective measures are desirable for consistent results and for ease of calculation, as much as signal-to-noise ratio serves as a reliable and simple measure for wideband speech coders. However, there are many complex factors that enter into the quality of speech signals, dealing with intelligibility and naturalness, as well as flexibility to simulate different voices and rates of speaking. In speech coding, we can compare the output and input speech signals, e.g., sample-by-sample in their waveforms for a signal-to-noise ratio; however, in text-to-speech synthesis, we do not have access to ideal speech waveforms for comparison, in the wide range of possibilities for different utterances and speaking styles. Thus evaluating synthetic speech is almost exclusively a subjective process. Even here however it is not clear what texts to use, nor what questions to ask listeners. We wish to have highly intelligible speech and a signal that sounds like clear human speech. To evaluate intelligibility, subjects are usually asked to indicate what words or sentences they hear, when listening to synthetic speech, either without the aid of a list of choices or from a closed list of textual possibilities. In the past, tests have been designed and used to evaluate specific phonemes, often in terms of their phonetic features, to discover how well a synthesizer conveys these features. For example, the Dynamic Rhyme

Test examines consonants (e.g., a choice between "foot" and "put" would test how well the synthesizer handles the fricative-stop distinction).

A flaw in such procedures is that they test very selectively only certain phonetic distinctions and in very limited environments, which are not realistic for practical applications. Recently, new evaluation procedures have been proposed [25],[29]. They seek to examine the performance of text-to-speech systems in terms of both the individual components of the systems, as well as their overall synergy. They differ from earlier methods in including the examination of consonant clusters, longer units than words (e.g., sentences and paragraphs), and intonation. The most recent work [31] looks at paradigms that detect concatenative speech units that cause the worst synthesizer performance, that judge transcription of proper names, and use the mean-opinion-score for more global evaluation.

11 CONCLUSION

Use of speech synthesis is becoming widespread, as the cost and size of computer memory decreases and as more special-purpose signal processing hardware is developed. Limited-vocabulary voice response systems will predominate in the near future since they yield the highest-quality speech. But the large number of applications for text-to-speech systems will lead to quality improvements that will increase their popularity. These improvements will come from further research into better models of intonation and vocal tract excitation as well as a clearer understanding of the speech production dynamics of the vocal tract and the processes of speech perception.

There has been a trend recently toward taking advantage of cheaper memory in terms of using very large inventories of speech units, to overcome the coarticulation and intonation problems. For example, one recent system stores more than 100 000 phone units, from many different phonetic contexts (spoken by one speaker over the course of several weeks, to ease the fatigue) [10]; the uncompressed memory for the spoken speech was 360 megabytes. This trend may well be an indication of the immediate future for TTS; indeed it follows closely that of recent automatic speech recognition (ASR) methods. The most widely accepted ASR method is via hidden Markov models (with neural networks becoming the second most used technique). Both of these stochastic methods involve quite simple network models, but require massive amounts of training and memory to accommodate the large amount of variability in the way speakers talk. While TTS need not model such speaker variability (e.g., modeling one speaker well can suffice for many TTS applications), it must handle well the large amount of variability (even within one speaker's voice) across the many different phonetic contexts met in normal speech. Thus we will likely see more TTS research that uses automatic training, requiring increased computer memory and computation. Eventually, however, increased understanding of how humans produce and perceive speech will yield more efficient TTS, and some combination of stochastic and knowledge-based methods will approach synthetic speech quite similar to that of humans.

Acknowledgment: Work supported by the Canadian Natural Sciences and Engineering Research Council (NSERC)].

REFERENCES

[1] J. Allen (1992) "Overview of text-to-speech systems," in *Advances in speech signal processing* (S. Furui & M. Sondhi, eds. - Marcel Dekker: New York), 741–790.

[2] D. Bigorgne et al, "Multilingual PSOLA Text-to-Speech System," (1993) in *Proceedings of the IEEE Conf. on Acoustics, Speech, and Signal Processing*, II-187–190.

[3] B. Caspers & B. Atal (1983) "Changing pitch and duration in LPC synthesized speech using multi-pulse excitation," J. Acoust. Soc. Am. **73, S1**.

[4] W. Cooper & J. Sorenson (1981) *Fundamental frequency in sentence production* (Springer-Verlag: New York).

[5] T. Dutoit (1994) "High quality text-to-speech synthesis: A comparison of four candidate algorithms," Proc. IEEE Int. Conf. ASSP, vol. I, 565–568.

[6] G. Fries (1994) "Hybrid time- and frequency-domain speech synthesis with extended glottal source generation," Proc. IEEE Int. Conf. ASSP, vol. I, 581–584.

[7] S. Grau, C. d'Alessandro, & G. Richard (1993) "A speech formant synthesizer based on harmonic + random formant-waveforms representations," Proc. Eurospeech-93, 1697-1700.

[8] S. Gupta & J. Schroeter (1993) "Pitch-synchronous frame-by-frame and segment-based articulatory analysis by synthesis," J. Acoust. Soc. Am. **94**, 2517– 2530.

[9] J. 't Hart & A. Cohen (1973) "Intonation by rule: a perceptual quest," J. Phonetics **1**, 309–327.

[10] A. Hauptmann (1993) "SpeakEZ: A first experiment in concatenation synthesis from a large corpus," Proc. Eurospeech-93, 1701–1704.

[11] S. Hertz (1991) "Streams, phones, and transitions: Toward a phonological and phonetic model of formant timing," J. Phonetics **19**, 91–109.

[12] J. Holmes (1983) "Formant synthesizers – cascade or parallel?" Speech Comm. **2**, 251–273.

[13] I. Karlsson (1992) "Consonants for female speech synthesis," Proc. Intern. Conf. on Spoken Language Processing, 491–494.

[14] I. Karlsson & L. Neovius (1993) "Speech synthesis experiments with the GLOVE synthesizer," Proc. Eurospeech-93, 925–928.

[15] J. Kerkhoff & L. Boves (1993) "Designing control rules for a serial pole-zero vocal tract model," Proc. Eurospeech-93, 893–896.

[16] D. Klatt (1976) "Linguistic uses of segmental duration in English: Acoustic and perceptual evidence," J. Acoust. Soc. Am. **59**, 1208–1221.

[17] D. Klatt (1980) "Software for a cascade/parallel formant synthesizer," J. Acoust. Soc. Am. **67**, 971–995.

[18] D. Klatt (1987) "Review of text-to-speech conversion for English," J. Acoust. Soc. Am. **82**, 737–793.

[19] D. Klatt & L. Klatt (1990) "Analysis, synthesis, and perception of voice quality variations among female and male talkers," J. Acoust. Soc. Am. **87**, 820–857.

[20] J. Koreman, L. Boves, and B. Cranen (1992) "The influence of linguistic variations on the voice source characteristics" Proc. Intern. Conf. on Spoken Language Processing, 125–128.

[21] M. Macchi, M. Altom, D. Kahn, S. Singhal, & M. Spiegel (1993) "Intelligibility as a function of speech coding method for template-based speech synthesis," Proc. Eurospeech-93, 893–896.

[22] E. Moulines & F. Charpentier (1990) "Pitch synchronous waveform processing techniques for text-to-speech synthesis using diphones," *Speech Communication*, vol. 9, 453–467.

[23] J. Olive (1975) "Fundamental frequency rules for the synthesis of simple English sentences," J. Acoust. Soc. Am. **57**, 476–482.

[24] S. Palmer and J. House (1992) "Dynamic voice source changes in natural and synthetic speech," Proc. Intern. Conf. on Spoken Language Processing, 129–132.

[25] L. Pols (1992) " Quality assessment of text-to-speech synthesis by rule," in *Advances in speech signal processing* (S. Furui & M. Sondhi, eds. – Marcel Dekker: New York), 387–416.

[26] Y. Sagisaka (1990) "Speech synthesis from text," *IEEE Communications Magazine*, vol. 28, no. 1, 35–41,55.

[27] Y. Sagisaka, N. Kaiki, N. Iwahashi and K. Mimura (1992) "ATR v-Talk speech synthesis system," Proc. Intern. Conf. on Spoken Language Processing, 483–486.

[28] H. Sato (1992) "Speech synthesis for text-to- speech systems," in *Advances in speech signal processing* (S. Furui & M. Sondhi, eds. – Marcel Dekker: New York), 833–853.

[29] R. van Bezooijen and L. Pols (1990) "Evaluating text-to-speech systems: some methodological aspects," *Speech Communication*, vol. 9, 263–270.

[30] J. van Santen (1993) "Timing in text-to-speech systems," Proc. Eurospeech-93, 1397–1404.

[31] J. van Santen (1993) "Perceptual experiments for diagnostic testing of text-to-speech systems," *Computer Speech and Language*, vol. 7, 49-100.

[32] Proc. ESCA Workshop on Speech Synthesis (1990) Autrans, France.

[33] C. Benoit, G. Bailly, and T. Sawallis, eds. (1992) **Talking Machines: Theories, Models and Applications**, Elsevier, North-Holland.

PART 5

APPLICATIONS OF MODELS

15

MICROPHONE ARRAY FOR HANDS-FREE VOICE COMMUNICATION IN A CAR

Stephen Oh and Vishu Viswanathan

Texas Instruments, Inc.
Dallas, Texas
USA

ABSTRACT

We present the results of our research on developing a speech acquisition and enhancement system so that a speech recognizer can reliably be used inside a noisy automobile environment, for the digital cellular telephone application. Our research results have demonstrated that a beamforming method with a microphone array is a reliable approach for hands-free voice dialing application in the car noise environment. Two beamforming algorithms were investigated in our research: delay-and-sum beamformer and generalized sidelobe canceler. Performance evaluation results of these two algorithms are presented for speech database collected in real cars under different noise conditions. Signal-to-noise ratio and speech recognition error rate were used as performance measures in these evaluations.

1 INTRODUCTION

Beamforming algorithms have been widely used for radar and sonar applications [17],[27]. These algorithms use an array of sensors for processing the spatially and temporally correlated signals and have been shown to be quite effective in reducing the effects of interferences such as noise processes and jammers transmitted from other hostile agents. Beamforming algorithms involving an array of microphones, therefore, hold promise for robust acquisition and enhancement of speech signal in noisy environments and in situations where there are competing talkers (jammers) in addition to the desired (target) talker. Speech applications of microphone-array-based beamforming include speech acquisition in an auditorium [1],[9],[10],[19], teleconferencing [1],[10], and voice communication in a noisy environment [5],[14],[15],[20]. Our interest in microphone array processing is to achieve reliable voice communication in a noisy automobile environment, for the digital cellular telephone application [20].

The noise process inside an automobile consists of a variety of components: road noise, mechanical vibration noise, reverberation, window noise, fan noise, car radio, other talkers, turn signal, engine noise, noise generated from nearby cars, etc; some of these noise components, such as car radio and fan noise, are controllable in that they can be suppressed easily. Others, such as road noise and noise due to passing vehicles, are not controllable and are also time-varying.

Hands-free cellular telephony requires dialing by voice with a speech recognition system. It is known that in an automobile, a speech recognition system performs poorly with a single microphone input unless the microphone is located close to the driver's mouth as in the case of a head-mounted or hand-held (handset) microphone; however, these methods are not desirable from safety, legal, or human-factors considerations. A driver's reaction to road events takes longer time while using a hand-held cellular telephone. Also, safety regulations in some countries prohibit using hand-held cellular telephones in road vehicles while driving. To address these important considerations, our approach uses an array of microphones mounted inside the car and produces enhanced speech from the multi-microphone speech data using beamforming algorithms.

Speech enhancement using a microphone array and beamforming has been previously investigated by a number of researchers. Flanagan et al. [10] investigated computer-steered microphone arrays for sound transduction in large rooms and for teleconferencing. Alvarado and Silverman [2] have investigated the optimal spacing of microphones of a linear array. Claesson et al. [5] have implemented a broadband adaptive array on a multiprocessor digital signal processing (DSP) system for use in a hands-free mobile radio telephone. Farrell et al. [9] have shown that beamforming with microphone arrays significantly reduces the effects of noise and reverberation in a room environment. Goulding and Bird [14] have discussed the properties of noise fields in automobile interiors and compared the performances of the adaptive noise cancelation method and the beamforming method. They claim that a simple delay-and-sum beamformer is effective in increasing the SNR and reducing the reverberant distortion of speech. Grenier [15] has also investigated beamforming with a microphone array in car environments. He reports that the microphone array is able to improve the speech recognition scores as compared to a single microphone. The authors [20] have investigated a variety of issues for the use of a microphone array in a car environment such as microphone types, adaptation of a speech recognition system, and highpass filtering. This chapter is based primarily on the authors' own work first reported in [20].

2 BEAMFORMING ALGORITHMS

Recently, beamforming algorithms have been used for speech enhancement. Beamforming approach for an automobile application is, however, different from that for the radar or sonar application since the automobile application is a near-field problem with a wideband signal (i.e., speech). Previous work dealing with radar and sonar applications has been carried out primarily for far-field and narrowband signals. If the minimum distance between the target and the microphones is greater than twice the total array aperture, it is reasonable to assume the far-field condition. Actual measurements in a car, however, support only a near-field condition. For handling the near-field condition, we assume that the acoustic signal amplitude attenuates proportionately with the square of the distance (spherical spreading).

2.1 Conventional Beamforming Algorithm

The conventional beamforming algorithm, also known as the delay-and-sum beamformer (DS-BF), is described in this section. Figure 1 shows the structure of the DS-BF. The microphone array has M microphones. Each microphone output contains the target (i.e., desired) signal and interferences such as noise and jammer (e.g., another talker) signals. If the delays between the target and the microphones are known or can be estimated, they can be compensated to synchronize the target signals for all channels; the interference signals will, however, remain incoherent. Let $s(t)$ be the target speech signal. The received signal at the i-th sensor, $x_i(t)$, can then be expressed as

$$x_i(t) = a_i s(t - \tau_i) + n_i(t) \quad i = 1, 2, \ldots, M \tag{15.1}$$

where a_i is the signal attenuation coefficient, τ_i is the propagation delay of the signal caused by the distance between the source (target) and the i-th sensor, $n_i(t)$ is the sum of noise and jammer signals, and M is the number of microphones. The DS-BF output is given by

$$y_{DS}(t) = \sum_{i=1}^{M} w_{di} x_i(t + \tau_i) = \mathbf{w}_d^T \mathbf{x}_t = \mathbf{w}_d^T \left[\mathbf{a} s(t) + \mathbf{n}_t\right] \tag{15.2}$$

where the coefficient w_{di} is the weight for the i-th channel signal, $\mathbf{x}_t = (x_1(t+\tau_1),\ldots,(x_M(t+\tau_M)))^T$, $\mathbf{w}_d = (w_{d1}, w_{d2},\ldots,w_{dM})^T$, $\mathbf{a} = (a_1,\ldots,a_M)^T$, and $\mathbf{n}_t = (n_1(t), n_2(t),\ldots,n_M(t)).^T$ As mentioned above, we assume that the signal delay values, τ_i, are known or can be estimated. The weight coefficients $\{w_{di}\}$ are generally selected so as to trade off the relationship between array beamwidth and average sidelobe level [16], [27].

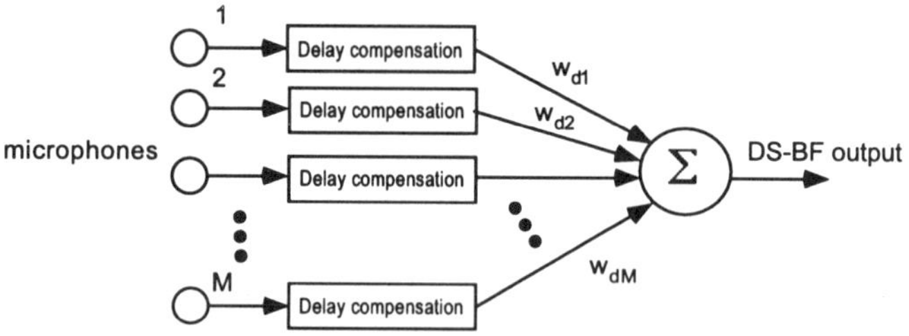

Figure 1 The structure of the delay-and-sum beamformer.

Delay Compensation

The signal delay at the i-th microphone can be computed as

$$\tau_i = \sqrt{(px_s - px_i)^2 + (py_s - py_i)^2 + (pz_s - pz_i)^2}/\nu \tag{15.3}$$

where $(px_s,\ py_s,\ pz_s)$ is the source location, $(px_i,\ py_i,\ pz_i)$ is the location of the i-th microphone, and ν is the speed of sound ($\nu = 340$ meters per second). Since the signal delay at each sensor is generally not an integer multiplication of the sampling time, the delay-compensated signal value, $x_i(t+\tau_i)$, is obtained from $x_i(t)$ through an interpolating filter [24].

Ideally, the delays of the received signals should be known exactly a priori. However, this is not possible in a real application since, as seen from equation (1.3), the signal delays are dependent on driver's position, height, etc. Moreover, it is likely that the driver's position is changing during talking. One solution for this problem is to use a source location estimation method [25]. At each time instance, the source location is estimated and tracked, and hence the time delays can be updated. However, it requires heavy computations to estimate the source location at each time instance, which may not be desirable in an automobile application because of the cost involved. Another alternative is to fix the values of signal delay at the expense of less accurate delay compensations. Less accurate delay compensations result in a smaller signal-to-noise ratio (SNR) improvement provided by the microphone array over a single microphone. However, if the source is located inside the main beam of the DS-BF, the degradation of SNR gain will not be severe in an automobile application since the source location changes only slightly. The performance degradation from delay compensation errors is discussed in detail in the next section.

Signal-to-Noise Ratio Gain

In an automobile environment, the dominating interference is the background ambient noise. Therefore, SNR is one of the important measures to use for evaluating how efficient a beamforming method is in reducing the effects of noise. SNR gain is also a useful way to measure how much improvement a beamforming algorithm provides over the single microphone signal. In this chapter, SNR gain is defined as the difference between the SNR in dB of the beamformer output and the SNR in dB of the output of a single microphone, which in our case is the center microphone of the array.

Let us define the SNR of the signal at the i-th microphone as

$$SNR_i = \frac{\text{Mean Square Value of } y_i(t)}{\text{Mean Square Value of } n_i(t)} \tag{15.4}$$

where $y_i(t) = a_i s(t - \tau_i)$. Equation (1.4) becomes

$$SNR_i = \bar{s}^2 \frac{a_i^2}{\rho_i} \tag{15.5}$$

where $\bar{s}^2$ is the mean square value of the source signal and ρ_i is the mean square value of the noise process $n_i(t)$. If the SNR of the DS-BF output is also defined in the same manner as in equation (1.4), the SNR of the DS-BF becomes

$$SNR_{DS} = \bar{s}^2 \frac{\mathbf{w}_d^T \mathbf{a}\mathbf{a}^T \mathbf{w}_d}{\mathbf{w}_d^T \mathbf{Q} \mathbf{w}_d} \tag{15.6}$$

where $\mathbf{Q} = E[\mathbf{n}_t \mathbf{n}_t^T]$. SNR gain can then be defined as

$$g_{SNR} = 10 \cdot \log_{10} \left[\frac{SNR_{DS}}{SNR_c} \right] \tag{15.7}$$

where SNR_c indicates the SNR of the center microphone.

Example: The SNR gain of the DS-BF can be computed under the ideal condition in which $a_i = 1$, $\rho_i = \rho$, $w_i = 1/M$ for all i, and the noise process is spatially uncorrelated and has a zero mean. Equations (1.5) and (1.6) then become

$$SNR_i = \bar{s}^2/\rho; \qquad SNR_{DS} = \bar{s}^2/(\rho M)$$

since $\mathbf{Q} = \rho\mathbf{I}$. Therefore, the SNR gain of the DS-BF is $10\log_{10} M$ dB. This means that the SNR gain increases with the number of microphones.

The SNR gain of the DS-BF may, however, not be sufficient for the car application as only a small number of microphones may be used for practical reasons. As a consequence, it is necessary to consider an adaptive beamforming algorithm to achieve larger SNR gains than provided by the DS-BF.

2.2 Adaptive Beamforming Algorithm

If the signal, noise, and interference characteristics are known and fixed, the channel weights of the DS-BF can be designed for optimal performance. For our application, however, this is not practical since the required information (such as noise statistics) changes with time. Therefore, it is necessary to have a beamforming technique which adapts to the noise and interference, and cancels out their effects. This can be done by using an adaptive scheme with a beamforming algorithm [11]. Also as mentioned above, out goal in using an adaptive beamformer is to achieve a large SNR gain with a small number of microphones.

Generalized Sidelobe Canceler

Among many types of adaptive beamforming algorithms, we selected to use the generalized sidelobe canceler (GSC), which is a linearly constrained minimum variance beamformer [16]. The GSC has advantages over other linearly constrained beamforming methods. The main advantage of the GSC is that it provides a convenient way of assessing the extent of the performance degradation caused by the steering and gain errors in the sensor array. The GSC is also simple to implement compared to other beamforming methods.

The structure of the GSC is shown in Figure 2, where a thick, dark line is used to denote multi-channel data, with the label indicating the number of channels. The top path is basically a conventional DS-BF beamformer. The lower path in Figure 2 is known as the sidelobe canceling path or adaptive noise canceling path. It consists of a blocking matrix, **B**, and a set of adaptive weights in an

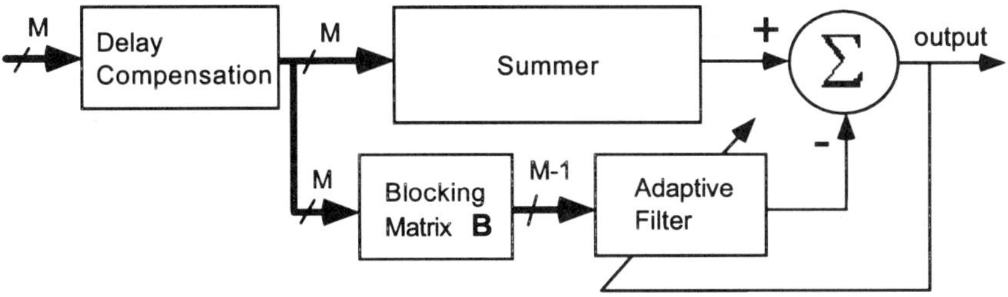

Figure 2 The structure of the Generalized Sidelobe Canceler

adaptive filter. The adaptive weights are chosen to minimize the output power, which is equivalent to minimizing the noise content of the GSC output.

Assuming that the signal delays are compensated exactly, the signal at the i-th channel after delay compensation can be written as

$$\begin{aligned} z_i(t) &= x_i(t+\tau_i) \\ &= a_i s(t) + n_i(t) \end{aligned} \tag{15.8}$$

where a_i is the attenuation of the signal. In vector notation, equation (1.8) becomes

$$\mathbf{z}_t = \mathbf{a}s(t) + \mathbf{n}_t \tag{15.9}$$

where $\mathbf{z}_t = (z_1(t), \ldots, z_M(t))^T$, $\mathbf{a} = (a_1, \ldots, a_M)^T$, and $\mathbf{n}_t = (n_1(t), \ldots, n_M(t))^T$

From equation (1.9), it is easy to see that there exists a vector so that its inner product with $\mathbf{z}_t$ does not contain the signal component $s(t)$. A blocking matrix can then be formed from a set of such vectors. In fact, the blocking matrix can be any matrix that satisfies the following two conditions:

- The inner product of a row vector and $\mathbf{a}$ defined in equation (1.9) should be zero.
- Row vectors of the matrix should be linearly independent.

For example, if $a_i = 1$ for $i = 1, \ldots, M$, the simplest form of blocking matrix can be found as

$$\mathbf{B} = \begin{bmatrix} 1 & -1 & 0 & \ldots & 0 \\ & 1 & -1 & & \vdots \\ \vdots & & \ddots & & \\ 0 & \ldots & 0 & 1 & -1 \end{bmatrix} \tag{15.10}$$

In this case, the output of the blocking function consists of pairwise differences of its input data. The output of the blocking matrix can then be written as

$$\tilde{\mathbf{z}}_t = \mathbf{B}\mathbf{z}_t \tag{15.11}$$

where the output vector $\tilde{\mathbf{z}}_t$ has $(M-1)$ elements and does not contain any desired signal components in this ideal case.

Let us define the coefficients of the adaptive filter at the k-th tap as

$$\mathbf{w}_a(k) = (w_{k,1}, w_{k,2}, \ldots, w_{k,M-1})^T \quad k = 0, 1, \ldots, N-1 \tag{15.12}$$

where N is the number of taps. The output of the lower path, which is referred to as the adaptive noise canceler output, is given by

$$y_{ANC}(t) = \sum_{k=0}^{N-1} \mathbf{w}_a^T(k)\tilde{\mathbf{z}}_{t-k} = \sum_{k=0}^{N-1} \mathbf{w}_a^T(k) \cdot \mathbf{B}\mathbf{z}_{t-k} \tag{15.13}$$

The output of the GSC beamformer can then be written as

$$\begin{aligned} y_{GSC}(t) &= y_{DS}(t) - y_{ANC}(t) \\ &= \mathbf{w}_d^T \mathbf{z}_t - \sum_{k=0}^{N-1} \mathbf{w}_a^T(k)\mathbf{B}\mathbf{z}_{t-k} \\ &= \sum_{k=0}^{N-1} (\mathbf{w}_d^T \delta_k - \mathbf{w}_a^T(k)\mathbf{B})\mathbf{z}_{t-k} \end{aligned} \tag{15.14}$$

where $y_{DS}(t)$ is defined in equation (1.2) and δ_k is 1 if $k=0$ and 0 otherwise. The coefficients of the adaptive filter are updated using the LMS (Least Mean Squares) algorithm as follows:

$$w_{ai}^{(t)}(k) = w_{ai}^{(t-1)}(k) + \mu \cdot y_{GSC}(t) \cdot \tilde{z}_i(t-k) \tag{15.15}$$

where $\tilde{z}_i(t-k)$ is the filter input at the i-th channel and k-th tap, and μ is the step size affecting the convergence rate.

Note that it is possible to have a frequency shaping filter in front of the summer in the upper path of Figure 2 to provide a desired frequency shaping for its output [4]. In this chapter, however, we do not consider this option, i.e., we assume that there is no frequency shaping filter.

Potential Problems of an Adaptive Beamformer

An adaptive beamformer such as the GSC, however, has a few drawbacks. If the target signals are not completely blocked out at the adaptive noise canceler (ANC) input (thereby causing target signal leakage), the adaptive filter will then adapt to the characteristics of the target signal; this results in (partial) cancelation of the target signal, which will be referred to as signal cancelation hereafter. When signal cancelation occurs, the beamformer output suffers from a significant signal component reduction; as a consequence, the beamformer output can lead to a negative SNR (i.e., it contains more noise than the desired signal). Possible causes of signal leakage include (1) uncertainty of the source (target) location, (2) reverberations of the signal, (3) inaccurate delay compensation, (4) unknown and unequal microphone characteristics, (5) inaccurate blocking matrix, and (6) quantization errors caused by the acquisition equipment.

Another drawback of an adaptive beamformer is that it may continuously modify the spectral characteristics of the beamformer output speech, thereby causing inconsistencies between training and test data, even if both involve the same noise condition. Speech recognition performance can degrade under this situation.

To minimize these drawbacks, we use two approaches. First, we incorporate the near-field condition, which is appropriate for the car application as noted at the beginning of Section 2. The difference or blocked signals in the lower branch in Figure 2 are computed after compensating for the attenuation over the respective target-to-sensor distance in accordance with the spherical spreading model. Second, an effective adapt-and-stop strategy is used in our GSC implementation: If the noise is consistent in characteristic for a short period of time (e.g., 5 seconds), it is possible for the adaptive filter to adapt fully on the noise components before the target signal starts. The ANC filter adaptation thus occurs over a certain period of time when only noise is present; the filter is held fixed after this initial adaptation period. This method is found to improve the SNR effectively without introducing any serious signal cancelation or spectral distortion. One way of instrumenting this adapt-and-stop strategy for the cellular car phone application is to require the driver to push a button (located, for example, on the steering column) indicating his intention to use the phone. This action triggers the ANC filter adaptation; after a few seconds, the adaptation stops and a green light comes on signaling the driver to start dialing with voice.

Signal cancelation can be reduced further by using a blocking matrix that is designed to minimize signal leakage. A blocking matrix, such as defined in equation (1.10), assumes that signal components of all channels are the same, which may not be true in practice. As a consequence, the GSC suffers from signal cancelation. A blocking matrix may be designed empirically beforehand to compensate for differences in the microphone characteristics [12].

In our investigation of GSC, we listened to the output of the adaptive noise canceler (output of the adaptive filter in Figure 2) as an effective way of monitoring speech signal leakage. If this ANC output does not have any audible speech signal, it is safe to assume that there is no signal leakage and hence no signal cancelation.

3 VALIDATION ON LABORATORY DATA

Substantial SNR gains of beamforming methods have been reported in the literature using simulated data and assuming ideal conditions. It is important to use real data to be able to estimate SNR gains that are actually achievable. We initially collected speech and noise data sets using the Brown University microphone array system [2],[26] and evaluated the DS-BF and GSC beamformers using these data sets. The locations of the microphone array, the target, the jammer, and the noise sources were determined based on actual measurements inside a car. Two noise sources were used at left and right front window locations, respectively. The actual noise data had been recorded previously inside a car running at 60 miles per hour (MPH), with the air conditioner on and the windows half-open. With the Brown University set-up, target and noise signals were collected separately so that we can generate noisy speech data with desired SNR values by using appropriate gain settings on the noise. An evaluation of the DS-BF and the GSC beamformers with a linear 7-microphone array was conducted under three SNR conditions (omni-directional microphones were used in this study): 0 dB, 5 dB, and 10 dB; the SNR values given correspond to the center microphone in the array. Figure 3 provides the measured SNR data. Notice that DS-BF provides about 5 dB SNR gain over a single microphone (the center microphone in the array), while GSC provides about 13 dB gain. Detailed results from speech recognition testing with Brown University data are not presented here for the following reason. To keep the data collection effort small, we had collected 10-digit sequences from one male and one female speaker, with only four distinct digits: one, four, five, and nine. For this data, the speaker-independent recognition algorithm with system adaptation (see Section 5) performed almost perfectly for both speakers and over all three SNR conditions, even for a single microphone. The recognition task involving only four distinct digits was too easy.

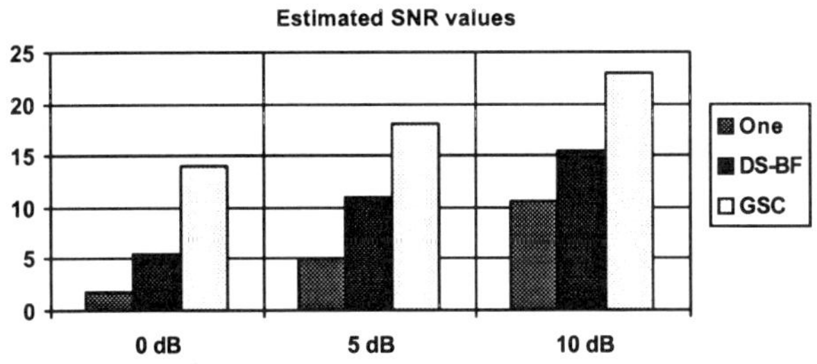

Figure 3 Estimated SNR values for a single microphone, DS-BF, and GSC when the input SNR is 0, 5, or 10 dB.

4 DATA COLLECTION IN A CAR

We collected multi-microphone speech data in two phases. Phase 1 data collection was designed for a pilot study involving one car and one type of microphones. Phase 2 data collection was performed using two different cars and three types of microphones. We describe below the details of our data collection including hardware set-up, number of talkers, noise conditions, speech material, types of microphones, and types of cars.

Data Collection Set-up

A hardware system was developed for speech data collection inside a car. This system consists of (1) seven equally spaced microphones (5 cm apart) in a linear array and one head-mounted close-talking microphone, (2) an eight-channel Sony DAT (Digital Audio Tape) recorder, and (3) interface between the Sony DAT recorder and a Sun SPARCstation for transferring digitized multichannel data. All collected data were processed on the SPARCstation. A DAT recorder was chosen for our data collection system because of its portability. For Phase 1, we collected speech data from one male talker; for Phase 2 we collected speech data from 12 talkers: 6 males and 6 females.

The microphone array was constructed from a strip of Plexiglas with properly sized holes to house the individual microphones. Figure 4(a) shows a photograph of the microphone array, and Figure 4(b) shows the array mounted on the driver-side visor.

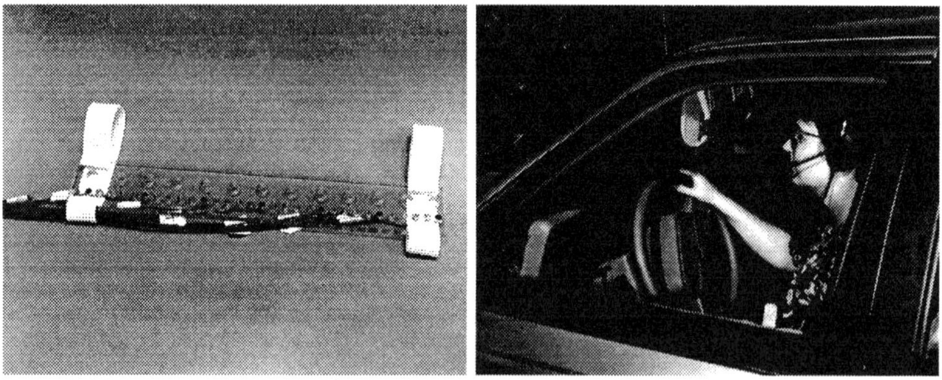

Figure 4 Photographs: (a) The microphone array and (b) the microphone array mounted on the driver-side visor. The head-mounted close-talking microphone shown here was used to serve as a performance reference.

Noise Conditions

For Phase 1 data collection, we included three idling-car conditions and two driving conditions. The three idling conditions used are: window closed and fan off (w0f0); window closed and fan on (w0f1); and window open and fan off (w1f0). The reason for including the idling conditions w0f0 and w0f1 is to provide speech training data for system adaptation of the speech recognition system (see Section 5.4). Such a training requirement by the driver in an idling condition provides adequate noise level, without the danger associated with training in a moving car. For the w1f0 condition, the car was parked right near the North Central Expressway at Dallas and was exposed to a high level of noise. This condition is extremely noisy and most people would consider it annoying; we included this condition in our pilot data collection to see how effective the microphone array solution is even in an extreme environment. The two driving conditions used are highway driving at 55 MPH and city or local driving. The local driving condition included stops at signal lights, acceleration, breaking, and turning with turn signal on. The speed range used in local driving was up to 40 MPH. The talker was seated at the front, passenger side for all five conditions because of safety and legal considerations.

For Phase 2 data collection, we used the following three noise conditions: idling car with fan on, local driving, and highway driving. We used two different settings for the fan speed. High fan speed (noisier) was set for the idling condition and low fan speed was used for local and highway driving conditions. We used the high fan speed for the idling condition (which is used for system adaptation of the speech recognition system) as the associated higher noise level leads to better system adaptation (see Section 5.6).

Speech Material

We used sixty randomly generated 10-digit numbers that were chosen to assimilate the actual phone numbers, for data collection at each noise condition. Digits consisted of 11 words: "zero," "one" to "nine," and "oh." During data collection, the talker had freedom to say "oh" or "zero" and was asked to talk as naturally as possible without any artificial constraints.

Automobiles and Microphones Used

The type of automobile used for the data collection affects the noise level inside the car. The more expensive cars generally produce significantly lower noise levels (higher SNRs). In our Phase 1 study, we collected data from a relatively noisy, compact car (Dodge Spirit), using an array of omni-directional microphones (Panasonic WM-60AT). For Phase 2, we collected data from two quieter cars: Lincoln Towncar and Nissan Maxima. The full data set for 12 talkers and 3 noise conditions was collected using the Lincoln Towncar and an array of uni-directional microphones (Panasonic WM-55A103). For comparison purposes, we collected three smaller data sets from only two talkers (one male, one female): (1) Nissan Maxima and uni-directional microphones; (2) Lincoln Towncar and omni-directional microphones (Panasonic WM-60AT); (3) Lincoln Towncar and noise-canceling microphones (Panasonic WM-55D103).

5 DATA ANALYSIS AND RESULTS

In this section, we present and discuss in detail various data analyses and the associated results. We processed the car data using the two beamforming algorithms, DS-BF and GSC, and compared against each other and against a single microphone, which is the center microphone in the array. Initially, we used SNR as the performance measure in these comparisons (see sections

5.1 and 5.2). Extensive speech recognition tests were then performed. Unless stated otherwise, the recognition error rates given are word (or digit) error rates (in contrast to sentence error rates). For voice dialing of 10-digit telephone numbers, the term sentence error refers to a mis-recognized telephone number; this error occurs when one or more digits in the telephone number get mis-recognized.

5.1 SNR Results

The data collected in the car was processed using the two beamforming algorithms, DS-BF and GSC, for each of the five Phase 1 noise conditions. SNRs were computed for the center microphone in the array, for DS-BF output, and for GSC output. (Recall from Section 4 that all Phase 1 data were collected using omni-directional microphones.) Both single microphone and DS-BF cases produce equivalent SNRs ranging from -2 dB to 12 dB; GSC provides SNR gains ranging from 2 dB to 9 dB over a single microphone.

5.2 Uncertainty of the Source Location

As mentioned above, it is difficult to locate the source (talker) position exactly. Therefore, with the Brown University laboratory database, we examined how much the performance of the beamformers deteriorates when the estimated source position involves an error. This test was done in the following manner: we measured the source position (talker's mouth) from the microphone array for the data collection and assumed this to be the correct location. Let (px_s, py_s, pz_s) be the correct location of the source. We divided the surrounding neighborhood of the correct location into 25 regions with the z-coordinate fixed : from $(-10 + px_s, -10 + py_s, pz_s)$ centimeters to $(10 + px_s, 10 + py_s, pz_s)$ centimeters, with 5 centimeter grid spacing. The distance of the center microphone of the array from the talker was about 30 centimeters. Beamformer outputs were generated based on these 25 erroneous source locations.

Results from the above test show that SNR does not deteriorate much due to the source location uncertainty, but the speech recognition performance degrades more significantly. However, the performance degradation is relatively small if the source location does not deviate by more than 5 centimeters. If the estimated source location is inside the main beam, the performance does not degrade severely.

5.3 Speech Recognition

A speech recognition algorithm was employed to evaluate the performance of the microphone array. This recognition algorithm was developed in Texas Instruments' Speech Research Laboratory, using the Hidden Markov Model (HMM) [22]. An earlier version of this algorithm is described in [7]. It is a speaker-independent continuous digit recognition algorithm and was trained on speech data collected over long-distance telephone lines [28].

For the cellular telephone application in an automobile, however, it may make sense to use a speaker-dependent speech recognition system because of expected higher performance and secure access. For consumer acceptance, however, the required training must be short. A practical solution is to use a speaker-independent algorithm and train (or adapt) the observation probabilities of the HMM model with a very limited number of training utterances collected from the given user. This training paradigm is sometimes referred to as "speaker adaptation." In our case, as noted above the speaker-independent algorithm was trained on data collected over long-distance telephone lines. Thus, the observation probabilities are to be adapted to account for the variations in such factors as the transducer (microphone array plus beamforming) and background noise as well as to correspond to the given speaker. For lack of a better term, we refer to this process as "system adaptation." Results presented in the next section show that the system adaptation step leads to substantial gains in the recognition performance.

As mentioned earlier, system adaptation of the speech recognizer should be done with speech data collected in a parked car condition for the driver's safety reason. The adaptation procedure in a real application can be envisioned as a driver sitting inside a car and speaking a few sentences or digit strings.

5.4 Performance Gain via System Adaptation

Here, we compare the performance of the speaker-independent case (SI) with the performance of the system-adapted case (SA), using Phase 1 data. Recall from the previous section that system adaptation requires the driver to provide a small amount of training speech in the car. For a number of reasons including safety and user acceptance, this training is best performed in an idling condition (i.e., in a parked car). We used 20 utterances collected in the w0f0 condition for system adaptation. In recognition testing for the system-adapted case, we used the remaining 40 utterances for w0f0 and all 60 utterances for each of the other four Phase 1 conditions.

Table 1 Performance comparison between speaker independent and system adapted speech recognition (Error rates in percent).

Sensor System	w0f0		w0f1		w1f0	
	SI	SA	SI	SA	SI	SA
One Mike	26.5	0.6	52	2.3	64.7	60.4
DS-BF	23.8	0.2	58.3	0.8	58.3	23.1
GSC	7.3	0.2	10.2	0.5	19.5	11.8

Table 1 shows the performance comparison of system-adapted and speaker-independent systems. All three sensor systems perform substantially better with system adaptation under each of the three Phase 1 idling-car conditions. We expect similarly impressive performance gains due to system adaptation in local and highway driving conditions. Since system adaptation leads to superior recognition performance, it was included as a standard feature in all our subsequent tests.

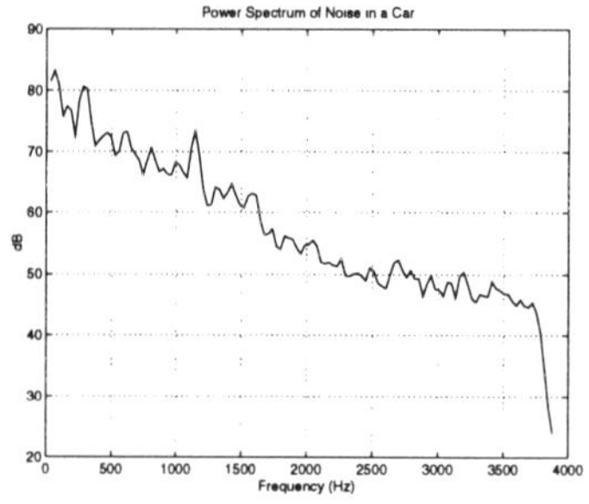

Figure 5 Power spectrum for the automobile noise (highway driving).

5.5 Highpass Filtering

Spectrograms of speech data collected in local and highway driving conditions (Phase 1) show the presence of strong low-frequency noise components below about 500 Hz. Figure 5 shows a typical spectrum of car noise during highway driving. The low-frequency noise components are likely the result of mechanical vibrations of a moving car and of road noises. We investigated the effect of applying a 500-Hz highpass filter to individual microphone signals prior to beamforming. The recognition error rates obtained with and without highpass filtering are plotted in Figure 6 for w1f0 and for local and highway driving conditions. The use of highpass filtering produces, in general, significant improvements in recognition performance. We note, however, for the low-noise conditions w0f0 and w0f1, highpass filtering has negligible effect.

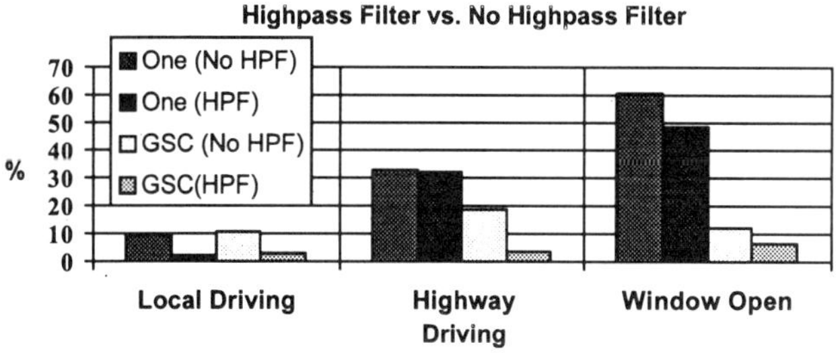

Figure 6 Word-error rates of system-adapted speech recognition on outputs of one microphone and GSC, with and without highpass filtering.

5.6 Selecting Training Database

System adaptation yields the best performance gain when the training speech is collected in the same condition as that used for testing. However, this is not feasible in practice, as noted above. We performed system adaptation in w0f1 and compared the results with those obtained previously using system adaptation in w0f0. Table 2 shows this comparison for a single microphone and for GSC. Clearly, w0f1 is a better condition to perform system adaptation than is w0f0. The reason for this result is that with all three test conditions being noisy, it is better to perform system adaptation in a noisier condition (w0f1 is noisier than w0f0). Even w0f1 is relatively noise-free for both beamformers DS-BF and GSC. In fact, we have achieved further performance gains for DS-BF and GSC, with system adaptation done using one-microphone signal in w0f1 as this signal is noisier than the respective beamformer output (see Section 5.8).

5.7 Microphone Types

Figure 7 shows speech recognition performance (in percent sentence error rate) of the DS-BF outputs, for each of the three microphones we compared: omni-directional, uni-directional, and noise-canceling. For this comparison, we used

Table 2 Performance comparison for different training sets of system-adapted speech recognition (Error rates in percent).

Training	local			highway			window open		
condition	One	DS	GSC	One	DS	GSC	One	DS	GSC
w0f0	2.2	7.1	2.5	32	51	6.4	49	14	10
w0f1	2.7	2.0	2.6	8.9	8.5	3.9	37	5.4	6

Phase 2 speech data collected from two speakers (one male and one female). Clearly, uni-directional microphones produce the best performance in each of the three test conditions shown in Figure 7. The main beam of the microphone array is quite narrow for noise-canceling microphones, making the performance of the DS-BF output sensitive to the location of the talker. This explains the poor performance of noise-canceling microphones. Data analysis for GSC indicates that the signal cancelation is the largest for noise-canceling microphones and the smallest for omni-directional microphones.

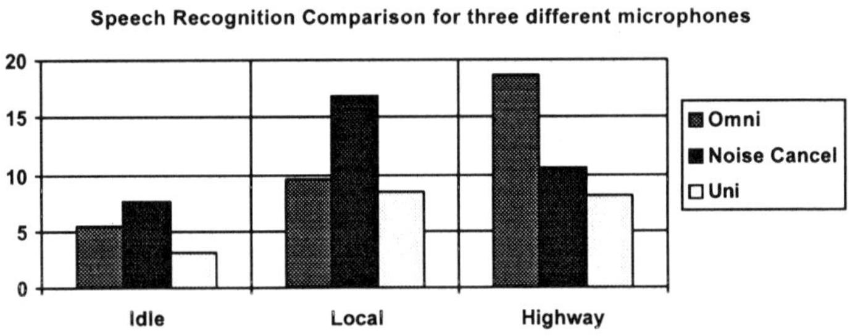

Figure 7 Performance comparison (sentence error rate) of the DS-BF output for omni-directional microphones, noise-canceling microphones, and uni-directional microphones. System adaptation and highpass filtering were used.

5.8 Overall Performance

In this section, we provide two sets of speech recognition performance comparisons of the three sensor systems: a single microphone, DS-BF, and GSC. For both sets, we used a 500-Hz highpass filter and performed system adaptation using training data collected under the w0f1 condition. For the first set of comparisons shown in Figure 8, we used omni-directional microphones and the Phase 1 speech data collected from one male speaker. Clearly, GSC produces substantially lower recognition error rates than does a single microphone and performs generally better than DS-BF. However, for the two-speaker (one male, one female) Phase 2 data, GSC actually produces slightly worse performance than DS-BF, quite similar to the result reported below.

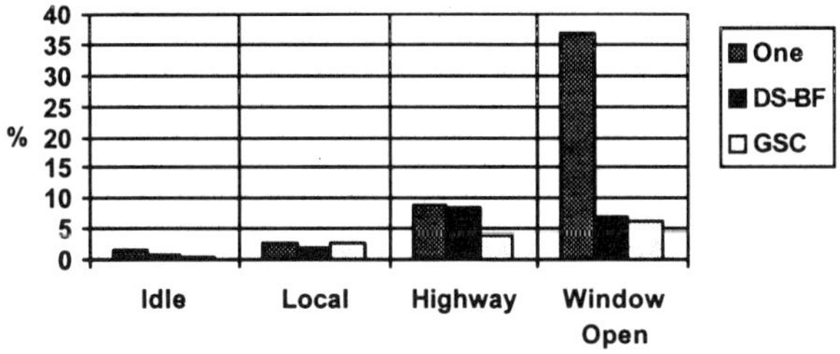

Figure 8 Speech recognition word error rate performance for four different noise conditions for omni-directional microphones and one speaker.

The second set of comparisons is shown in Figure 9. For this set, we used the Phase 2 data collected with uni-directional microphones from 10 speakers, 5 male and 5 female. (We excluded two of the 12 speakers from data analysis because of certain data collection problems.) From Figure 9, we see that DS-BF provides the best overall performance and significantly better performance than does GSC in highway driving.

The major reason for the poor performance of GSC with uni-directional microphones is that the desired signal leaks into the adaptive noise canceling path (see Figure 2), which leads to partial signal cancelation at the GSC output. The main beam of the microphone array is narrower for uni-directional microphones than for omni-directional microphones. This explains why partial signal cancelation in GSC is more likely to occur for uni-directional microphones than for omni-directional microphones.

We investigated several ways of improving the performance of the GSC beamformer consisting of uni-directional microphones. One that worked well is described next. Recall that we performed system adaptation using the GSC output for the training condition w0f1. This GSC output was, however, relatively noise-free, with a high SNR, compared to the center microphone signal. We, therefore, decided to use the center microphone signal in w0f1 for system adaptation, instead of using the GSC output. Unfortunately, at the time we tried this experiment we had access to speech data of only sever speakers (three male, four female). The sentence error rate for GSC in highway driving was 13.7% with GSC output in w0f1 used for system adaptation. When we used

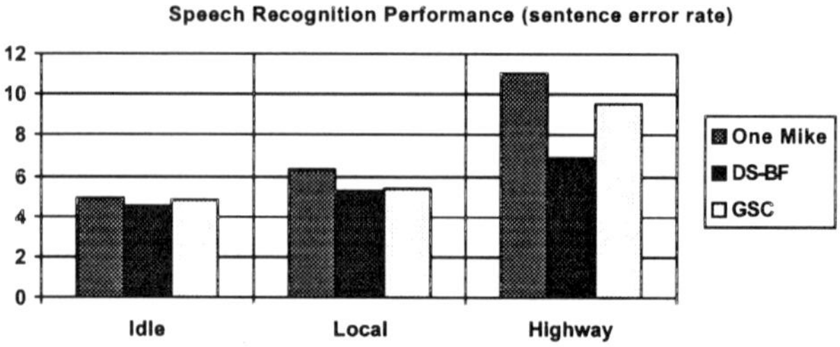

Figure 9 Overall speech recognition performance (sentence error rate) for three noise conditions.

the center microphone signal for system adaptation, this error rate decreased to only 8.5%. We repeated this test for the idling and local driving conditions. The sentence error rate for GSC decreased from 5% to 4.3% for the idling condition and from 6% to 4.8% for local driving. We believe that this method of using noisier training data for system adaptation would have produced better overall GSC performance than shown in Figure 9; as mentioned above, we did not have access to the data from the other three speakers and hence could not run this experiment on their data. (Use of the center microphone signal for system adaptation helped improve the performance of DS-BF but by only a small amount.) Notwithstanding this improved result, we still believe that the issue of partial signal cancelation in GSC must be addressed to ensure robust performance.

5.9 Other Test Results

The results given in this section were obtained using Phase 1 speech data collected with omni-directional microphones.

Spectral Subtraction vs. Beamforming

We investigated the effect of using a single-channel speech enhancement algorithm; for this experiment, we used the well-known spectral subtraction algorithm [3]. Both 500-Hz highpass filtering and system adaptation in w0f1 (enhanced speech was used in training) were employed. For a single microphone, using speech enhancement reduced the recognition error rate from 8.9% to 6% for highway driving and from 37.4% to 23% for w1f0. These values for GSC are 3.9% → 1.5% for highway driving and 6.3% → 3.6% for w1f0. For highway driving, the error rate drops from 8.5% to 3.7% for DS-BF. It is interesting to note that GSC without speech enhancement is still superior to a single microphone with speech enhancement. Furthermore, the spectral subtraction method introduces audible distortions commonly referred to as "musical noises." Results from informal listening tests show a clear preference for speech from GSC without speech enhancement over the enhanced single microphone signal. This preference should be preserved, if not strengthened, under speech coding, which is required in digital cellular systems.

Close-Talking Microphone

The head-mounted close-talking (noise-canceling) microphone produces only 0.5% word recognition error in all Phase 1 conditions. This result indicates the need for further research as there is more room for improvement of the microphone-array-based beamforming approach.

Use for Speech Coding

A brief evaluation test was conducted comparing the performance under speech coding. This test is important because the various digital cellular systems will all use speech coding, as noted above. As speech coder, we used the US Federal Standard 1016 4.8 kbits/s CELP (Code Excited Linear Prediction) algorithm. Informal listening tests show that the speech quality of the CELP coder output is substantially better for GSC than for DS-BF and single microphone cases. Specific speech quality improvements due to GSC include lower background noise level and more natural-sounding voice; the CELP output speech sounds rough for the other two cases.

6 SIGNAL CANCELATION IN ADAPTIVE BEAMFORMING

As discussed in Section 5, the adaptive beamformer GSC has the potential of providing substantial SNR gains over the DS-BF and a single microphone. However, for practical applications the problem of partial signal cancelation in GSC must be resolved satisfactorily. This problem has been investigated for narrowband beamforming by a number of researchers as part of robust adaptive beamforming [6],[8],[18],[29]. More recently, an approach involving calibration of the sensor array has been proposed for wideband beamforming for dealing with some of the factors responsible for signal cancelation such as source location error and unequal sensor characteristics [12],[23]. The method proposed in [12] uses a short segment of speech as the calibration signal and computes from the measured microphone data an optimal blocking matrix that minimizes leakage of the calibration signal into the noise canceling path of GSC. Simulation results provided in [12] show that in the presence of array sensor gain and delay errors, the calibrated GSC produces substantially better performance than the uncalibrated GSC. Even with such calibration, the signal cancelation problem still remains in a reverberant environment [12]. Finally, while writing this chapter the authors became aware of another promising approach that deals with source location errors by combining explicit source (or target) tracking with adaptive beamforming [13]. The approach involves an iterative algorithm in which the estimated source signal (i.e., the beamformer output) is used to adaptively track the correct source position, which in turn is fed back to the beamformer to produce a better source signal estimate. It is not clear to the authors as to how well this approach will carry over to the problem of real-time hands-free voice communication in a car. Suffice it to say that further research is warranted to resolve satisfactorily the problem of signal cancelation in adaptive beamforming.

7 CONCLUSIONS

For hands-free voice communication in an automobile environment, we have presented an effective speech acquisition and enhancement approach that uses a small array of microphones and a beamforming algorithm for processing the microphone outputs. Although beamforming has been widely used in the past in dealing with narrowband and far-field signals, our investigation deals with wideband speech input from a driver in the near-field of a microphone array. Two beamforming algorithms, the non-adaptive DS-BF and the adaptive GSC, were evaluated on multi-speaker speech data collected in real cars under several noise conditions, using as performance measures SNR and word or sentence error rate of an HMM-based continuous speech recognition system. Major results of these evaluations are as follows:

- System adaptation of the speech recognition models using a small amount of training speech collected in a parked car condition leads to dramatic increases in speech recognition accuracy, for all three sensor systems: a single microphone and the two 7-microphone beamformers, DS-BF and GSC.

- A highpass filter with a 500-Hz cutoff at the output of each microphone produces significant improvements in speech recognition performance, for all three sensor systems.
- Uni-directional microphones lead to better speech recognition performance than do either omni-directional microphones or noise-canceling microphones, for DS-BF. This result is, in general, not true for GSC because of partial signal cancelation occurring with uni directional microphones.
- Both DS-BF and GSC produce better overall speech recognition performance than does a single microphone. Because GSC is vulnerable to partial signal cancelation in some situations, DS-BF is a more reliable approach.

Finally, as GSC has the potential of providing substantial performance gains over DS-BF and a single microphone, we recommend further research into resolving the problem of signal cancelation in GSC.

REFERENCES

[1] J. B. Allen, D. A. Berkley, and J. Blauert, "Multimicrophone signal-processing technique to remove room reverberation from speech signals," *J. Acoust. Soc. Amer.*, vol.62, pp. 912-915, Oct. 1977.

[2] V. M. Alvarado and H. F. Silverman, "Experimental results showing the effects of optimal spacing between elements of a linear microphone array," in *Proc. IEEE Int. Conf. Acoust., Speech, Signal Processing*, pp. 837-840, Apr. 1990.

[3] S. Boll, "Suppression of acoustic noise in speech using spectral subtraction," *IEEE Trans. Acoust., Speech, Signal Processing*, vol. ASSP-27, pp. 113-117, Apr. 1979.

[4] K. M. Buckley, "Spatial/spectral filtering with linearly constrained adaptive arrays," *IEEE Trans. Acoust., Speech, Sig. Processing*, vol. ASSP-35, pp. 249-266, Dec. 1987.

[5] I. Claesson, S. E. Nordholm, B. A. Bengtsson, and P. Eriksson, "A multi-DSP implementation of a broad-band adaptive beamformer for use in a

hands-free mobile radio telephone," *IEEE Trans. Vehicul. Technol.*, vol. 40, pp. 194-202, Feb. 1991.

[6] H. Cox, R. M. Zeskind, and M. M. Owen, "Robust adaptive beamforming," *IEEE Trans. Acoust., Speech, Signal Processing*, vol. ASSP-35, pp.1365-1375, Oct. 1987.

[7] G. Doddington, "Phonetically sensitive discriminants for improved speech recognition," in *Proc. IEEE Int. Conf. Acoust., Speech, Signal Processing*, pp. 556-559, May 1989.

[8] M. H. Er and A. Cantoni, "A unified approach to the design of robust narrow-band antenna array processors," *IEEE Trans. Antenn. Propagat.*, vol. AP-38, pp. 17-23, Jan. 1990.

[9] K. Farrell, R. J. Mammone, and J. L. Flanagan, "Beamforming microphone arrays for speech enhancement," in *Proc. IEEE Int. Conf. Acoust., Speech, Signal Processing*, pp. 285-288. Mar. 1992.

[10] J. L. Flanagan, J. D. Johnston, R. Zahn, and G. Elko, "Computer-steered microphone arrays for sound transduction in large rooms," *J. Acoust. Soc. Amer.*, vol. 78, pp. 1508-1518, Nov. 1985.

[11] O. L. Frost, "An algorithm for linearly constrained adaptive array processing," in *Proc. IEEE*, vol. 60, pp. 926-935, Aug. 1972.

[12] G. Fudge and D. Linebargar, "A calibrated generalized sidelobe canceller for wideband beamforming," *IEEE Trans. Signal Processing*, vol. 42, pp. 2871-2875, Oct. 1994.

[13] S. Gazor, S. Affes, and Y.Grenier, "Robust adaptive beamforming via target tracking," *Ecole Nationale Supérieure des Télécommunications*, Paris, France, Submitted to IEEE Trans. Signal Processing, Dec. 1993.

[14] M. Goulding and J. S. Bird, "Speech enhancement for mobile telephony," *IEEE Trans. Vehicul. Technol.*, vol. 39, pp. 316-326, Nov. 1990.

[15] Y. Grenier, "A microphone array for car environments," in *Proc. IEEE Int. Conf. Acoust., Speech, Signal Processing*, pp. 305-308. Mar. 1992.

[16] L. J. Griffiths and C. W. Jim, "An alternative approach to linearly constrained adaptive beamforming," *IEEE Trans. Antenn. Propagat.*, vol. AP-30, pp. 27-30, Jan. 1982.

[17] S. Haykin, *Array Signal Processing*, Englewood Cliffs, NJ: Prentice-Hall, 1985.

[18] N. K. Jablon, "Adaptive beamforming with the generalized sidelobe canceller in the presense of array imperfections," *IEEE Trans. Antenn. Propagat.*, vol. AP-34, pp. 996-1012, Aug. 1986.

[19] W. Kellermann, "A self-steering digital microphone array," in *Proc. IEEE Int. Conf. Acoust., Speech, Signal Proccessing*, pp. 3581-3584, May 1991.

[20] S. Oh. V. Viswanathan, and P. Papamichalis, "Hands-free voice communication in an automobile with a microphone array," in *Proc. IEEE Int. Conf. Acoust., Speech, Signal Processing*, pp. 281-284. Mar. 1992.

[21] B. I. Pawate, and G. Doddington, "Implementation of a hidden Markov model-based layered grammar recognizer," in *Proc. IEEE Int. Conf. Acoust., Speech, Signal Processing*, pp. 801-804, 1989.

[22] J. Picone, "Continuous speech recognition using hidden Markov models," *IEEE Acoust., Speech, Signal Processing Magazine*, pp. 26-41, Jul. 1990.

[23] J. W. Pierre and M. Kaveh, "Wideband sensor array processing using a laboratory array testbed," in *Proc. 35th Midwest Symp. on Circuits, Syst.*, Aug. 1992.

[24] R. Pridham and R. A. Mucci, "A novel approach to digital beamforming," *J. Acoust. Soc. Amer.*, vol. 63, pp. 425-434, Feb. 1978.

[25] R. O. Schmidt, "Multiple emitter location and signal parameters estimation," *IEEE Trans. Antenn, Propagat.*, vol. AP-34, pp. 276-280, Mar. 1986.

[26] H. F. Silverman, "Some analysis of microphone arrays for speech data acquisition," *IEEE Trans. Acoust., Speech, Signal Processing*, vol. ASSP-35, pp. 1699-1712, Dec. 1987.

[27] B. D. Van Veen and K. M. Buckley, "Beamforming: A versatile approach to spatial filtering," *IEEE Acoust., Speech, Signal Processing Magazine*, pp. 4-24, Apr. 1988.

[28] B. Wheatley and J. Picone, "Voice across america: Toward robust speaker-independent speech recognition for telecommunications applications," *Digital Signal Processing: A Review Journal*, vol. 1, pp. 45-63, Apr. 1991.

[29] B. Widrow, K. M. Duvall, R. P. Gooch, and W. C. Newman, "Signal cancellation phenomena in adaptive antennas: Causes and cures," *IEEE Trans. Acoust., Speech, Signal Processing*, vol. ASSP-30, pp. 469-478, May 1982.

16

THE PITCH MODE MODULATION MODEL AND ITS APPLICATION IN SPEECH PROCESSING

Michael A. Ramalho
Richard J. Mammone

Bellcore
331 Newman Springs Road
Red Bank, New Jersey 07701-5699, USA [1], *and*
CAIP Center, Rutgers University
Piscataway, New Jersey, 08855-1390, USA [1]

PRECIS: *The techniques currently used for speech coding or enhancement critically depend upon some form of statistical stationarity, either in the speech signal, the noise signal, or both, in order to accomplish the coding or enhancement. Virtually all speech processing techniques utilize a speech model to reduce the amount of information necessary to characterize the speech signal. Although the speech signal is known to be highly redundant, it is also non-stationary. This non-stationarity requires that the parameters of these models be extracted from short duration signal segments, where the stationarity assumption in the models is not seriously violated. Unfortunately, the use of short speech frames makes the estimation of the model parameters difficult and sometimes obscures the very redundancy the model was based on. The use of a longer frame size is desirable for many signal processing techniques that require increased frequency domain resolution.*

1 INTRODUCTION

The Pitch Mode Modulation Model (PMMM) developed in this chapter is a *long-term* speech model which accommodates non-stationarity in normal speech and, therefore, can process longer speech segments. The PMMM components of voiced speech are found to vary in time synchrony via a specific time warping function, called the beta function. Independent estimates of this beta function can be estimated from every observable pitch harmonic. This beta function

[1] We gratefully acknowledge Rome Laboratories, Contract No. F30602-91-C-0120 for their support of this work.

redundancy is used to *non-linearly time-warp* the speech signal such that the quasi-periodic speech waveform becomes periodic. After this *non-linear* processing, the speech (or other quasi-periodic signal) can be processed by standard *linear* techniques for enhancement or other signal processing ends. To date, the primary application for the PMMM has been speech enhancement for Speaker Identification (SI) systems.

The organization of this chapter is as follows. The PMMM is introduced in Section 2, and the spectral domain effect of its formulation is addressed in Section 3. PMMM parameter estimation specifics are addressed in Section 4. Application of the PMMM to a speaker identification problem is performed in Section 5. Other PMMM synthesis and other applications are discussed in Section 6. Summary and conclusion follow in Section 7.

2 THE PITCH MODE MODULATION MODEL

The Pitch Mode Modulation Model [13, 14, 15, 16] can be interpreted as a generalization of the Fourier Series representation for quasi-periodic signals. The PMMM is a decomposition of a speech signal, $s(t)$, into amplitude modulated (AM) and phase modulated (PM) sinusoidal components. A signal $s(t)$ can be represented as:

$$s(t) = \sum_{i=1}^{N} a_i(t)\cos(\phi_i(t)) = \sum_{i=1}^{N} Re\{a_i(t)e^{j\phi_i(t)}\} \qquad (16.1)$$

where $a_i(t)$ and $\phi_i(t)$ are the amplitude and phase modulation functions, respectively, of the ith harmonic component. For a given sound, the N function pairs $\{a_i(t), \phi_i(t)\}$ possess redundant information. This redundancy is due to the synchronous variation of all the components. This synchronous behavior can be understood by examining the PMMM form for the phase of the ith quasi-harmonic component:

$$\phi_i(t) \approx i\,\bar{\omega}_1[t + \beta(t) + \Theta_i] = i\,\bar{\omega}_1\tau + (i\bar{\omega}_1\Theta_i)\ \forall\ i, \qquad (16.2)$$
$$t \in (t_a, t_b],\ \tau \in (\tau_a, \tau_b],\ \beta(t_a) \triangleq 0,$$

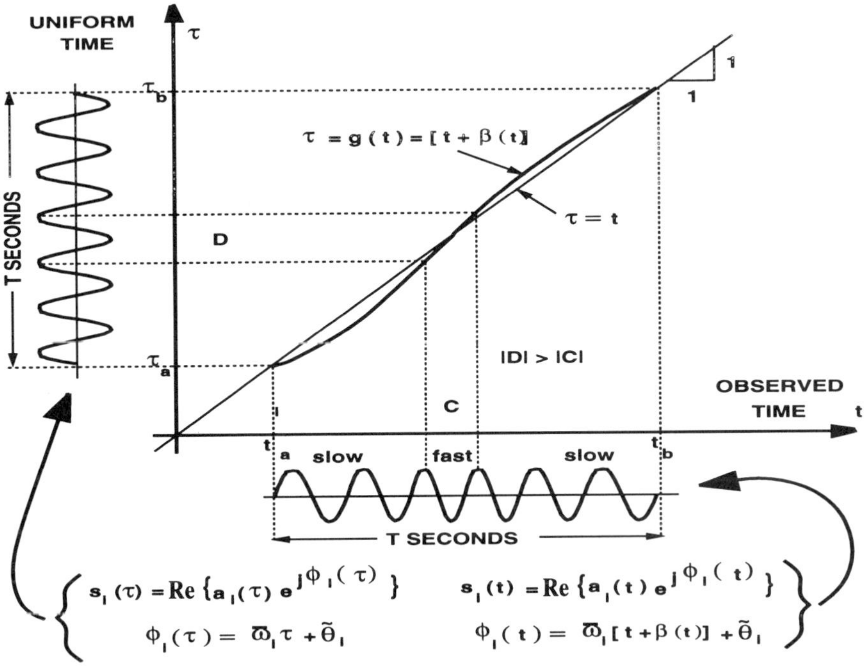

Figure 1 The Mapping $\tau = [t + \beta(t)]$

where $\bar{\omega}_1$ is the average fundamental (pitch) frequency for a time interval $(t_a, t_b]$, and Θ_i is the phase offset. The time warping function $\beta(t)$ is such that the mapping $\tau = g(t) = [t+\beta(t)]$ is a *non-linear* warping of the measured quasi-periodic signal into a periodic signal in τ with period $(2\pi)/\bar{\omega}_1$. This is illustrated in Figure 1.

Note that the time warping function $\beta(t)$ is *defined* as the *deviation* of the mapping $\tau = g(t)$ from the line $\tau = t$ during time interval $(t_a, t_b]$, and that $\beta(t_a) \stackrel{\Delta}{=} \beta(t_b) \stackrel{\Delta}{=} 0$. The variation of $\beta(t)$ is exceedingly small for voiced speech; typical values for $\max|\beta(t)| \approx 0.001*(t_b - t_a) \sim 0.005*(t_b - t_a)$ for voiced frame sizes $(t_b - t_a)$ of $35 \sim 200$ milliseconds. In other words, the plots of $\tau = t$ and $\tau = [t + \beta(t)]$ for voiced speech would be virtually indistinguishable if plotted together on the same scale.

Each PMMM component can be considered as a phase modulated sinusoid with carrier frequency $i\ \bar{\omega}_1$ and a phase deviation about that carrier. The spectral domain effect of a coherent $\beta(t)$ time warping on each pitch harmonic is a broadening of the bandwidth of the higher order quasi-harmonics. This important effect is detailed in the following section.

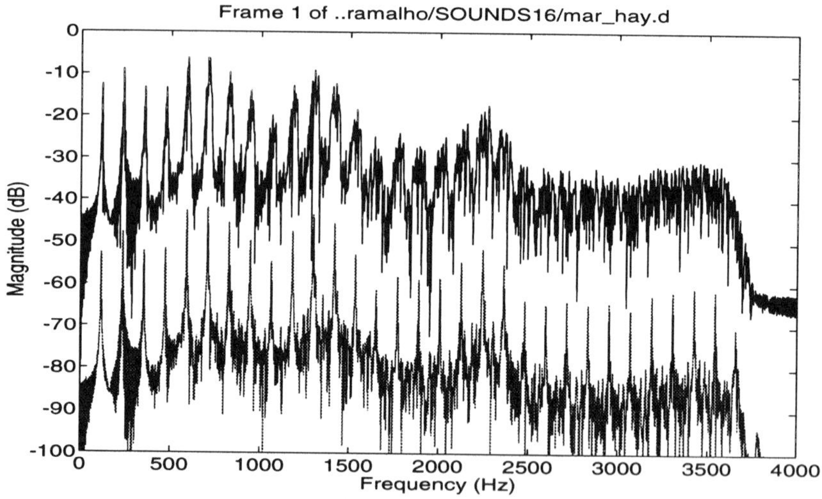

Figure 2 "How are you": Spectra of Original and De-Warped Spectra

As a demonstration of the power of PMMM processing, a a 4096 sample portion of the utterance "How *are* you?" was *de-warped* (i.e., made more periodic) according to the PMMM procedure to be described in Section 4. The portion of the utterance in *italics* was contained in the 4096 samples, and it was sampled at 16 kHz (256 msec of speech). The utterance was spoken naturally, with no effort to maintain pitch or speaking timbre. Figure 2 presents two spectra; the top spectra is from the original 4096 speech sample of the utterance, and the bottom is from the same 4096 sample of speech after the PMMM de-warping. The bottom spectra is plotted 40 dB below the top spectra for comparison purposes. Clearly seen is the sharpening of the pitch harmonics toward that of a line spectra[2]. The authors are unaware of any other long-term speech model that has accomplished a 256 msec de-warping as dramatic as this.

The present embodiment of PMMM parameter extraction is described in Section 4. In this section the PMMM phase modulation form, $\phi_i(t)$ will be justified. Prior to describing PMMM parameter extraction, however, it is instructive to analytically quantify the effect that a *coherent*, $\beta(t)$ induced modulation has on the spectrum of a voiced speech signal. This is the goal of the following section.

[2]For a perfect line spectra, the component amplitudes would be required to be constants (i.e. no amplitude modulation) and the 4096 samples would need to be an exact multiple of the fundamental period.

3 SPECTRAL IMPLICATIONS OF $\beta(T)$ MODULATION

This section addresses the effects that a *coherent*, $\beta(t)$ induced, phase modulation has on the spectrum of a voiced speech signal. To do this, we first assume that the PMMM form for the phase holds exactly for all quasi-harmonics, that is:

$$\phi_i(t) = i\,\bar{\omega}_1[t + \beta(t) + \Theta_i] \quad \forall\, i,\ t \in (t_a, t_b]. \tag{16.3}$$

This conjecture is then examined in the frequency domain to determine the spectral effect of the $\beta(t)$ induced modulation.

Figure 1 illustrates yet another way to interpret the mapping $\tau = g(t) = [t + \beta(t)]$. The phase modulation term of Equation 16.3, $(i\bar{\omega}_1\beta(t))$, can be viewed as the phase modulation function necessary to modulate each harmonic component of a purely periodic harmonic series of sinusoids into the set of quasi-harmonic sinusoids that are the PMMM decomposition of speech. To see this, the instantaneous frequency of the ith quasi-harmonic component (i.e., the time derivative of the phase function $\phi_i(t)$ [3]) is given by:

$$\omega_i(t) = \frac{d\phi_i(t)}{dt} = i\,\bar{\omega}_1 + i\,(\bar{\omega}_1\frac{d\beta(t)}{dt}). \tag{16.4}$$

Borrowing from communications theory lexicon, the first term can be interpreted as a carrier frequency $(i\ \bar{\omega}_1)$ and the second term as the frequency modulation about that carrier[4]. Since the $(\bar{\omega}_1\frac{d\beta(t)}{dt})$ part of the frequency modulation is common to all quasi-harmonics, the pitch harmonics with higher

[3] Strictly speaking, there are many definitions for the instantaneous frequency for both amplitude and frequency modulated signals. Most definitions approach, in some sense, the time derivative of the phase function for small amplitude modulation in the time interval considered. The small AM assumption will be justified in Section 6. For more information concerning instantaneous frequency definitions see [7, 3]. For details on *discrete-time* instantaneous frequency definition and computation see [19].

[4] It is recognized in communications literature that one cannot tell from a given waveform whether it has been phase or frequency modulated. The essential difference is the nature of the dependency on the modulating signal. The following analysis follows from a frequency modulation viewpoint.

center frequency carriers (higher harmonic number i) have a correspondingly higher instantaneous frequency deviation.

Phase Modulation (PM), unlike amplitude modulation, is a *nonlinear* form of modulation. It is not possible to evaluate the spectra of any PMMM component for an arbitrary $\beta(t)$ because the superposition principle does not apply [18, 20]. In AM modulation, if one knows the spectra for the modulating signal, then the spectra of the modulated signal is a frequency translated version of this spectra (due to the modulation property of the Fourier Transform). If the AM modulating signal can be decomposed into a *linear sum* of elementary functions (e.g., a harmonic or power series), the spectrum of the modulated signal would be a frequency shifted version of the *linear sum* of their respective spectra. No such decomposition for the purposes of estimating PM spectra is possible. Theoretically, one would have to perform an analysis for every $\beta(t)$ one would expect to encounter.

In order to gain analytical and mathematical insight to this type of modulation, we are forced to consider a specific form for the $\beta(t)$ modulation in which a closed form is known to exist – the case of sinusoidal modulation. Thereafter, the results for this case will be heuristically generalized in a completely analogous manner to the case of Frequency or Phase Modulation in communications theory.

3.1 Analysis for Sinusoidal $\beta(t)$

Consider the specific case where the modulation function $\beta(t)$ is sinusoidal. From communications theory, we know that if $\beta(t)$ is sinusoidal with frequency ω_{mod}, (i.e. $\beta(t) = \alpha \sin(\omega_{\text{mod}} t)$), then the resultant spectra would have 'sidebands' spaced symmetrically on either side of the pitch 'carrier' $(i\bar{\omega}_i)$ at frequency separations ω_{mod}, $2\omega_{\text{mod}}$, $3\omega_{\text{mod}}$, etc. [20, 18]. This is illustrated in Figure 3.1 where the magnitude of the sidebands are labeled J_1, J_2, etc.. For the case of coherent $\beta(t)$ modulation, consider a quasi-harmonic component with $\beta(t) = \alpha \sin(\omega_{\text{mod}} t)$. The resulting quasi-harmonic component is given by:

$$s_i(t) = a_i(t)\text{Re}[e^{j\, i\bar{\omega}_1(t+\Theta_i)} e^{j\, i\bar{\omega}_1 \alpha \sin(\omega_{\text{mod}} t)}]. \qquad (16.5)$$

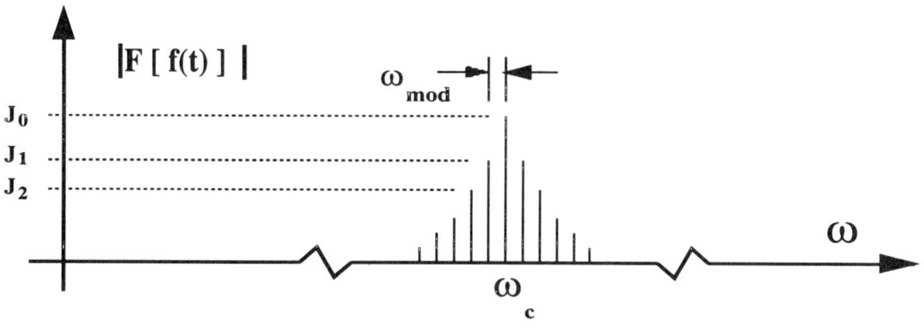

Figure 3 Illustrative magnitude line spectra for FM signals with sinusoidal modulation

We will find it useful to define $\psi = i\bar{\omega}_1\alpha$ as the modulation index. The right-most complex exponential of the previous equation can be expanded into its Fourier series representation as:

$$e^{j\, i\bar{\omega}_1\alpha\, \sin(\omega_{\mathrm{mod}}t)} = e^{j\psi\, \sin(\omega_{\mathrm{mod}}t)} = \sum_{n=-\infty}^{\infty} J_n(\psi) e^{jn\omega_{\mathrm{mod}}t}. \tag{16.6}$$

$J_n(\psi)$ is the Fourier Series coefficient for a given ψ and is computed by:

$$J_n(\psi) = \frac{1}{T_{\mathrm{mod}}} \int_{-T_{\mathrm{mod}}/2}^{T_{\mathrm{mod}}/2} e^{j\psi \sin(\omega_{\mathrm{mod}}t)} e^{-jn\omega_{\mathrm{mod}}t} dt, \tag{16.7}$$

where T_{mod} is the period of ω_{mod}. Making a change of variables $\xi = \omega_{\mathrm{mod}}\, t = (2\pi/T_{\mathrm{mod}})\, t$, we get:

$$J_n(\psi) = \frac{1}{2\pi} \int_{-\pi}^{\pi} e^{j(\psi \sin\xi - n\xi)} d\xi\ . \tag{16.8}$$

This integral equation occurs in many physical problems and has been tabulated extensively. It is a function of n and ψ, denoted by $J_n(\psi)$, and is called the Bessel function of the first kind (signified by the "J") of order n and argument ψ.

Rewriting Equation 16.5 by using the expansion of Equation 16.6 results in:

$$s_i(t) = a_i(t)\mathrm{Re}\left[e^{j\, i\bar{\omega}_1(t+\Theta_i)} \sum_{n=-\infty}^{\infty} J_n(\psi)e^{jn\omega_{\mathrm{mod}}t}\right], \tag{16.9}$$

$$= a_i(t) \sum_{n=-\infty}^{\infty} J_n(\psi)\cos(\,[i\bar{\omega}_1 + n\omega_{\mathrm{mod}}]t + i\bar{\omega}_1\Theta_i)\,, \tag{16.10}$$

the last step being allowed because the Bessel functions, $J_n(\psi)$, are real. Furthermore, due to the fact that $J_n(\psi) = J_{-n}(\psi)$ for n even, and $J_n(\psi) = -J_{-n}(\psi)$ for n odd, we can further expand Equation 16.10 as:

$$\begin{aligned} s_i(t) &= a_i(t)\,\{\; J_0(\psi)\;[\cos(i\bar{\omega}_1 t + i\bar{\omega}_1\Theta_i)] \qquad (16.11)\\ &+ J_1(\psi)\;[\cos([i\bar{\omega}_1 + \omega_{\mathrm{mod}}]t + i\bar{\omega}_1\Theta_i) - \cos([i\bar{\omega}_1 - \omega_{\mathrm{mod}}]t + i\bar{\omega}_1\Theta_i)]\\ &+ J_2(\psi)\;[\cos([i\bar{\omega}_1 + 2\omega_{\mathrm{mod}}]t + i\bar{\omega}_1\Theta_i) + \cos([i\bar{\omega}_1 - 2\omega_{\mathrm{mod}}]t + i\bar{\omega}_1\Theta_i)]\\ &+ J_3(\psi)\;[\cos([i\bar{\omega}_1 + 3\omega_{\mathrm{mod}}]t + i\bar{\omega}_1\Theta_i) - \cos([i\bar{\omega}_1 - 3\omega_{\mathrm{mod}}]t + i\bar{\omega}_1\Theta_i)]\\ &+ \;\;\ldots\;\}. \end{aligned}$$

This is what we set out to prove. Namely, that if $\beta(t)$ is of the form of a sinusoid of frequency ω_{mod}, the resultant spectra would have 'sidebands' symmetrically on either side of the pitch 'carrier' ($i\bar{\omega}_i$) at frequency separations ω_{mod}, $2\omega_{\mathrm{mod}}$, $3\omega_{\mathrm{mod}}$, etc. We now see that the number of these sidebands is infinite, and the magnitude of a given sideband is given by the solution of the Bessel function of the first kind for the specific modulation index, $\psi = i\bar{\omega}_1\alpha$, and order, n. This is in sharp contrast to Narrowband AM modulation. If the amplitude of a constant frequency sinusoid of frequency $\omega_{carrier}$ was modulated at a frequency ω_{mod} ($\omega_{\mathrm{mod}} \ll \omega_{carrier}$), the resultant spectra would only have *one* sideband no matter how large the modulation. AM is called a *linear* form of modulation, while angle modulation (FM or PM) is inherently *nonlinear*.

Although the PM introduced by $\beta(t)$ theoretically introduces an infinite number of sidebands, the magnitudes of the higher-order sidebands become negligible and, for all practical purposes, the power is contained within a finite bandwidth. To show this, Figure 4 shows a plot of the coefficients $J_n(\psi)$ plotted as a function of the modulation index ψ.

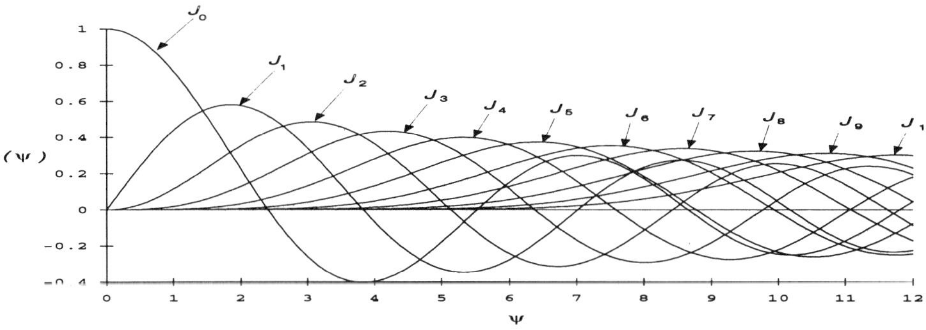

Figure 4 $J_n(\psi)$ as a Function of Modulation Index ψ

For small frequency modulation indices, $\psi = i\bar{\omega}_1\alpha \leq 0.2$, the modulation is approximated well as a *linear* form of modulation. This is because the only significant $J_n(\psi)$ are $J_0(\psi)$ and $J_1(\psi)$, i.e. only one significant sideband. This case is referred to as Narrowband Frequency Modulation in communications literature.

However, for large modulation indices the modulation becomes increasingly nonlinear, and the number of sidebands of significant energy grows. The case of $\psi = i\bar{\omega}_1\alpha > 0.2$ is referred to as Wideband Frequency Modulation in communications literature. Depending on the modulation index $\psi = i\bar{\omega}_1\alpha$, the low harmonic index pitch harmonics may be approximated well by narrowband PM, while for increasingly higher pitch harmonics the modulation is increasingly wideband. This will be illustrated shortly.

Two more points should be made in regard to the Bessel functions, $J_n(\psi)$, and angle modulation in particular. The Bessel functions are normalized such that $J_0^2(\psi) + 2J_1^2(\psi) + 2J_2^2(\psi) + 2J_3^2(\psi)... = 1$ for any ψ. This equality states that the sum of the power in the sidebands (whose power is proportional to their $J_n^2(\psi)$, $n \neq 0$) plus the power in the carrier ($J_0^2(\psi)$) equals the power of an unmodulated carrier ($J_0^2(0) = 1$ and $J_n^2(0) = 0$, $\forall\, n \neq 0$), or

$$J_0(0) = 1 = \left[J_0^2(\psi) + 2\sum_{n=1}^{\infty} J_n^2(\psi) \right]. \tag{16.12}$$

Thus the power which was concentrated at a particular frequency of an unmodulated sinusoid is "spread out" over the sidebands where $J_n^2(\psi)$ is significant[5].

To illustrate the power of PMMM-based de-warping in estimating the quasi-harmonic component parameters, the table below presents the expected sideband magnitudes for a PMMM component $s_i(t)$ and warping of the form $\beta(t) = \alpha \sin(\omega_{\text{mod}} t)$. The pitch frequency used for this example is 100 Hz., the magnitude of the warping function, α, is set to a typical value of 0.1%, and the modulation frequency, ω_{mod}, is assumed to be small compared to the pitch ($\omega_{\text{mod}} << \bar{\omega}_1$).

$$\textbf{CASE:}\quad s_i(t) = \cos\left[\, i\bar{\omega}_1(t + \alpha \sin(\omega_{\text{mod}} t) + \Theta_i) \,\right]$$

FIRST HARMONIC (100Hz) CASE: $i\bar{\omega}_1\alpha = 0.628$ $(i = 1, f_0 = 100\text{Hz}, \alpha = 0.001)$	FIFTEENTH HARMONIC (1500Hz) CASE: $i\bar{\omega}_1\alpha = 9.42$ $(i = 15, f_0 = 100\text{Hz}, \alpha = 0.001)$	
$J_0^2(0.628) = 0.99$	$J_0^2(9.3) = 0.16$	$2J_6^2(9.3) = 0.14$
$2J_1^2(0.628) = 0.11$	$2J_1^2(9.3) = 0.20$	$2J_7^2(9.3) = 0.30$
$2J_2^2(0.628) = 0.00$	$2J_2^2(9.3) = 0.20$	$2J_8^2(9.3) = 0.31$
$\Uparrow$ 99%*Energy*	$2J_3^2(9.3) = 0.11$	$2J_9^2(9.3) = 0.24$
	$2J_4^2(9.3) = 0.27$	$2J_{10}^2(9.3) = 0.14$
	$2J_5^2(9.3) = 0.12$	$\Uparrow$ 98 % Energy

For these values, the first harmonic undergoes relatively narrowband phase modulation (only one significant sideband) and the fifteenth (at only 1500 Hz) undergoes wideband modulation with 10 significant sidebands! Figure 5 diagrammatically shows this case and what were to happen if the PMMM warped signal was de-warped with the correct $\beta(t)$. The power of the fifteenth harmonic spread out over the 10 sidebands about 1500Hz would become compressed at 1500 Hz resulting in a 10 dB magnitude gain in the magnitude!

Magnitude gain of this order is possible for suitably long frame sizes. In fact, if the original and PMMM de-warped magnitude spectra of Figure 2 were plotted on top of one another, the harmonic peaks of the de-warped spectra would be approximately 10 dB higher than the original spectra in the 3500 Hz range. This magnitude gain is also useful for extracting the PMMM model parameters in the presence of noise. The PMMM de-warping procedure does not signif-

[5] See [20] for detailed analysis and [11] for proof of the Bessel function identity in Equation 16.12.

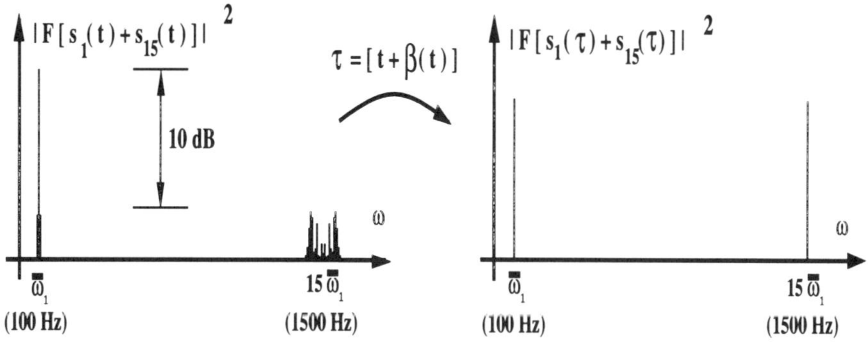

Figure 5 Illustration of PMMM Magnitude Gain

icantly change the magnitude response for the broadband noise components of signals, yet the de-warping dramatically compresses the bandwidth of the higher order components - literally pulling them out from the broadband noise floor. This sometimes makes the estimation of these component magnitudes possible, where they were previously hidden beneath the broadband noise.

The Short Term Fourier Transform (STFT) magnitude of voiced and mixed-voice speech show the trend of increasing harmonic bandwidth with increasing harmonic number. One may expect that the spectral shapes should be somehow similar since the identical form of the modulation (up to multiplication by the harmonic number) is identical for all harmonics. Dissimilar harmonic spectra have lead many speech coding researchers to (incorrectly) surmise that the excitation for these frequency bands of dissimilar spectra were created by a different excitation than the other bands[6]. Because the $\beta(t)$ modulation is a *nonlinear* form of modulation, we *should not* assume similar spectral shapes, as superposition arguments do not hold. We should instead *expect* significant visual differences in the individual harmonic spectra[7].

The main point of this discussion on spectra is to note that it is unreasonable to expect similar spectral shapes for the case of quasi-harmonics that have been modulated in a wideband manner. The above analysis considered the case of sinusoidal $\beta(t)$ and made some observations about what we should expect for the spectra of $s(t)$. Fortunately, frequency modulation practices in communications show that these results, called tone modulation in communications,

[6] For example, the Multiband Excitation Coder [6].

[7] To further confuse things, it is well recognized that similar spectra can be created by different forms of modulation, and that significantly different spectra can be formed by seemingly similar modulation forms [10, 21].

and observations can be heuristically extended to the case of an arbitrary $\beta(t)$ modulation (see [18, 20]).

4 PMMM PARAMETER ESTIMATION ALGORITHM

The PMMM processing exploits the fact that a subset of independent time warping estimates can be intelligently combined to provide a robust estimate of $\beta(t)$. Finding $\beta(t)$ is key in the PMMM paradigm and is tantamount to demodulating each quasi-harmonic. The bandwidths for the higher harmonic number quasi-harmonics may overlap in frequency, as was shown in Figure 2. This makes demodulation of these harmonics in one step impossible and iterative techniques to refine the $\beta(t)$ estimate are required.

Immediately below we present an example to justify the previous statement that the PMMM form for the phase function holds approximately for all quasi-harmonic components of voiced speech. In the subsection following this, the numerical specifics for the PMMM parameter extraction are presented.

4.1 Beta Function Extraction Example

As a simple example of the statements already made, Figure 6 presents the magnitude of a 4096 point FFT of an vowel /æ/, as in the word "bat", sampled at 16 kHz. This particular utterance is unusually long (256 milliseconds) and spoken with an effort to maintain a steady pitch. As a result, all quasi-harmonics are able to be demodulated without resorting to the iterative techniques to follow in Section 4.3. Although not typical in this regard, it is shown here to illustrate the redundancy of voiced speech.

Figure 7 presents the $\beta_i(t)$ estimates for the first ten components of the same utterance as the Short Term Fourier Transform (STFT) of Figure 6. The time warping functions, $\beta_i(t)$ $i = 1...10$, estimated from the components of this utterance are consistent, that is, they tend to cluster about an average $\beta(t)$. A representative time warping function, $\beta(t)$, can be obtained from this cluster of $\beta_i(t)$'s such that Equation 16.2 holds approximately for all quasi-harmonic components. Note that the *deviation* from uniform time necessary to make the quasi-harmonics periodic is very small, about 0.2% (about eight samples out of the 4096 sample duration).

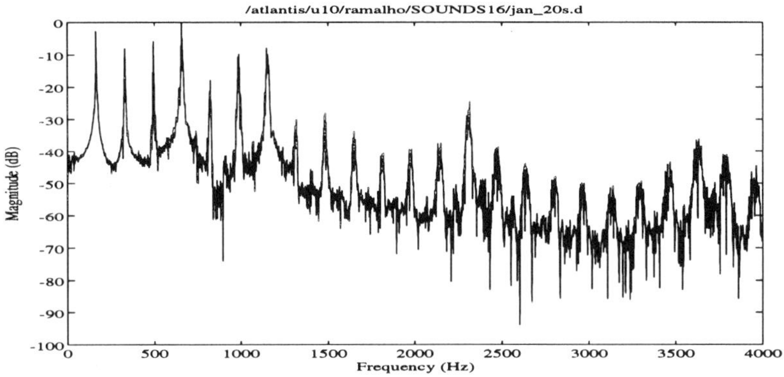

Figure 6 STFT Magnitude of vowel /æ/ in "bat" (4096 samples @ 16kHz sampling)

The beta function extraction methodology exploits the fact that $\beta_i(t)$'s that are associated with the wanted quasi-harmonic sound will tend to cluster. In the procedure that follows, all Fourier domain peaks are demodulated as potential harmonic components and $\beta_i(t)$'s are calculated. Then a subset of those $\beta_i(t)$'s that cluster will be classified as components belonging to the wanted quasi-harmonic sound. A representative $\beta_i(t)$ is created from this subset and is used to de-warp the sound.

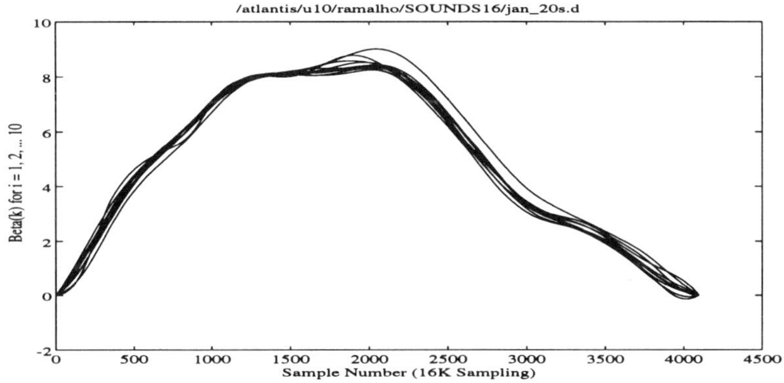

Figure 7 $\beta_i(t)$ Estimates for $i = 1, 2, \ldots 10$

4.2 Beta Function Extraction Procedure

Independent $\beta_i(t)$ estimates are found for each peak in the Fourier spectrum by solving: $\phi_i(t) = \bar{\omega}_i[t+\beta_i(t)+\Theta_i]$ for $\beta_i(t)$. Each $\phi_i(t)$ estimate is calculated from the complex-valued analytic signal, obtained by single-side bandpass filtering about a Fourier domain spectral peak. Specifically, a frame of speech is first windowed using a Hamming window and a Fast Fourier Transform (FFT) is computed. Then the FFT is single-side bandpass filtered about the positive frequency component peak and an Inverse FFT is performed to obtain the analytic signal $S_i(k) = a_i(k)e^{j\phi_i(k)}$ (with an upper case S). The phase estimates $\phi_i(k)$ from only the center half of the Hamming windowed frame are used[8]. The average frequency, $\bar{\omega}_i$, of the component is calculated as:

$$\bar{\omega}_i = \frac{1}{|t_b - t_a|}\int_{t_a}^{t_b} \dot{\phi}_i(t)dt \tag{16.13}$$

where $\dot{\phi}_i(t)$ is the derivative of $\phi_i(t)$ with respect to time. Numerically, the derivative is approximated by a first difference[9] and the integration is performed using the trapezoidal rule.

A $\beta_i(k)$ is found by solving the equation $\phi_i(k) = \bar{\omega}_i[k + \beta_i(k) + \Theta_i]$ for $\beta_k(k)$ subject to the constraint $\beta_i(k_{\text{begin}}) = 0$[10]. The $\beta_i(k)$ so constructed from the $\phi_i(k)$ will also satisfy the constraint $\beta_i(k_{\text{end}}) = 0$.

It has been found via experimentation that the time warping estimates can be improved if more than one estimate is made for the Fourier domain peak considered. To do this, we form three different analytic signals from the same Fourier domain peak, each one using a slightly different bandwidth. The largest bandwidth used, in number of FFT frequency bins, is taken to be the $2L$, where L is the first local minimum of:

[8] Using synthetic signals it has been found that this method of obtaining the *phase* of a known frequency modulated signal was superior to most other methods and is relatively straightforward to implement. Later, when we are interested in estimating the *amplitude* of a harmonic, a different windowing strategy will be employed.

[9] For high noise environments a method for obtaining a robust $\beta(t)$ from noisy $\phi(t)$ estimates has been developed (see [15]).

[10] The discrete time index is k and the sample indices k_{begin} and k_{end} are used to denote the first and last sample in a given $\phi_i(k)$ frame.

$$BW(L,k) = \left[\left(\frac{1}{psd(k)} \sum_{l=-L}^{L} psd(k-l) \right) - 1 \right] /2L \tag{16.14}$$

as L is increased from 1 and $psd(k)$ is the magnitude of the FFT bin peak considered. The two other bandwidths used were integer approximations of $4L/3$ and $2L/3$. The time warping estimates so obtained should be substantially similar to each other if the peak was a pitch harmonic. It has been found that if this peak was a spurious noise peak, the time warping estimates will not be similar to each other – because noise is poorly modeled as a frequency modulated signal. If the three $\beta(t)$ estimates are not substantially similar, the $\beta(t)$ for this Fourier domain peak is discarded. If the three $\beta(t)$ estimates are substantially similar, the $\beta(t)$ estimates for this spectral peak are averaged and kept as a potential $\beta_i(t)$.

A 3-class K-Means clustering [5] algorithm was designed to obtain a representative $\beta(t)$ from the $\beta_i(t)$'s. Two classes are initialized with the $\beta_i(t)$ from the most positive and from the most negative extremes. The remaining class is initiated with a vector of all zeros. After the K-means algorithm converges, the cluster with the most elements in it is chosen as the representative $\beta(t)$ class. To further improve the $\beta(t)$ estimate, a tightly clustered subset of M of the $\beta_i(t)$'s are averaged to obtain the representative $\beta(t)$ using the following algorithm. $\mathcal{L}_1$ distances from the class mean are calculated, sorted with ascending distance into a vector d, and the vectors with the largest distances are eliminated until $\left[2\sum_{i=1}^{M-1} d(i)\right]/(M-1)$ is no longer less than $d(M)$. The $\beta(t)$ is then taken as the average of the M time warping functions $\beta_i(t)$'s.

Figure 8 presents an example of the beta clustering strategy. This figure presents the beta function estimates for all the pitch peaks of a 2048 sample segment of the utterance "How are you?", sampled at 16 kHz. Previous to generating these estimates, the signal had been iteratively de-warped ten times to make the utterance as periodic as possible. This allows for *all 29* of the pitch peaks from 0 to 3500 Hz to be demodulated. Some of the pitch peaks demodulated have signal power 40 dB down from the peak signal. As a result, some beta function estimates are impaired by the noise, resulting in a noisy beta function estimate. The beta functions have been classified into three classes, the beta functions depicted in these classes are plotted with different line characteristics. For example, the class with the largest positive beta extremes are plotted as dotted lines. Clearly seen is the "middle" cluster of consistent beta estimates. The final beta function estimate used for this utterance is the av-

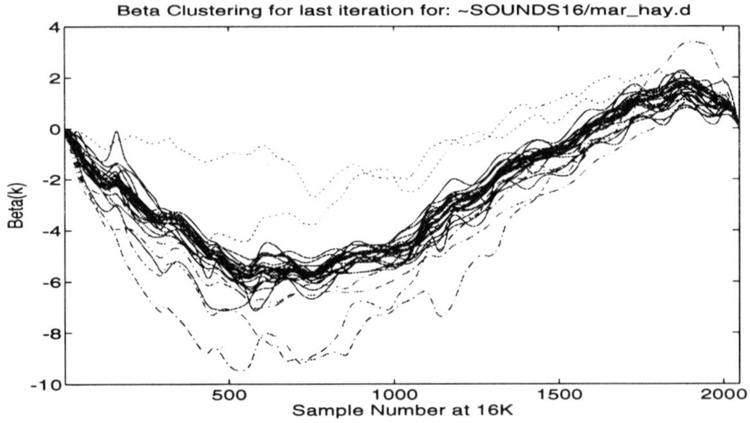

Figure 8 Beta clustering example for 2048 sample segment of "How are you?"

erage of the M beta functions using the algorithm above. For this particular case, M was equal to 9 and the $\beta_i(k)$'s corresponded to harmonic numbers 7, 8, 9, 15, 16, 24, 25, 27, and 28 of the 29 possible harmonics in the 0 to 3500 Hz range.

The $\phi_i(t)$'s that correspond to this subset of M time warping functions are, with high confidence, a proper subset of the desired N components of the desired quasi-harmonic sound. The average frequencies of those M quasi-harmonics, $\bar{\omega}_i$, are found to be harmonically related to a very high accuracy. Experimentation with the M average frequency estimates, $\bar{\omega}_i$, has shown that if $\bar{\omega}_A$ is the average radian frequency of the Ath harmonic component and $\bar{\omega}_B$ is that of the Bth harmonic, then:

$$\frac{\bar{\omega}_A}{A} = \frac{\bar{\omega}_B}{B} \pm X\% \tag{16.15}$$

where X is dependent on the frame size. X is approximately 1% for 32 millisecond frame sizes and decreases (i.e., improves) for larger frames. For a wholly voiced 256 millisecond utterance analyzed in Figure 2, X was less than 0.1%. Knowing this property, algorithms can be designed that compute the common integral factor of these M component frequencies and one such algorithm can be found in [15].

The ability of this technique to obtain $\beta(t)$ in the presence of realistic noise marks a significant difference to other, independently derived, speech research which use time-warping principles to make quasi-periodic pitch harmonics periodic [1, 4, 17]. If, for example, narrowband noise or other frequency selective impairments contaminate the speech signal, an accurate estimate of $\beta(t)$, and subsequently the pitch, are still obtained from the quasi-harmonics not so impaired. This robustness of the pitch estimate is significant, as the performance of many speech coding [1, 9] and enhancement techniques [8] are critically dependent on accurate fundamental frequency tracking. Also note that the $\beta(t)$ processing above allows for the identification of pure tones since their $\beta_i(t) = 0$.

4.3 Iterative Parameter Estimation

An iterative technique for obtaining $\beta_i(t)$ estimates is required because the spectra of the higher order quasi-harmonics typically overlap in frequency. The strategy used is the following.

A $\beta(t)$ is estimated from the components that can be demodulated and relation $\tau = [t + \beta(t)]$ is applied to de-warp $s(t)$ to become more periodic. After this de-warping, the pitch harmonics should have become more sharpened (their deviation from their harmonic carrier frequencies should be lessened) and additional pitch peaks may now be able to be demodulated. A $\delta\beta(t)$ is estimated that, when added to the previous $\beta(t)$, serves to make the quasi-harmonics more periodic. This process is repeated until all pitch harmonics are of uniform bandwidth.

A block diagram of this processing is shown as the solid blocks and lines of Figure 9. The solid lines and blocks of Figure 9 depict the iterative procedure to obtain the necessary PMMM parameters for a given frame j: the time-warping function, $\beta_j(k)$, and the average fundamental frequency, $\bar{\omega}_{1_j}$.

Since the $\beta(t)$ estimation is iterative, it is not required to converge to the optimal $\beta(t)$ in one step. For clean 8 KHz speech, it is often possible for the iterative process to converge in no greater than three iterations. For highly mixed-voice or for noisy environments, it may be advisable to purposely slow the convergence process in order to help assure the correct $\beta(t)$ result[11].

The de-wapring procedure, in discrete time, is an interpolation processes. The de-warping procedure interpolates new signal samples at the new (discrete time)

[11] See [15] for details.

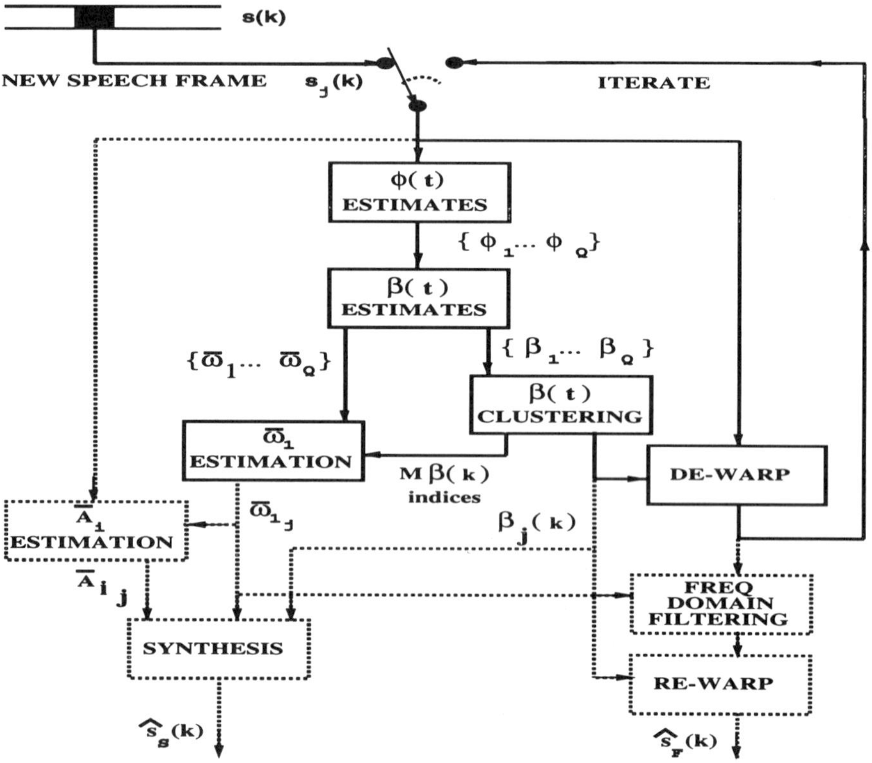

Figure 9 PMMM Iterative Parameter Estimation Block Diagram

uniform time sample points τ_k[12]. Computer simulations have shown that an interpolation "granularity" to the nearest $1/50^{\text{th}}$ of a sample yields an interpolation error (after warping) which is flat and less than 60 dB below the peak signal magnitude. Many such iterative warpings, as inferred by Figure 9, would lead to an unacceptable error since the warping errors are cumulative. In practice, the de-warping is accomplished by warping the *original* speech frame by the combination (i.e., sum) of all the warpings to the present iteration. This procedure, with $1/50^{\text{th}}$ interpolation granularity, insures that all interpolation errors are insignificant.

[12] The uniform time base sample index is denoted by τ_k.

5 PMMM SPEECH ENHANCEMENT RESULTS

Presented here are simulation results for PMMM-based enhancement for a speaker identification (ID) application. For this application, the de-warped speech frame is enhanced via frequency domain comb filtering and then *re-warped* to re-introduce the quasi-periodicity present in the original speech. The comb filter utilized Tukey windows ($\alpha = 0.5$) of bandwidth $\approx$ 31 Hz and centered on the harmonics of $\bar{\omega}_{1_j}$. These operations are shown as the dotted lines and blocks in Figure 9 and the resultant speech is denoted $\hat{s}_F(k)$.

After this PMMM enhancement, the speech is classified by a cepstrum-based speaker identification system [2]. This system includes pre-emphasis by $H(z) = 1 - 0.95z^{-1}$, 12-order LPC-based cepstrum generation, and LVQ clustering. A 30 ms Hamming window is applied every 10 ms. The speech corpus for this application is a subset of the DARPA TIMIT database that included 20 speakers, 10 male and 10 female, from the New England dialect. The speech was sampled at 8 KHz. Three types of noise were considered: a white Gaussian noise, babble noise and a low-pass colored noise. The babble noise was created by summing the speech of 10 male and 10 female speakers of the TIMIT database after silence removal and signal power normalization. The colored noise was created by filtering white noise by a cascade of two low-pass filters designed by the Hamming window technique with orders of 3 and 5 and cutoff frequencies of 50 and 1500 Hz respectively.

Non-Enhanced Speech: Figure 10 presents the probability of correct identification, $P_{correct}$, against SNR for the 20 speaker identification task for the three types of noise considered. When trained on clean speech, the performance in white noise degrades much more quickly than for the colored or babble noise as the noise level of the respective noise sources are increased. White noise at a level of 25 dB SNR, while subjectively annoying, does not impair human listeners in identifying speakers.

In order to improve the robustness with respect to SNR for white noise, the same system was trained with 25 dB of white noise added to the training set. These results are shown in Figure 11. While the white noise performance at and near 25 dB improves to 100% and the babble noise at low SNR improves, the performance in colored noise, high SNR babble noise and clean testing data degrades. This performance characteristic has been noted previously [12]. The inadequacy comes from the fact that the noise characteristics are seldom known *a priori.*

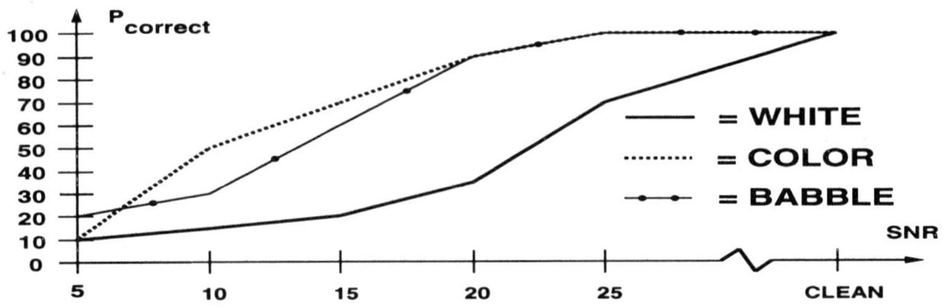

Figure 10 Training on Clean Speech (No enhancement)

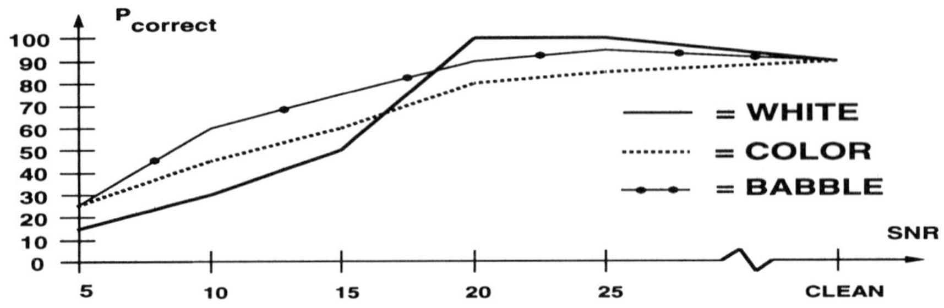

Figure 11 Training on 25 dB White Noise (No enhancement)

PMMM Enhanced Speech: Figure 12 presents the performance of the system trained with enhanced versions of the clean speech and tested with the enhanced versions of the testing set with their respective noise sources. The performance of the white noise and the babble testing set improves dramatically over that of Figure 10. Additionally, there is a small performance improvement for the colored noise as well as 100% correct identification for clean speech.

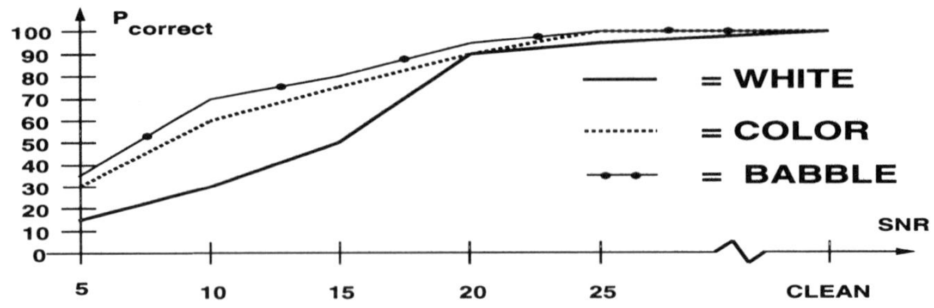

Figure 12 Training on Enhanced Clean Speech

6 PMMM-BASED SPEECH SYNTHESIS

Speech enhancement via synthesis from PMMM parameters is also possible. For this case, the quasi-harmonic component amplitudes, $a_i(t)$, are calculated from analytic signals formed with Tukey windows about the harmonics of $\bar{\omega}_{1_j}$ from the de-warped speech. For enhancement, the de-warped analytic signal bandwidth used is approximately 31 Hz. For speech coding application in low noise environments, a wider bandwidth may be used. The resultant synthesized speech signal is denoted $\hat{s}_F(k)$ in Figure 9 and was synthesized on a frame by frame basis according to:

$$\hat{s}_{S_j}(k) = \sum_{i=1}^{N} a_{i,j}(k) \cos(i\,\bar{\omega}_{1_j}[k + \beta_j(k)] \; + K_{i,j}). \tag{16.16}$$

where j is the frame index. The constants $K_{i,j}$ are adjusted such that the cosine argument is smooth across frames - no attempt was made to recreate the original relative phase terms across components (i.e., the $i\,\bar{\omega}_1 \Theta_i$ terms in Equation 16.2). Experimentation has found that the $a_i(t)$ are low frequency waveforms and synthesis with $a_i(t)$ bandlimited to 31 Hz results in a the synthesized signal $\hat{s}_S(k)$ sounding identical to the comb-filter enhanced signal $\hat{s}_F(k)$. Thus, for synthesis applications, the $a_{i,j}(k)$ can simply be formed by interpolation from the 31 Hz bandlimited component amplitudes.

PMMM synthesis (and comparisons to of it to other harmonic-based speech coding techniques), and PMMM application to other speech applications (e.g., time-scale and pitch-scale modification of speech, speaker separation, etc.) is contained in [15].

7 SUMMARY AND CONCLUSION

The Pitch Mode Modulation Model is a long-term speech model that is able to model the non-stationarity found in normal voiced and mixed voice speech. The PMMM components are found to vary in time synchrony with a specific time warping function, $\beta(t)$, which causes a coherent phase modulation of the components. Its major attributes as a long-term model are:

- No specific parameterization of the time warping function (e.g., a time series) was used as a model constraint in order to estimate $\beta(t)$.
- No specific parameterization of the amplitude of the PMMM components (e.g. constants) was used as a model constraint.
- No sensitivity of the PMMM technique to *a priori* pitch estimates, as none are used.
- The PMMM is the first long-term model able to associate, on the harmonic-level, quasi-harmonic speech components and disassociate them from non-speech components.
- The PMMM is the first mathematical formulation that describes the spectral effect of a coherent time warping of PMMM components using phase modulation principles.
- A robust pitch estimate is obtained *after* PMMM processing, using signal components *known* to be correlated (in time modulation characteristic) to other speech components.
- The PMMM is the first technique to obtain and qualitatively describe a spread spectrum-like processing gain for amplitude estimation of PMMM components.

PMMM enhancement was demonstrated in a text independent speaker identification application and was shown to reduce the sensitivity of the system to mismatches in the noise environment between training and testing data. Other PMMM synthesis applications were also discussed.

REFERENCES

[1] L. B. Almeida and J. M. Tribolet, "Nonstationary Spectral Modeling of Voiced Speech", *IEEE Trans. Acoust., Speech, Signal Processing*, vol. ASSP–31, no. 3, pp. 664–677, June 1983.

[2] K. T. Assaleh, R. J. Mammone, and J. L. Flanagan, "Speech Recognition Using the Modulation Model", *Proc. ICASSP–93*, vol. 2, pp. 664–667, 1993.

[3] B. Boashash, editor, *Time-Frequency Signal Analysis, Methods and Application*, New York, NY: John Wiley & Sons, 1992.

[4] D. Chazan, Y. Stettiner, and D. Malah, "Optimal Multi-Pitch Estimation Using the EM Algorithm for Co-Channel Speech Separation", *Proc. ICASSP–93*, vol. 2, pp. 728–731, 1993.

[5] A. Gersho and R. M. Gray, *Vector Quantization and Signal Compression*, Boston, MA.: Kluwer Academic Publishers, 1992.

[6] D. W. Griffin and J. S. Lim, "The Multiband Excitation Vocoder", *IEEE Trans. Acoust., Speech, Signal Processing*, vol. 36, no. 8, pp. 1223–1236, August 1988.

[7] G. Jones and B. Boashash, "Instantaneous Frequency, Instantaneous Bandwidth and the Analysis of Multicomponent Signals", *Proc. ICASSP–90*, vol. E, pp. 2467–2470, 1990.

[8] J. S. Lim, *Speech Enhancement*, Englewood Cliffs, NJ: Prentice-Hall, 1983.

[9] R. J. McAulay and T. F. Quatieri, "Speech Analysis/Synthesis Based on Sinusoidal Representation", *IEEE Trans. Acoust., Speech, Signal Processing*, vol. ASSP–34, no. 4, pp. 744–754, August 1986.

[10] A. Papoulis, "Random Modulation: A Review", *IEEE Trans. Acoust., Speech, Signal Processing*, vol. ASSP-31, no. 1, pt. 1, pp. 96–105, February 1983.

[11] L. A. Pipes, *Applied Mathematics for Engineers and Physicists*, New York, NY: McGraw-Hill, 1958.

[12] L. R. Rabiner and B-H. Juang, *Fundamentals of Speech Recognition*, Englewood Cliffs, NJ: Prentice-Hall, 1993.

[13] M. A. Ramalho and R. J. Mammone, "New Speech Enhancement Techniques Using the Pitch Mode Modulation Model", Rutgers University, Piscataway, NJ, CAIP Technical Report, CAIP–TR–155, February 12, 1993.

[14] M. A. Ramalho and R. J. Mammone, "New Speech Enhancement Techniques Using the Pitch Mode Modulation Model", *Proc. of the* 36th *Midwest Symposium on Circuits and Systems*, Detroit, MI, August 16–18, 1993.

[15] M. A. Ramalho, *The Pitch Mode Modulation Model with Applications in Speech Processing*, doctoral dissertation, Rutgers University, New Brunswick, NJ, January 1994.

[16] M. A. Ramalho and R. J. Mammone, "A New Speech Enhancement Technique with Application to Speaker Identification", *Proc. ICASSP-94*, vol. 1, pp. 29–32, 1994.

[17] Y. Stettiner, D. Malah, and D. Chazan, "Estimation of a Long-Term Model for Accurate Representation of Voiced Speech", *Proc. ICASSP-93*, vol. 2, pp. 534–537, 1993.

[18] F. G. Stremler, *Introduction to Communications Systems*, Reading, MA: Addison-Wesley Publishers, Series in Electrical Engineering, 1982.

[19] M. Sun and R. J. Sclabassi, "Discrete-Time Instantaneous Frequency and Its Computation", *IEEE Trans. Signal Processing*, vol. 41, no. 5, pp. 1867–1880, May 1993.

[20] H. Taub and D. L. Schilling, *Principles of Communications Systems*, New York, NY: McGraw-Hill, 1986.

[21] R. J. Webster, "Spectral Line Profiles Generated by Deterministic Frequency Modulation", *IEEE Trans. Signal Processing*, vol. 39, no. 4, pp. 1012–1017, April 1991.

17

AUDITORY MODELS AND HUMAN PERFORMANCE IN TASKS RELATED TO SPEECH CODING AND SPEECH RECOGNITION

Oded Ghitza

AT&T Bell Laboratories, Acoustics Research Department,
Murray Hill, New Jersey 07974

ABSTRACT

Auditory models that are capable of achieving human performance in tasks related to speech perception would provide a basis for realizing effective speech processing systems. Saving bits in speech coders, for example, relies on a perceptual tolerance to acoustic deviations from the original speech. Perceptual invariance to adverse signal conditions (noise, microphone and channel distortions, room reverberations) and to phonemic variability (due to non-uniqueness of articulatory gestures) may provide a basis for robust speech recognition. In this paper we describe a state-of-the-art auditory model that simulates, in considerable detail, the outer parts of the auditory periphery up through the auditory nerve level. Speech information is extracted from the simulated auditory nerve firings, and used in place of the conventional input to several speech coding and recognition systems. The performance of these systems improves as a result of this replacement, but is still short of achieving human performance. The shortcomings occur, in particular, in tasks related to low bit-rate coding and to speech recognition. Since schemes for low bit-rate coding rely on signal manipulations that spread over durations of several tens of ms, and since schemes for speech recognition rely on phonemic/articulatory information that extend over similar time intervals, it is concluded that the shortcomings are due mainly to a lack of perceptually related integration rules over durations of 50–100 ms. These observations suggest a need for a study aimed at understanding how auditory nerve activity is integrated over time intervals of that duration. We discuss preliminary experimental results that confirm human usage of such integration, with different integration rules for different time-frequency regions depending on the phoneme-discrimination task.

Reprinted, with permission, from the IEEE Transactions on Speech and Audio Processing, Vol. 2, No. 1, 115-132, January 1994.

1 INTRODUCTION

In building speech-processing systems, we make the axiomatic assumption that the speech waveform contains the entire information produced by the human speech production mechanism. In a particular speech-processing system, however, only part of this information is used, depending on the task for which the system is designed. For speech recognition tasks, for example, only phonemic information is needed. For speech coding tasks, on the other hand, information associated with the quality of speech is also required.

Traditionally, the task-related information is extracted from the speech waveform (or its Fourier representation) using statistical inference methods. This paper discusses an alternative way to extract this information that relies on processing principles derived from properties of the auditory system. The premise of this approach is that such processing principles can provide a basis for realizing effective speech processing systems. Saving bits in speech coders, for example, relies on perceptual tolerance to acoustic deviations from the original speech. Perceptual invariance to adverse signal conditions (noise, microphone and channel distortions, room reverberations) and to phonemic variability (due to non-uniqueness of articulatory gestures) may provide a basis for robust speech recognition.

The potential advantages that can be gained by utilizing auditory models for speech processing depend on how accurate the models are in mimicking human performance. And building such accurate models depend on the amount of knowledge we have about the auditory system. This knowledge is acquired by combining data that has been collected in both psychophysical and physiological studies of the auditory system. Studies of cochlear mechanics and studies of mechanical to neural transduction in the cochlea provide insight into the processing of sounds in the pre auditory-nerve stages of the auditory periphery. Studies of the population response of single auditory nerve fibers in the cat to speech-like signals provide a rich source of information concerning the principles by which such sounds are encoded in the auditory nerve. In contrast, little is known, at present, about the functional operation of auditory nuclei beyond the auditory nerve.

Current research activity in auditory modeling is mostly devoted to the study of the auditory periphery. Several papers have been published that examine how the response of the cochlea may be processed to provide a relevant representation of the speech signal. Each study utilizes a computational model to simulate either the direct firing activity or another related representation of the cochlear output. However, the manner in which this information is processed differs among the studies, reflecting differences in the structural properties of the central processor hypothesized by each study. These structural properties can be described by using a two-component characterization: the place/non-place component, that indicates if the central processor utilizes explicit knowledge about the fibers' tonotopic place of origin in the cochlear partition, and the rate/temporal component, that indicates whether the central processor uses instantaneous firing rate measurements alone or higher-order firing statistics (e.g., the interspike interval statistics). Using this two-component criterion, the following categories are traditionally used: (1) place/rate category, where the central processor possesses explicit knowledge of place and uses only instantaneous rate information [7, 24, 30], (2) place/temporal category, where place information is used together with detailed temporal information of local neural responses [14, 31, 32, 8], and (3) non-place/temporal category, where place information is omitted altogether and the only sources of information are the temporal properties of the global neural response [1, 9].[1]

This paper focuses on a model that belongs to the last category. Detailed modeling of the auditory periphery (up through the auditory nerve level) is used to simulate the firing activity of the auditory nerve. The model consists of 190 cochlear channels, distributed from 200 Hz to 7000 Hz according to the frequency-position relation suggested by Greenwood [15]. Each channel comprises Goldstein's nonlinear model of the human cochlea [13], followed by an array of five level-crossing detectors that simulate the auditory nerve fibers innervating one inner hair cell. At this junction of the model the speech sound is represented as a 950-dimensional point process (190 channels times 5 levels per channel) that simulates the firing activity of the auditory nerve fibers.

One way to achieve human performance is to process the simulated auditory nerve firing activity according to principles derived from properties of post auditory-nerve nuclei. Unfortunately, such information is not available at present. Our approach, therefore, is to process the simulated auditory nerve firing patterns according to principles that are motivated by observed properties of actual auditory nerve response. In Section 2, a representation of this kind is described as an Ensemble Interval Histogram (EIH). Conceptually, the EIH is a measure of the spatial (tonotopic) extent of coherent activity across the simulated auditory nerve. This measure signifies different physical properties of the acoustic stimulus depending on frequency: Information at low frequencies is represented in terms of the extent of coherent firing activity that is

[1] For an excellent overview of auditory models, the reader is referred to the theme issue "Representation of Speech in the Auditory Periphery", edited by S. Greenberg, *Journal of Phonetics*, Volume 16, No. 1, January 1988.

phase-locked to the underlying resolved components, while information at high frequencies is represented as the extent of coherent instantaneous rate of the firing activity driven by a wideband, unresolved (in frequency) temporal event. This representation is in accordance with observed properties of auditory nerve response, which show a diminishing degree of phase locking of the neural firings to the driving stimulus, as the frequency of the stimulus increases (e.g., [18]).

As pointed out before, the advantages that can be gained by utilizing auditory models for speech processing depend on how accurate the models are in mimicking human performance. In Sections 3 and 4 we evaluate the capability of the EIH to achieve human performance. Using psychophysical results as a reference we examine the extent to which speech processing systems that use the auditory model mimic human performance, in tasks related to speech recognition and to speech coding. In Section 3 we evaluate the EIH in the context of a speech recognition task (after [10]). We first identify a psychophysical experiment that addresses such a task, the Diagnostic Rhyme Test, or DRT [33] and present data we have collected on how accurate human listeners are in performing the task. Next, we simulate the DRT procedure. The auditory periphery is replaced by the EIH and the higher auditory elements by an automatic speech recognition system. Errors are displayed in terms of a distribution among six phonemically distinctive features. The error patterns of the simulated procedure are then compared with those of the human subjects, to provide a quantitative measure on how close the model performance is to the actual human performance. The evaluation process shows that performance improves by replacing a conventional speech representation by the EIH, but is still short of achieving human performance.

In Section 4 we evaluate the EIH in the context of speech coding. Following the methodology used in Section 3, we first identified a psychophysical experiment that addresses a speech coding task (the Mean-Opinion-Score, or MOS, a test which is widely used to assess quality of speech coders) and collected data on how human listeners score the quality of synthetic speech produced by different speech coding systems. Then, we use the EIH as a basis for a system aiming at predicting the MOS. Since the EIH is computed from simulated auditory nerve responses modeled in considerable detail, we hypothesize that it contains only perceptually-relevant speech information. If this hypothesis holds, perceptual differences between the original and the coded speech should be realistically reflected in the EIH domain. To examine this hypothesis, we mimicked the MOS test by measuring the L_2-norm of the difference between the EIHs of the original and the coded speech. For comparison purposes we also examined the L_2-norm in the cepstral domain. The comparison shows that EIH provides better MOS predictions for coders with a bit-rate of 16 kb/s and above, but underrates the quality assessment of CELP-type of coders (i.e., coders with a bit-rate of 8 kb/s and below).

The evaluation process described in Sections 3 and 4 demonstrates that performance is improved by replacing conventional speech representation methods by the auditory model, but is still short of achieving human performance. The evaluation process indicates that shortcomings occur, in particular, in tasks related to low bit-rate coding and to speech recognition. In Section 5.1 we discuss possible reasons for this inadequacy. Since schemes for low bit-rate coding rely on signal manipulations that spread over durations of several tens of ms, and since schemes for speech recognition rely on phonemic/articulatory information that extend over similar time intervals, it is concluded that the shortcomings are due mainly to the inappropriate manner by which we currently measure a distance between two speech segments. A typical way to represent a speech segment is in terms of an ordered sequence of observations, sampled at a uniform rate. (The observations can be expressed, for example, in terms of EIH or in terms of Fourier power spectrum). Hence, to measure the distance between two speech segments we must define a distance metric between two **sequences** of observations. This, in turn, requires a definition of a distance metric between two **single** observations. In the experiments of Sections 3 and 4, the distance between two single EIH observations is measured in the L_2-norm sense. The distance between two sequences of EIH observations is measured as the accumulated sum of distances between corresponding single observations along a path that best matches the sequences in the sense of minimizing the accumulated sum. The experimental results of Sections 3 and 4 suggest that both, the L_2-norm (of the difference between two single EIH observations) and the accumulated sum (for two sequences of EIH observations) are inadequate measures for the purpose of predicting human performance. They also imply a need for a study aimed at understanding how auditory nerve activity is integrated over durations of 50–100 ms. In Section 5.2 we discuss preliminary experimental results that confirm human usage of such integration, with different integration rules for different time-frequency regions depending on the phoneme-discrimination task.

2 THE ENSEMBLE INTERVAL HISTOGRAM (EIH) REPRESENTATION

This section describes the EIH model. The model comprises two stages, one that models the pre auditory-nerve section and one that stands for the post auditory-nerve section of the auditory periphery. The pre auditory-nerve section has been modeled in considerable detail, guided by the physiological structure of the auditory periphery. In contrast, the post auditory-nerve section is represented in an heuristic manner since little is known, so far, about the operation of auditory functions associated with that part of the auditory periphery. In Section 2.2, we review the pre auditory-nerve portion of the auditory periphery. The considerations in modeling this part of the auditory periphery are discussed next: the model for the mechanical displacement of the basilar membrane is described in Section 2.3, and the model for the mechanical-to-neural transduction is described in Section 2.4. The second stage of the EIH model is discussed in Section 2.5.

Before describing the EIH model, let us briefly describe the structure of the mammalian auditory system.[2] This will put into perspective the state of our knowledge and what part of it is being utilized in our current models.

2.1 A Brief Description of the Auditory Pathway

The auditory system is symmetrically organized around a midline between the left and the right sides (Figure 1). The most peripheral part at each side consists of the **external and middle ears**, the **cochlea**, and the **auditory nerve**. This portion of the auditory system will be described in detail, in Section 2.2. The auditory nervous system, which receives its inputs from both left and right auditory nerves, consists of several neural nuclei that can be grouped into three major parts — the auditory brainstem, the auditory midbrain, and the auditory cortex. Each nucleus can be divided into different zones, characterized by their morphological structure, neurophysiological response, and their input/output mappings. The main information flow is along an ascending pathway that begins at the external ears and ends in the auditory cortex. In parallel, there is an information flow in a descending pathway that begins at the auditory cortex and can be traced all the way down to the middle ears.

Figure 1 shows the major nuclei of the mammalian auditory system. The auditory nerve at each side projects to its corresponding **cochlear nucleus** which is the first nucleus in the auditory brainstem. The projections of various divisions in each cochlear nucleus form two parallel branches, one on the same

[2] For an introductory survey of the auditory system the following references are recommended: Pickles [27] and Kiang and Peake [19].

side of the cochlear nucleus and the other across the midline. These projections, on both sides, proceed either directly or indirectly via other brainstem nuclei (the **trapezoid body** and the **superior olivary complex**), to the auditory midbrain nuclei (the **lateral leminiscus** and the **inferior colliculus**). The inferior colliculus provides the major ascending path to the **medial geniculate body** which, in turn, provides inputs to the **primary auditory cortex**.

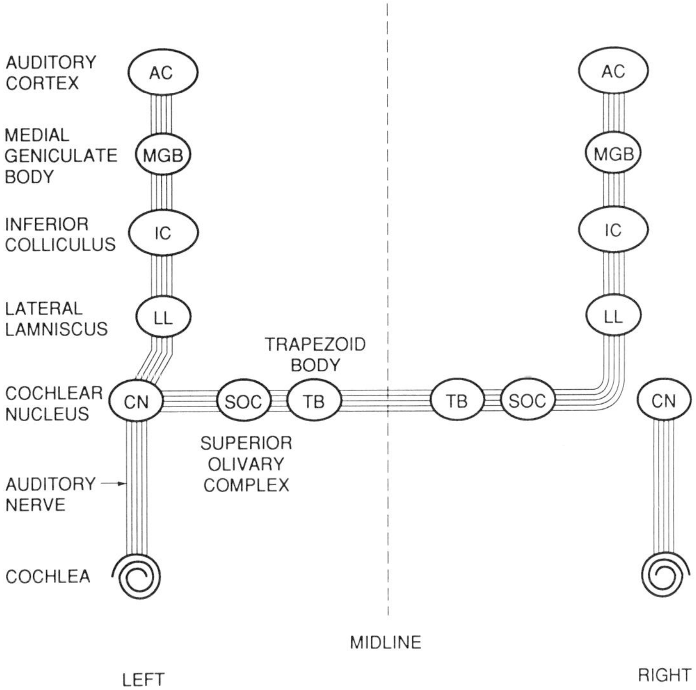

Figure 1 Major nuclei of the mammalian auditory system.

2.2 Physiological Basis for the EIH Model

A diagram of the pre auditory-nerve portion of the auditory periphery is shown in Figure 2(a). This part of the auditory periphery comprises three distinct parts: the outer ear, the middle ear, and the inner ear. The outer ear consists of the **pinna** (the ear surface surrounding the canal in which sound is funneled) and the **external canal**. Sound waves are guided through the outer ear to the middle ear, which consists of the **eardrum** (which moves due to the sound pressure) and a mechanical transducer comprises the **hammer**, the **incus** and the **stapes** (which conveys the motion of the eardrum into mechanical vibrations along the inner ear). The inner ear consists of the **cochlea**, which is a fluid-filled chamber partitioned by the basilar membrane, and the **auditory nerve**. The mechanical vibrations at the entrance of the cochlea (a 2 1/2 turn, snail-like tube, shown as a straight tube in Figure 2(b)) excites the fluid inside the cochlea and cause the basilar membrane to vibrate at places associated with the frequencies of the input acoustic wave. Distributed along the basilar membrane (in a dense but discrete manner) are sensors called **inner hair cells** (IHC) that are innervated by the auditory nerve fibers and transform the mechanical displacement of the basilar membrane into firing activity at the nerve fibers.

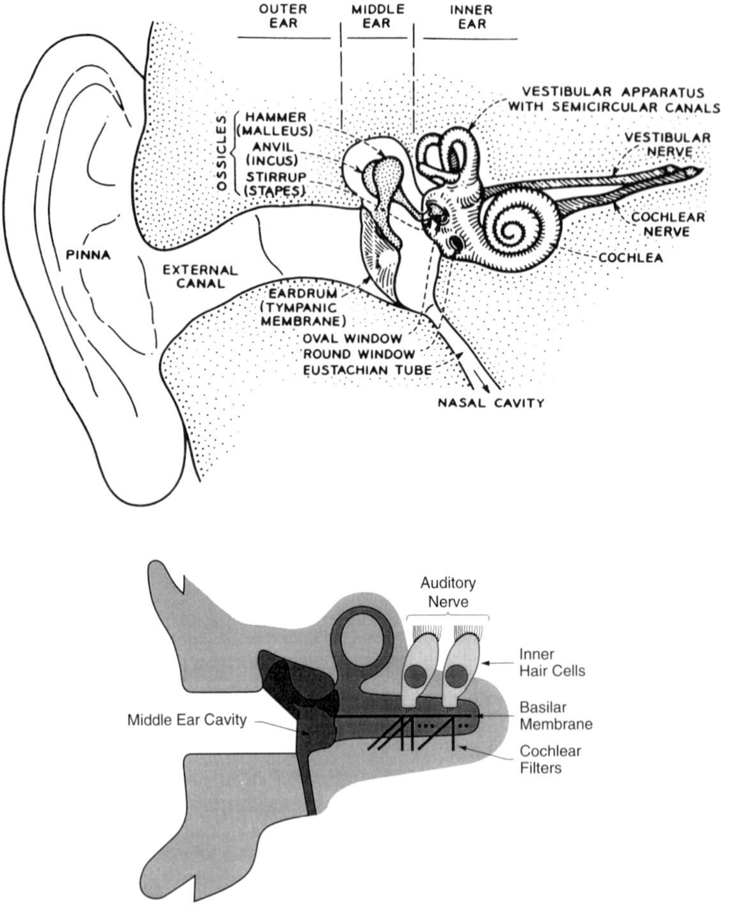

Figure 2 (a) A physiological model of the pre auditory-nerve parts of the auditory periphery. (b) A simplified view of (a). The 2 1/2 turn, snail-like shape of the cochlea is shown as a straight tube.

The mechanical displacement of the basilar membrane, at any given place, can be viewed as the output signal of a band-pass filter whose frequency response has a resonance peak at a frequency which is characteristic of the place. This resonance frequency is called **characteristic frequency**, CF. The log of CF is approximately proportional to distance along the membrane, and the distribution of the inner hair cells (IHCs) along the cochlear partition is essentially uniform. (There are some 4000 IHCs along the basilar membrane). The displacement of the basilar membrane is reflected in the AC component of the IHC receptor potential. The transformation from mechanical motion to receptor potential involves several nonlinearities, the most relevant to the present discussion being the half-wave rectification which is a consequence of the unidirectional depolarization of the IHC. Each IHC is innervated by approximately ten auditory-nerve fibers, whose spontaneous activity ranges between 0 and 100 discharges per second. The spontaneous rate is highly correlated with the fiber diameter and the size of the synaptic region between the fiber and inner hair cell. The spontaneous discharge rate is also correlated with the threshold of response. For any given CF region, fibers with high spontaneous rate tend to have between 5 and 20 dB lower threshold than units with low rates of background activity. Occasionally, low-spontaneous units may have as much as 40–60 dB higher thresholds than high-spontaneous units of comparable CF [22, 23].

2.3 Modeling the Mechanical Displacement of the Basilar Membrane

The EIH model is schematically illustrated in Figure 3. Its first stage represents the auditory periphery up through the level of the auditory nerve. The middle ear is modeled as a high-pass filter, with a cutoff frequency at 1000 Hz and a slope of 20 dB/decade. The mechanical displacement of the basilar membrane is sampled by 190 IHC channels distributed from 200 Hz to 7000 Hz according to the frequency-position relation suggested by Greenwood [15]:

$$F = A(10^{ax} - 1) \tag{17.1}$$

where F is frequency in Hz, x is the normalized distance along the membrane (i.e., $0 < x < 1$, $x = 0$ is the apex) and the appropriate constants for the human cochlea are $A = 165.4$ and $a = 2.1$. The corresponding mechanical motion in each channel is simulated using a model of the human cochlea suggested by Goldstein [13]. The model is termed MBPNL, for "Multi Band Pass Non Linear" filter. It operates in the time domain and changes its gain and bandwidth with changes in the input intensity, in accordance with observed psychophysical

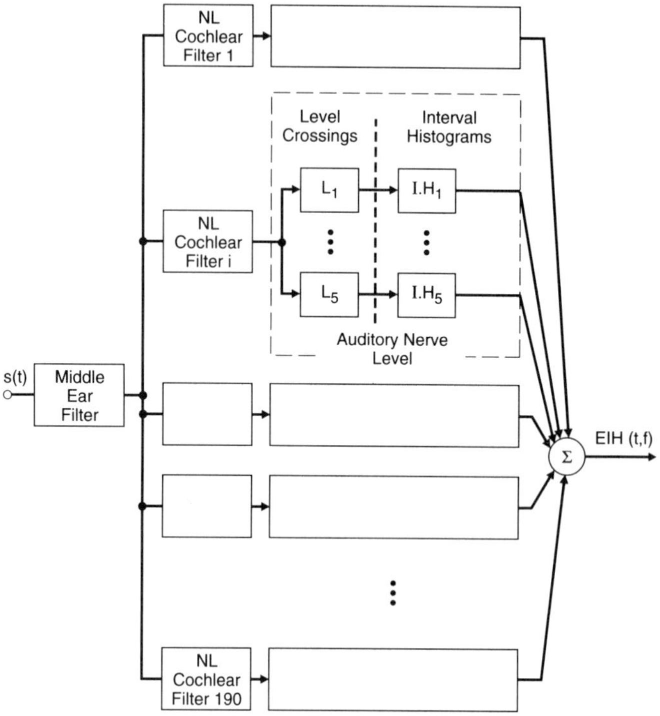

Figure 3 The Ensemble Interval Histogram (EIH) model.

behavior. The MBPNL model is shown in Figure 4. The lower signal processing path (H1-H2) is a compressive nonlinear filter that represents the sensitive, narrowband compressive nonlinearity at the tip of the basilar-membrane tuning curves. The upper signaling path (H3-H2) is a linear filter that represents the

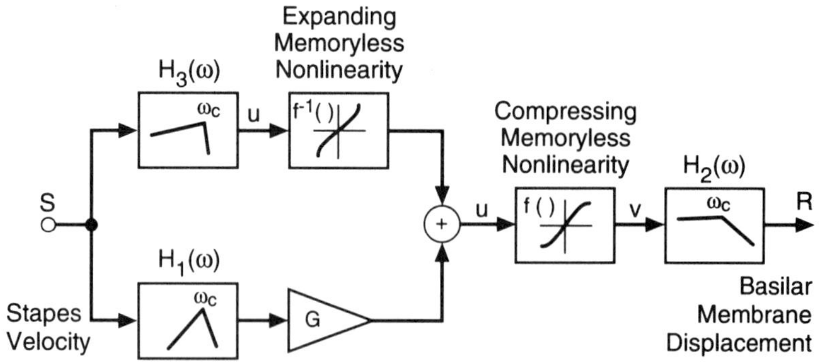

Figure 4 The MBPNL cochlear model (After [13]).

insensitive, broadband linear tail response of basilar-membrane tuning curves (after [13]).

The "iso-input" frequency response of an MBPNL filter at CF of 3400 Hz is shown in Figure 5. For an input signal $s(t) = A\sin(\omega_o t)$, with A and ω_o fixed, the MBPNL behaves as a linear system, with a fixed "operating point" on the nonlinear curves f and f^{-1} of Figure 4, determined by A. Figure 5 shows the iso-input frequency response of the system for different values of A: For a given A, a discrete "chirp" signal was presented to the MBPNL, with a slowly changing frequency. Changes in ω_o occurred only after the system reached steady-state, for a proper gain measurement. For a 0 dB input level ($A = 1$), the gain at CF is approximately 40 dB. As the input level increases the gain drops, the bandwidth increases and CF shifts. These observations are in accordance with psychophysical behavior.

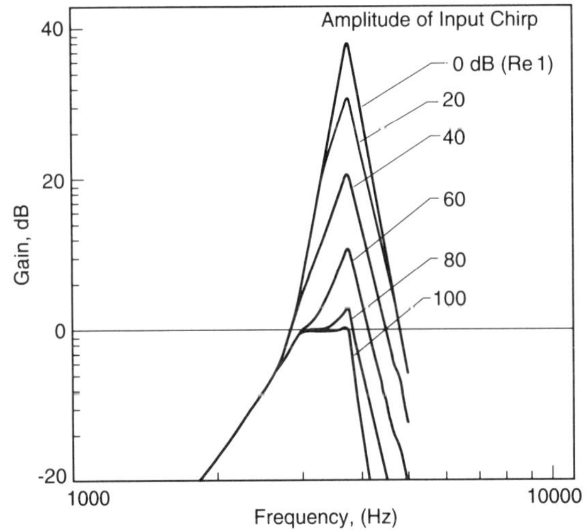

Figure 5 Iso-input frequency response of the MBPNL model at $CF = 3400$ Hz.

Figure 6 illustrates the high degree of overlap among the simulated cochlear channels. It shows the iso-input response at input level of 60 dB. Only one of every ten channels is drawn.

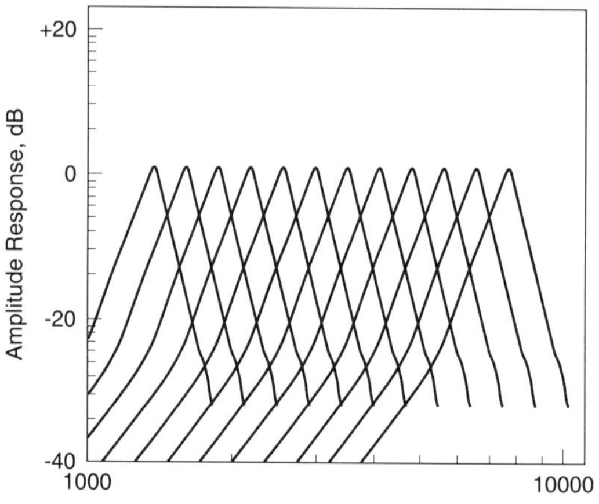

Figure 6 An illustration of the degree of overlap among the cochlear channels. Shown are the iso-input response at input level of 60 dB. Only one of every ten channels is drawn.

2.4 Modeling the Mechanical to Neural Transduction

The ensemble of nerve fibers innervating a single IHC is simulated with an array of level-crossing detectors at the output of each cochlear filter (i.e., each level-crossing detector is equivalent to a fiber of specific threshold). A neural firing is simulated as the positive-going level crossing. The detectors are distributed across a range of positive levels, to account for the half-wave rectification nature of the IHC receptor potential. The values assigned to the level j of every filter is a random Gaussian variable, with a mean, L_j, and a standard deviation, $\sigma_j = 0.2L_j$. The mean values, $\{L_j\}_{j=1}^{5}$, are uniformly distributed on a log scale over the amplitude range of the MBPNL output. The randomness in the values of the jth level across the cochlear channels simulates the fact that diameters and synapse-connection sizes of fibers innervating the same side of different IHCs along the cochlear partition have a certain amount of intrinsic variability (which is characteristic of most physiological systems). The model omits certain properties of the auditory-nerve function. By modeling the nerve-fiber firing activity as a point process produced by a level-crossing detector, the probabilistic nature of the neural firing mechanism is essentially neglected. Thus, the level crossings should be interpreted as the **combined** firing activity of a collection of fibers originating in different IHCs located in a range along the basilar membrane small enough to ensure similar cochlear tuning characteristics.

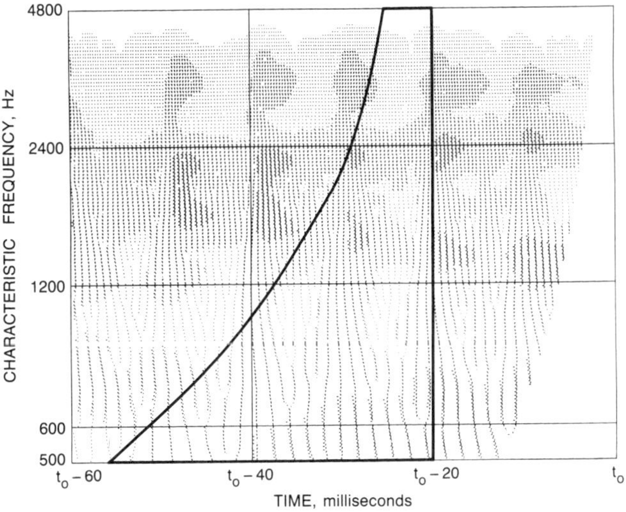

Figure 7 Simulated auditory-nerve activity for the first 60 ms inside the vowel [o] in the word "gob". The abscissa represents time and the ordinate represents the characteristic frequency of the IHC channels. Note the logarithmic scale of the characteristic frequency, which represents the place-to-frequency mapping at the basilar membrane. In the figure, a level-crossing occurrence is marked as a single dot, and the output activity of each level-crossing detector is plotted as a separate trace. Each IHC channel contributes five parallel traces, with the lower trace representing the lower-threshold level-crossing detector. If the magnitude of the filter's output-signal is low, only one level will be crossed, as is the case for the very top channels of the figure. However, for large signal magnitudes, several levels will be activated, creating a "darker" area of activity. The figure also illustrates how the length of the analysis window in each channel is related to its characteristic frequency (CF).

The outputs of the level-crossing detectors represent the discharge activity of the auditory-nerve fibers in terms of a 950-dimensional point process (190 channels times 5 levels per channel). Figure 7 shows simulated auditory-nerve activity for the first 60 ms inside the vowel [o] in the word "gob". The abscissa represents time and the ordinate represents the characteristic frequency of the IHC channels. Note the logarithmic scale of the characteristic frequency, which represents the place-to-frequency mapping on the basilar membrane. In the figure, a level-crossing occurrence is marked as a single dot, and the output activity of each level-crossing detector is plotted as a separate trace. Each IHC channel contributes five parallel traces, with the lower trace representing the lower-threshold level-crossing detector. If the magnitude of the filter's output signal is low, only one level will be crossed, as is the case for the very top channels of Figure 7. However, for large signal magnitudes, several levels will be activated, creating a "darker" area of activity.

2.5 Measure of Synchrony and Instantaneous Rate Across the Simulated Fibers

The 950-dimensional point process at the output of the level-crossing detectors serves as the input to the second stage of the EIH. With the purpose of achieving human performance, this point process, which simulates the auditory nerve firing activity, would have to be processed following principles derived from properties of post auditory-nerve nuclei. Unfortunately, such information is not available at present. Our approach, therefore, is to apply processing principles that are motivated by observed properties of actual auditory nerve response.

Measurements of firing responses of cats' auditory-nerve fibers (e.g., [29, 37, 3, 4, 5, 6]) show a significant difference between the properties of the firing patterns of low CF (say, below 1000 Hz) and high CF fibers. This difference is determined mainly by the mechanical properties of the basilar membrane. At low CFs, harmonics are resolved with high precision and neural discharges of auditory nerve fibers are phase locked to the underlying driving component. That is, synchrony (between neural discharges and the underlying driving component) is maintained. At high CFs, frequency resolution is poor and the phase-locking of the discharges is greatly reduced. The instantaneous rate of firing, however, conveys temporal information with fine time resolution.

It is also evident from these measurements that as the sound pressure level increases, more fibers fire **in coherence** with certain temporal properties of the stimulus waveform. For this reason, we suggest the spatial (tonotopic) extent of such coherent activity as a measure of the perceptual importance of the underlying temporal events. The extent of coherent activity signifies different properties of the underlying stimulus, depending on frequency. At low frequencies, it is the extent of coherent firing activity **phase-locked** to an underlying resolved component. At high frequencies, it is the extent of coherent **instantaneous rate** of the firing activity driven by a wideband, unresolved (in frequency) temporal event. Obviously, there is no distinct boundary between these auditory nerve regions. Rather, the change in properties is gradual (e.g., [18]).

In the model, the amount of coherent neural activity across the simulated fiber array is measured by determining the similarity in the short-term interval probability density functions of individual level-crossing detectors. An estimate of the interval probability density function of a given level can be obtained by computing a histogram of the intervals from the point process data produced by the level-crossing detector. Only intervals between successive upward-going level crossings are considered. Since we prefer an auditory representation in the frequency domain, the histogram of the **inverse intervals** is computed. This is accomplished by distributing the reciprocal of the intervals (i.e., converting the intervals into units of frequency) in a histogram consisting of successive bins, ranging from 0 to, say, 4000 Hz. (We shall discuss the considerations involved in choosing the bin allocation, the window length and the number of bins momentarily). The similarity between all individual interval probability density functions is measured by collecting the individual histograms into one **Ensemble Interval Histogram** (via summing the corresponding histogram bins across all levels and all channels). The resulting representation is the EIH. The tonotopic extent of a coherent neural activity generated by an underlying temporal event is encoded as the magnitude of the corresponding bin in the EIH.

To illustrate this point, consider the case where the input signal is $s(t) = A\sin(2\pi f_o t)$. First, consider the channel with CF equal to f_o (Figure 8(a)). For a given amplitude A, the cochlear filter output will activate only part of the level-crossing detectors, depending on the value of A. For a given detector, the time interval between two successive positive-going level crossings is $1/f_o$. Since the histogram is scaled in units of frequency, this interval contributes a count to the f_o bin. For the input signal in the illustrated example, all of the intervals are the same, resulting in a histogram where the magnitude of each bin, save one (f_o), is zero. As the signal amplitude A increases, more levels are activated. As a result, this cochlear filter contributes additional counts to the f_o bin of the EIH. Since the crossing levels are equally distributed on a log-amplitude scale, the magnitude of any EIH bin is related, in some fashion, to decibel units. However, this relation is not a straightforward one since there are several sources contributing counts to the f_o bin in a nonlinear manner. Figure 8(b) shows an input signal $s(t) = A\sin(2\pi f_o t)$ driving five adjacent cochlear channels. For the sake of simplicity, we assume that the filters are linear, with an amplitude response $|H_i(f)|$ and a phase response $\phi_i(f)$, i=1,2 ,..., 5. Due to the shape of the filters, more than one cochlear channel will contribute to the f_o bin. In fact, all the cochlear filters which produce $s_i(t) = A|H_i(f_o)|\sin(2\pi f_o t + \phi_i(f_o))$ will contribute to the f_o bin of the EIH, provided that $A|H_i(f_o)|$ exceeds any of the level crossing thresholds. In Figure 8(b) only cochlear filters 2, 3 and 4 are contributing non-zero histograms to the EIH. The number of counts is different for each filter, depending on the magnitude of $A|H_i(f_o)|$.

Two factors affect the properties of the interval histogram, the bin allocation over the frequency range and the choice of the window length. Motivated by the tonotopic organization along the auditory pathway we assign the bins according to the ERB-rate scale (shown in Figure 9, after Moore and Glasberg,

[26]), which is related to the psychophysical critical band assignment, or the Bark scale. In the following paragraph we will define the ERB-rate scale.

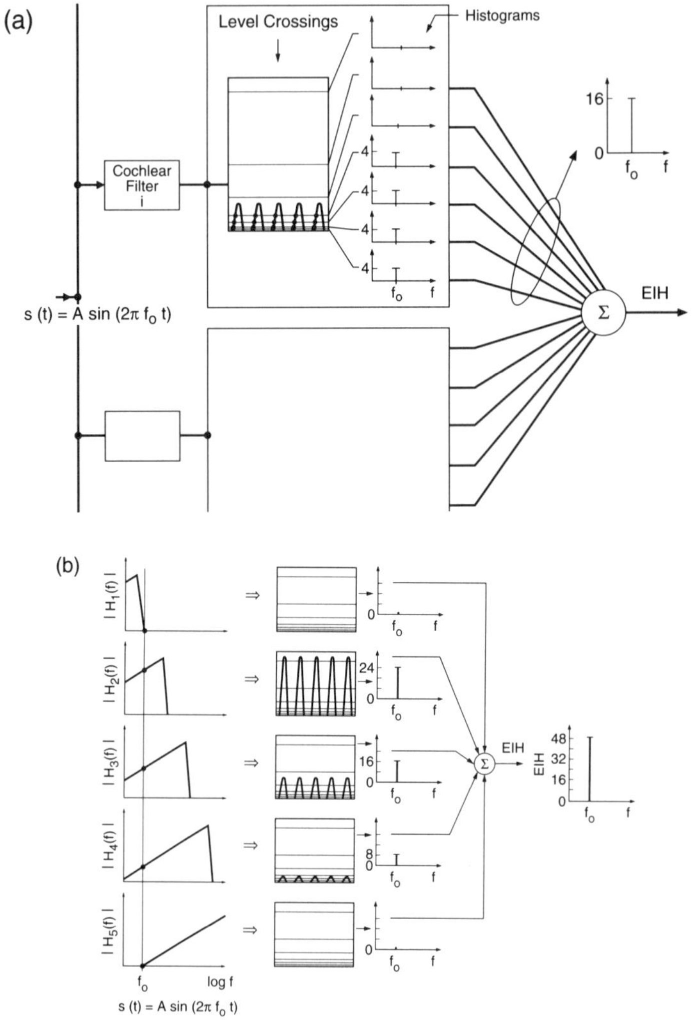

Figure 8 (a) The contribution of the i-th channel to the Ensemble Interval Histogram for an input signal $s(t) = A\sin(2\pi f_o t)$. Only the lowest four level-crossing detectors are contributing non-zero histograms to the ensemble. (b) The contribution of five successive channels $|H_i(f)|$, $i = 1, 2, \ldots, 5$, to the EIH, for an input signal $s(t) = A\sin(2\pi f_o t)$. Channel i contributes to the f_o bin of the EIH provided that $A|H_i(f_o)|$ exceeds any of the level-crossing thresholds.

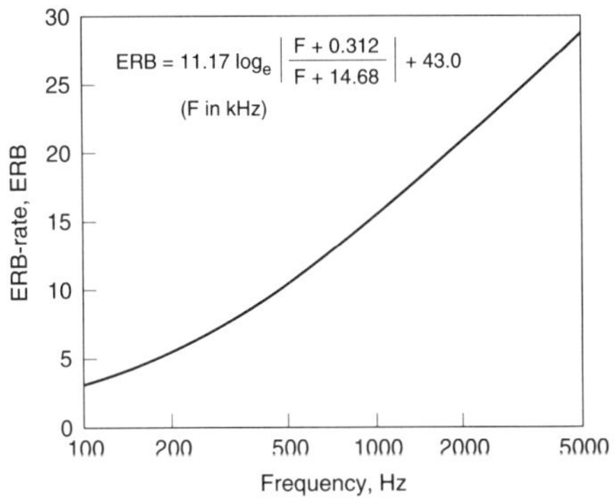

Figure 9 ERB-rate of the human auditory filters, as a function of CF. Suggested by Moore and Glasberg [26], based on psychophysical experiments. See text for a definition of ERB.

Let $|H(f)|$ be a unimodal frequency response of a filter, and let $|H(f_o)|$ be the maximum gain of the filter, at frequency f_o. The **Equivalent Rectangular Bandwidth** (ERB, in Hz) of $|H(f)|$ is defined as follows:

$$\mathrm{ERB} = \frac{\int |H(f)| df}{|H(f_o)|} \tag{17.2}$$

In words, ERB is the **bandwidth** of an hypothetical rectangular filter, with a gain of $|H(f_o)|$, such that the integral over its frequency response is equal to the integral over $|H(f)|$. Using psychophysical measurements of the ERB of human auditory filters, Moore and Glasberg [26] derived the following quadratic fit, as a function of the characteristic frequency of the auditory filter:

$$\mathrm{ERB} = 6.23F^2 + 93.39F + 28.52 \tag{17.3}$$

Where F is frequency in kHz. (Note that very similar ERBs can be derived from Eq. (17.1) of Greenwood, where an ERB at a given frequency corresponds to a constant distance of 0.85 mm on the basilar membrane [25]). Using the ERB of the auditory filter as a unit of measurement, Moore and Glasberg [26] suggested the **ERB-rate scale** which relate number of ERBs to the frequency interval $[0, F]$. This scale was obtained by integrating the reciprocal of Eq. (17.3), yielding:

$$\mathrm{ERB} - \mathrm{rate} = 11.17 \log_e \frac{F + 0.312}{F + 14.675} + 43.0 \tag{17.4}$$

Where F is frequency in kHz. (The constant 43.0 was chosen to make the number of ERBs = 0 when F = 0.) To summarize, Eq. (17.3) specifies the ERB of a human auditory filter at a given frequency F, and Eq. (17.4) determines the number of successive ERBs which covers the frequency range $[0, F]$. Using the ERB-rate scale, we quantized the frequency range [0,4000] Hz into 32 bins, which is roughly the number of ERBs in this frequency range. Momentarily, we will illustrate how the choice of bin allocation determine the properties of the EIH.

Another parameter that affects the properties of the interval histogram is the size of the observation window. Motivated, again, by the tonotopic organization along the auditory pathway, we set the window length to be inversely proportional to characteristic frequency. That is, at time t_0, intervals produced by a level-crossing detector located at a characteristic frequency CF_0 are collected over a window of length $\frac{10}{CF_0}$ that ends at time t_0 (see Figure 7).

Figure 10 illustrates the relationship between the bandwidth characteristics of the filters, the window length and the bin allocation. The figure is organized from left to right. It shows the response of two hypothetical cochlear filters, H_1 and H_2, to a pulse-train input, $s(t)$, with a pulse every 20 ms. The center frequency of H_1 is 100 Hz, and that of H_2 is 2000 Hz, with bandwidths of 30 Hz and 300 Hz, respectively. The bandwidths of the filters dictate the properties of their outputs. Thus, H_1, which resolves the frequency component at 100 Hz, results in a sinusoidal output $y_1(t)$. In contrast, the output of H_2 is wider in bandwidth and follows sharp temporal changes of $s(t)$. (In the limit, with the pulse-width approaching zero, $y_2(t)$ is the impulse response of the filter). In the example of Figure 10, the signal is sampled at a uniform rate, once every 5 ms, and only interval histograms of the **zero crossings** are considered. Because of our choice of window length, zero-crossings of $y_1(t)$ are collected over a 100 ms window, and zero-crossings of $y_2(t)$ are collected over a 5 ms window. Figure 10 shows the location of five successive windows, for $y_1(t)$ and $y_2(t)$. The interval histograms for these frames are shown in the right-hand-side of the figure. The figure shows typical interval histograms for two choices of bin allocation, linear (with, say, 128 bins over [0,4000] Hz and a fine frequency resolution) and ERB (with 32 bins over the same frequency range). In the case of linear bin allocation, the narrow-band signal $y_1(t)$ contributes identical intervals to the 100 Hz bin. And since the window length of H_1 is much longer than the frame rate, the interval histograms hardly change with time. In contrast, the wide-band signal $y_2(t)$ contributes intervals of different values, resulting in histograms that extend over several bins. And since the window length of H_2 is similar to the frame rate, the histograms change rapidly with time, demonstrating high temporal resolution. In the case of the ERB bin allocation, bins at low frequencies are narrow, resulting in fine frequency resolution, similar to the frequency resolution of the histograms with linear bin allocation. However, bins at high frequencies are wide, covering a filter bandwidth (e.g., 300 Hz at $CF = 2000$ Hz), and a frequency range of one ERB bin contains several linearly allocated bins. Therefore, the interval count at this ERB bin equals to the **sum** of intervals over all the linearly allocated bins at that frequency range. In other words, at time t_0, ERB bins at high frequencies contain the overall number of intervals collected over the window, irrespective of the shape of the interval pdf. Therefore, we view the changes in time at the high frequency bins as a measure of instantaneous rate.

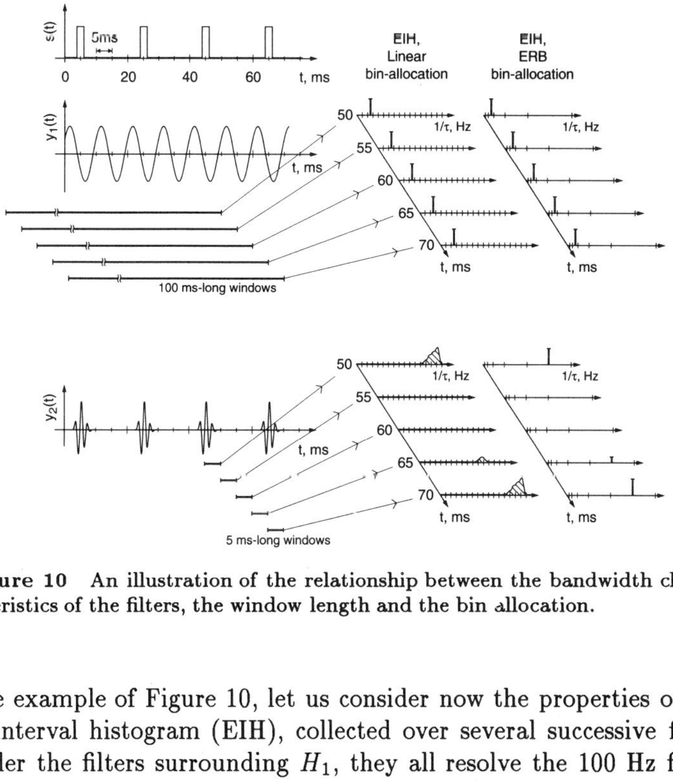

Figure 10 An illustration of the relationship between the bandwidth characteristics of the filters, the window length and the bin allocation.

Using the example of Figure 10, let us consider now the properties of the **ensemble** interval histogram (EIH), collected over several successive filters. If we consider the filters surrounding H_1, they all resolve the 100 Hz frequency component and contribute intervals to the same bin (located at 100 Hz). Therefore, the magnitude of the EIH at this bin can be viewed as a measure of the number of level-crossing detectors and the number of successive filters that are synchronized (or phase locked) to the 100 Hz underlying component. If we consider the filters surrounding H_2, they all result in output signals similar to $y_2(t)$ of Figure 10. The change with time of the corresponding EIH bin can, therefore, be viewed as a measure of the extent of coherent instantaneous-rate activity across the array.

In summary, choosing a bin allocation and a window length that are matched to the bandwidth characteristics of the cochlear filters provide a unified representation that exhibits fine frequency resolution at low frequencies (based on a measure of synchrony), and fine temporal resolution at high frequencies (based on a measure of instantaneous rate). There is no distinct boundary that signals the switch from one type of measurement to the other. Rather, the change in properties is gradual.

3 USING THE EIH IN A RECOGNITION TASK

In this section we examine the extent to which the EIH is capable of mimicking human performance in a task related to speech recognition. We confine ourselves to a recognition task with a minimal cognitive load since we are aiming at measuring the performance of the EIH, in isolation from the other parts of the evaluation system. We first identify a psychophysical experiment that addresses such a task, the Diagnostic Rhyme Test, or DRT [33] and collect data on how accurate the human listener is in performing the task. Next, we simulate the DRT procedure. The auditory periphery is replaced by the EIH and the higher auditory elements by an automatic speech recognition system especially designed to keep the errors due to the decision process to a minimum. Errors are displayed in terms of a distribution among six phonemically distinctive features. The error patterns of the simulated procedure are compared with those of the human, to provide quantitative measure on how close the model performance is to the actual human performance. The DRT is described in Section 3.1, the simulated DRT in Section 3.2 and the experimental results in Section 3.3.

3.1 Brief Description of Voiers' DRT

The psychophysical experiment that we selected is the one used in the standard Diagnostic Rhyme Test (DRT), suggested by Voiers [33]. In general, the DRT test attempts to evaluate how well phonemic information is perceived by a human listener. The test is divided into two parts, measuring the human performance (via psychophysical experiments) and deriving an intelligibility score. For our purposes only the first part, i.e. the data collection, is relevant. As we shall see momentarily, the test is appropriate for our needs for two reasons. First, from the acoustic point of view the database spans the speech subspace associated with initial diphones in a uniform manner. Hence, the performance of the EIH is examined in terms of its capability to represent all possible cells in that speech subspace. And second, it allows us to separate the effects of the auditory periphery from those due to cognition.

Voiers' DRT database covers initial diphones of spoken words of the Consonant-Vowel-Consonant (CVC) type. Table 1 shows the list of words used in the DRT. The list consists of 96 pairs of confusable words spoken in isolation. Words in a pair differ only in their initial consonants. The consonants are equally distributed among 6 phonemic distinctive features (16 word-pairs per feature) and among 8 vowels. The feature classification follows the binary system suggested by Jakobson, Fant and Halle [16].[3] The vowels are [ee] (like in peen) and [it] (like in bit) for high-front vowels, [eh] (like in zed) and [at] (like in fast) for high-back vowels, [oo] (like in zoo) and [oh] (like in note) for low-front vowels, and [aw] (like in boss) and [ah] (like in bond) for low-back vowels.

The psychophysical procedure is very carefully controlled. The listeners are well trained and are very familiar with the database, including the voice quality of the individual speakers. A One Interval Two Alternative Forced Choice (1I2AFC) paradigm is used. First, the subject is presented visually with a pair of rhymed words. Then, one word of the pair (selected at random) is presented aurally and the subject is required to indicate which of the two words was played. This procedure is repeated until all the words in the database have been presented. The errors can be displayed either in terms of a confusion matrix (between consonants), or as a distribution among the 6 phonemic distinctive features.

Two points are noteworthy. First, since it is a discrimination test, we can categorize the DRT as a speech recognition task. And second, due to the controlled nature of the test procedure, we assume that all cognitive information needed for the discrimination task is available to the listener prior to the aural presentation. (Of course, we also assume that the subject is indeed utilizing all this information). If this assumption is correct, an error in identifying the word is due mainly to inaccuracy in the internal auditory representation of the

[3]The six features are *Voicing, Nasality, Sustention, Sibilation, Graveness* and *Compactness*. The *Voicing* feature characterizes the nature of the source, being periodic or non-periodic. The *Nasality* feature indicates the existence of a supplementary resonator. The terms *Sustention* and *Sibilation* are due to Voiers. They correspond respectively to the continuant-interrupted and strident-mellow contrasts of Jakobson, Fant and Halle. Finally, *Graveness* and *Compactness* represent broad resonance features of the speech sound, related to place of articulation.

Table 1 Stimulus Words used in the DRT

VOICING *Voiced–Unvoiced*	**NASALITY** *Nasal–Ora2*	**SUSTENTION** *Sustained–Interrupted*
veal–feel	meat–beat	vee–bee
bean–peen	need–deed	sheet–cheat
gin–chin	mitt–bit	vill–bill
dint–tint	nip–dip	thick–tick
zoo–Sue	moot–boot	foo–pooh
dune–tune	news–dues	shoes–choose
voal–foal	moan–bone	those–doze
goat–coat	note–dote	though–dough
zed–said	mend–bend	then–den
dense–tense	neck–deck	fence–pence
vast–fast	mad–bad	than–Dan
gaff–calf	nab–dab	shad–chad
vault–fault	moss–boss	thong–tong
daunt–taunt	gnaw–daw	shaw–chaw
jock–chock	mom–bomb	von–bon
bond–pond	knock–dock	vox–box

SIBILATION *Sibilated–Unsibilated*	**GRAVENESS** *Grave–Acute*	**COMPACTNESS** *Compact–Diffuse*
zee–thee	weed–reed	yield–wield
cheep–keep	peak–teak	key–tea
jilt–gilt	bid–did	hit–fit
sing–thing	fin–thin	gill–dill
juice–goose	moon–noon	coop–poop
chew–coo	pool–tool	you–rue
Joe–go	bowl–dole	ghost–boast
sole–thole	fore–thor	show–so
jest–guest	met–net	keg–peg
chair–care	pent–tent	yen–wren
jab–dab	bank–dank	gat–bat
sank–thank	fad–thad	shag–sag
jaws–gauze	fought–thought	yawl–wall
saw–thaw	bond–dong	caught–taught
jot–got	wad–rod	hop–fop
chop–cop	pot–tot	got–dot

Table 2 Stimulus Words used in the DALT

VOICING *Voiced–Unvoiced*	NASALITY *Nasal–Oral*	SUSTENTION *Sustained–Interrupted*
teethe–teeth	screen–screed	seethe–seed
jib–gyp	ring–rig	dish–ditch
mood–moot	noon–nude	goof–goop
brogue–broke	moan–mode	both–boat
liege–leash	gleam–glebe	chief–cheep
ridge–rich	rim–rib	give–gib
prove–proof	tomb–tube	soothe–sued
loathe–loath	gloam–globe	jove–job (e)
led–let	hen–head	rev–reb
have–half	lam–lab	calve–cab
jaws–joss	brawn–broad	froth–fraught
hodge–hotch	nom–nob	slav–slob
peg–peck	gem–jeb	flesh–fletch
lathe–lath	fan–fad	path–pat
flaws–floss	long–log	frothe–fraud
fob–fop	bomb–bob	bosh–botch

SIBILATION *Sibilated–Unsibilated*	GRAVENESS *Grave–Acute*	COMPACTNESS *Compact–Diffuse*
peach–peak	sheave–sheathe	league–lead
kiss–kith	skim–skin	hick–hit
sues–soothe	rufe–ruth	fugue–feud
poach–poke	oaf–oath	bloke–bloat
breeze–breathe	neap–neat	creek–creep
bridge–brig	miff–myth	sling–slim
truce–truth	rube–rude	luke–loop
clothes–clothe	lobe–load	rogue–robe
bess–beth	deaf–death	mesh–mess
badge–bag	dab–dad	lag–lad
ross–wroth	shawm–shawn	dog–daub
notch–knock	top–tot	cock–cot
ledge–leg	web–wed	egg–ebb
mass–math	raff–rath	knack–nap
maws–mothe	trough–troth	gong–gone
bodge–bog	sauve–swathe	chock–chop

stimulus. Hence, the error list provided by the test reflects errors in the internal human auditory representation during the discrimination task.

3.2 Simulating the DRT (after [10])

In the simulation, the peripheral part of the auditory pathway is modeled by the EIH, and the cognitive process is replaced by an array of recognizers, one for each pair of words in the DRT database. The errors due to the recognition procedure should be kept to a minimum, so that the overall detected errors are due mainly to inaccuracies in the front-end representation.

The Recognition System

A recognizer in the array represents one DRT word-pair. For a test word (from a given word-pair), the recognizer makes a maximum likelihood decision between two hidden Markov (HMM) word models, one for each word in the pair. An HMM word-model is defined as a left-to-right phonemic sequence. Each state of the HMM word-model represents one phonemic unit. The recognizer used is an HMM recognizer with time-varying states, suggested by Ghitza and Sondhi [12]. In this recognizer, a state of the HMM represents one phonemic unit in terms of a time-varying mean sequence of ordered frames — a template — and a block covariance matrix that characterizes the intra-frame statistical dependence within the phonemic unit. The particular phonemic unit that was selected is a diphone. In this way the dynamic nature of coarticulation between two successive phonemes is represented accurately, and the ability to discriminate between the initial consonants of the DRT words is improved.

For every word in the DRT database, the HMM has a trivial transition matrix that allows only one deterministic transition (from the first diphone to the second). For such a simple transition matrix, the HMM model reduces to a modified version of the dynamic time warping (DTW) method of speech recognition (e.g., [36]). In this framework, a word-model for a C_iVC_f word is a concatenation of two state-templates C_iV and VC_f, and the decision between the two word models is a **minimum distance** decision. The input-word and the word-models are represented in terms of ordered sequences of frames (or observation vectors). The distances between the word-model sequences and the input-word sequence are measured as defined in the DTW method: we first define $d(\mathbf{a}, \mathbf{b})$ to be the distance between any two **single** observation vectors, $\mathbf{a}$ and $\mathbf{b}$ as:

$$d(\mathbf{a}, \mathbf{b}) = (\mathbf{a} - \mathbf{b})'\Sigma^{-1}(\mathbf{a} - \mathbf{b}) \tag{17.5}$$

Here, $'$ denotes vector transpose, and Σ is a covariance matrix whose (i,j)th entry is the covariance between the ith and the jth components of the observation vector. For $\Sigma = I$, $d(\mathbf{a}, \mathbf{b})$ is simply the L_2-norm of the difference between $\mathbf{a}$ and $\mathbf{b}$. In terms of d, we can define the distance between the **sequences** $\mathbf{O}^i$ and $\mathbf{O}^j$ by the usual DTW procedure. Let $\mathbf{o}^i_m$, $m = 1, 2, \ldots, M$ be the observations in sequence $\mathbf{O}^i$, and $\mathbf{o}^j_n$, $n = 1, 2, \ldots, N$ the observations in sequence $\mathbf{O}^j$. We define $D(\mathbf{O}^i, \mathbf{O}^j)$ as:

$$D(\mathbf{O}^i, \mathbf{O}^j) = \frac{1}{N} \min_{m(n)} \sum_{n=1}^{N} d(\mathbf{o}^j_n, \mathbf{o}^i_{m(n)}) \tag{17.6}$$

The mapping $m(n)$ is constrained such that $m(1) = 1$ and $m(N) = M$. Thus $D(\mathbf{O}^i, \mathbf{O}^j)$ is the average distance between corresponding observation vectors in the two sequences, after the sequence $\mathbf{O}^j$ has been optimally warped onto the sequence $\mathbf{O}^i$.

Simulating the One-Interval-Two-Alternative-Forced-Choice Paradigm

To eliminate effects due to variability among speakers, the simulation is done on a speaker-dependent basis. Every speaker provides two repetitions of the DRT word-list, one for training and one for testing. As a part of the training phase, a vocabulary of diphones is obtained for each speaker by segmenting the training part of the database manually. The vocabulary covers all the diphones that appear in the DRT word list. If several tokens of a particular diphone appear in the DRT word list, the diphone is represented by only one of these tokens.

The testing phase is a simulation of the 1I2AFC paradigm. For testing a particular word-pair, the recognizer is first loaded with the appropriate diphone-state models (drawn from the inventory) and transition matrices. This step simulates the effect of the visual presentation of the word-pair to the listener. Then, the two words are presented one at a time to the recognizer, analogously to the aural presentation to the listener. Based on the recognizer's phonemic transcription, it is decided whether or not the word was correctly recognized. This procedure is repeated until all the word pairs in the database have been scanned. The overall error list is displayed in terms of a distribution among the six phonemic features.

Figure 11 illustrates the simulation procedure. To test the word-pair *peak/teak*, for example, the state models for the diphones *pi, ti* and *ik* are drawn from the state vocabulary, along with the appropriate transition matrix that allows only the necessary transitions. The recognizer is then presented with the word *peak*, and produces a phonemic transcription which is either *p-i-k* or *t-i-k*. If the first transcription occurs, the result of the simulated discrimination task is considered to be correct. Otherwise, an error is registered. Next, the word *teak* is tested. Identical state models and transition rules are used, and the same sequence of steps is repeated. This concludes the test for this word-pair. To test the next word-pair (*moon/noon*, in Figure 11), the recognizer is loaded with new state models (*mu*, *nu* and *un*) and a new transition matrix, and the above procedure is repeated. Note that in testing the word-pair *peen/bean*, the state model for the diphone *pi* (which is required to model the word *peen*) is the same state model used previously for the word *peak*.

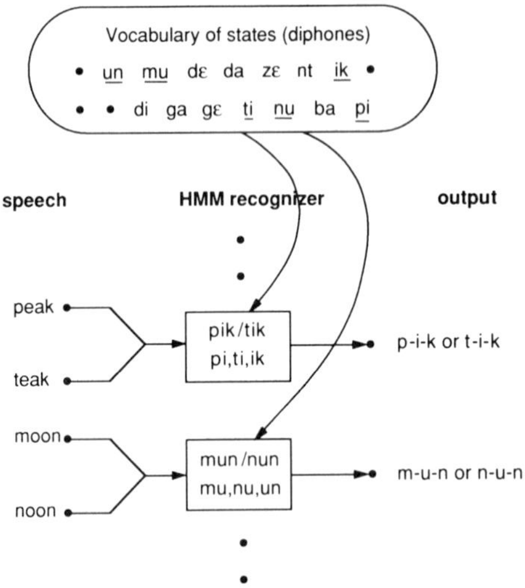

Figure 11 An illustration of the DRT simulation procedure. To test the word-pair *peak/teak*, for example, the state models for the diphones *pi*, *ti* and *ik* are drawn from the states vocabulary, along with the appropriate transition matrix that allows only the necessary transitions. The recognizer is then presented with the word *peak*, and produces a phonemic transcription which is either *p-i-k* or *t-i-k*. If the first transcription occurs, the result of the simulated discrimination task is considered to be correct. Otherwise, an error is registered. Next, the word *teak* is tested. Identical state models and transition rules are used, and the same sequence of steps is repeated. To test the next word-pair (e.g, *moon/noon*) the recognizer is loaded with new state models (*mu*, *nu* and *un*) and a new transition matrix, and the above procedure is repeated.

3.3 Experimental Results

As pointed out before, the DRT simulation is performed on a speaker-dependent basis — to eliminate errors due to between speaker variability. A single-speaker simulation procedure (training and testing) is repeated for every speaker in the database. The overall data is then analyzed, to find similarities in the error patterns across listeners, speakers and phonemic features.

Signal Conditions

Three male speakers were used, two from Voiers' database (speakers RH and CH). Each speaker provided two repetitions of the DRT word-list, one for training and one for testing. The signals were lowpass filtered to 3600 Hz and sampled at an 8-kHz rate. A "noisy" version of the testing repetitions was created by adding white noise to the original ("clean") signals. The signal-to-noise ratio (SNR) was 10 dB.[4] The noisy version was sent to Dynastat Inc. (a company established by Voiers) for the psychophysical evaluation. To comply with Dynastat's procedure, the processed words were recorded at a rate of a word every 1.3 seconds. For the recordings to sound continuous over time, we first set the variance of the white noise generator to a level that remained unchanged until all the words in the DRT word list had been recorded in sequence. To record a particular word, the speech signal was amplified (or attenuated) by a gain factor that was calculated in advance, to maintain the desired global SNR.

For training, a vocabulary of diphones was created from the clean repetition of the DRT word-list assigned for training. For testing, we used the same noisy versions that were sent to Dynastat Inc. .

Analysis Methods

We tested two speech representation methods, the EIH and the Fourier power spectrum. The auditory model for the EIH representation is described in Section 2. The model uses 190 cochlear channels and 5 thresholds per channel. An EIH observation was computed once every 10 ms. The interval statistics

[4] The SNR was defined using global measurements. First, the total energy, E_{Tot}, of the original (noise-free) word was computed. Then, the average energy per digital sample, E_{Samp}, was determined, by dividing E_{Tot} by the number of sample points in the signal. E_{Samp} was used to set the variance of a white noise generator to a level dependent on the desired global-SNR. Note that this definition of global-SNR overestimates the actual signal to noise ratio during the consonantal segments since the magnitude of the noise is largely dependent on the amplitude of the vocalic portion of each word.

at time t_0 are collected from all 950 (190 times 5) threshold detectors, using all simulated firing records which exist in the windows that end at time t_0 (see Figure 7). The length of each window is $\frac{10}{CF}$, where CF is the characteristic frequency of the cochlear channel. The interval statistics are collected into a 32-bin vector, where the bins divide the frequency range [0,4000] Hz according to the ERB-rate scale. The sum over the 32 bins was normalized to 1, to eliminate the effect of "loudness".

For the Fourier power spectrum, an 11th-order cepstral representation was computed every 10 ms. That is, the feature-vector under consideration is the Fourier **spectral envelope**. At time t_0, the cepstral coefficients are derived from the 10th-order LPC coefficients, computed from a 30 ms-long Hamming window centered at t_0. The first cepstral coefficient (c_0) was set to 0 and only the next 10 coefficients were used. In this way, the envelope is normalized in the sense that the average value of the LPC log spectrum is 0.

As for the recognition system, the distance measures d and D of Eqs. (17.5) and (17.6) were used, respectively. For d, we used $\Sigma = I$ (i.e., d is the L_2-norm of the difference vector). The same distance measures were used for both analysis methods. We shall discuss the implications of this choice for d and D in Section 5.

Results

The raw data that summarizes the outcome of one experimental run (i.e., one speaker) are organized in the form of a matrix with 12 rows and 16 columns. Each row represents a phonemic category (attribute present or attribute absent for each of the six phonemic features — see Table 1), and each column represents a word in the DRT word-list, associated with the corresponding row. For the simulated procedure, an entry in the matrix is a binary number, a 0 (for a correct answer) or 1 (for an error). For the psychophysical procedure, the value of a matrix element is an integer number between 0 and 8 and it indicates the number of listeners who made a mistake in identifying the corresponding word (8 is the number of listeners participating in the test). Calculating statistics across the columns, we computed the average error and the 95% confidence interval for every row (i.e., for every phonemic category we averaged across all vowels and all listeners). Then, we averaged across all speakers.

Figure 12 shows the resulting error distribution. The abscissa of every plot indicates the twelve phonemic categories: "vc" is for *Voicing*, "ns" for *Nasality*, "st" for *Sustention*, "sb" for *Sibilation*, "gv" for *Graveness* and "cm" for

Compactness. The "+" sign stands for attribute present and the "−" sign for attribute absent. The figure contains four plots, where the left-upper plot is a superposition of the other three plots, excluding the confidence intervals. Note that the line connecting the measurements is only for display purposes, to enable the reader to distinguish between error patterns that belong to a particular parameter value.

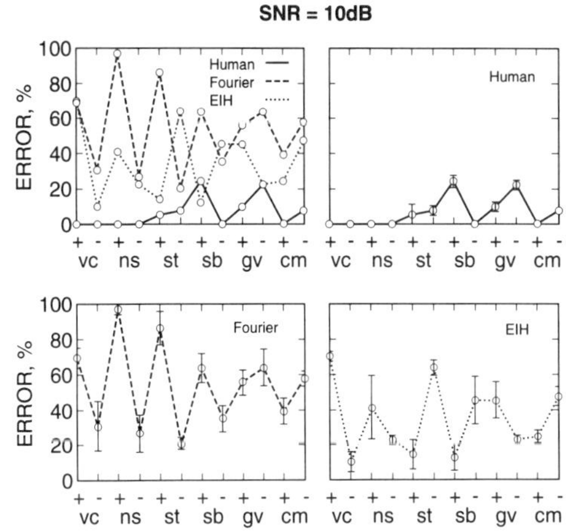

Figure 12 Distribution of errors made by the human listener, an EIH with an ERB bin allocation and a Fourier power spectrum, in a 10 dB signal-to-noise ratio. The left-upper plot is a superposition of the other three plots, excluding the confidence intervals. The abscissa of every plot indicates the six phonetic features: "vc" is for *Voicing*, "ns" for *Nasality*, "st" for *Sustention*, "sb" for *Sibilation*, "gv" for *Graveness* and "cm" for *Compactness*. The "+" sign stands for attribute present and the "−" sign for attribute absent. The line connecting the measurements is only for display purposes, to enable the reader to distinguish between error patterns that belong to a particular parameter value. The noise is additive and white, and the signal-to-noise ratio is defined using global measurements (see text). Note that an error value of 100% in a given phonemic category means that all 16 words in the category were mistakenly identified as their DRT counterparts.

Three points are noteworthy. First, the human observer performs much better than the EIH and Fourier power spectrum. Second, on the average EIH is more robust to noise than the Fourier power spectrum, in agreement with previous reports (e.g, [9]). And third, errors made by the Fourier power spectrum analyzer are mainly in the presence of Voicing, Nasality, Sustention and Sibilation, while errors made by the EIH are mainly in the presence of Voicing and the absence of Sustention. (We will show later (in Section 5.1) that other measures of interval statistics lead to improved results, and that the EIH performance can be brought closer to that of a human observer (see Figure 16).)

4 USING THE EIH TO PREDICT MEAN-OPINION-SCORE (MOS) OF SPEECH CODERS

In this section we examine the extent to which the EIH is capable of mimicking human judgement of speech quality. We follow the methodology used in Section 3: first, we identify a psychophysical experiment that addresses this task. Then, we collect data on how the human listener scores the quality of synthetic speech produced by different speech coding systems. Finally, we use the EIH as a basis for a system aiming at predicting the human scores.

The psychophysical experiment that was selected is the MOS test, which is widely used to assess the quality of speech coders. It is a subjective test that can be categorized as a rating procedure. Subjects are presented, once, with a speech sentence and are requested to score its quality using a scale of 5 grades. The grades (and their numerical aliases) are Excellent (5), Good (4), Fair (3), Poor (2) and Bad (1). The MOS is the mean score, averaged over the database and the subjects.

It is obviously interesting to find out how adequate EIH is in predicting MOS. Since EIH is computed from simulated auditory nerve responses modeled in considerable detail, we may hypothesize that it contains only perceptually-relevant speech information. If this hypothesis holds, perceptual differences between the original and the coded speech should be realistically reflected in the EIH domain. To examine this hypothesis, we mimicked the MOS test by measuring the L_2-norm of the difference between the EIHs of the original and the synthetic speech. For reference purposes we also compared the original and the coded speech in the cepstral domain, using the L_2-norm metric. (We are aware of numerous methods for an objective quality assessment of coded speech (e.g., [28, 20, 35]), but a detailed comparison is beyond the scope of this paper).

4.1 Database

We collected MOS scores for 18 different speech coding conditions. The assessed coders were a 64 kb/s μ-law PCM, two versions of a 32 kb/s ADPCM, a 16 kb/s LD-CELP, two versions of a 8 kb/s CELP and three versions of a 6.6 kb/s CELP.[5] The ADPCM and LD-CELP coders were tested in 4 tandem configurations.[6] The speech material contained sentences spoken by 4 male and 4 female speakers, each contributing 2 sentences.

4.2 Analysis Methods

For each coding condition, the coded waveform was first aligned (in time) with the original waveform. The EIH representation was computed as described in Section 3.3. The original and the coded speech signals were sampled synchronously. At time t_0, an L_2-norm of the difference between the normalized EIHs of the original and the coded speech was calculated first. This "frame" distance was than weighted by the "loudness" of the original speech at this time instant. (Our definition of loudness is the sum over the 32 bins, prior to normalization). The "EIH-distance" between a coded speech sentence and the original is defined as the mean value of the weighted frame distances over the entire sentence.

The cepstral representation was computed as described in Section 3.3. The original and the coded speech signals were sampled synchronously. At time t_0, an L_2-norm of the difference between the normalized cepstra of the original and the coded speech was calculated first. This "frame" distance was then weighted by the c_0 of the original. The "Cepstral-distance" between a coded speech sentence and the original is defined as the mean value of the weighted frame distances over the entire sentence.

[5] μ-law PCM and ADPCM are described by Jayant and Noll, [17]. LD-CELP is described by Chen *et al*, [2]. A description of CELP can be found in Kroon and Atal, [21].

[6] A tandem connection is a configuration used in communication networks, where several speech coders are connected in series. In a 2-tandem connection of ADPCM coders, for example, the synthetic speech produced by an ADPCM coder is passed through a second ADPCM coder. The synthetic speech produced by the second coder is the signal used for testing. See Jayant and Noll, [17].

4.3 Experimental results

Figures 13 and 14 summarize the results. The abscissa indicates the MOS and the ordinate the EIH-distance (Figure 13) or the Cepstral-distance (Figure 14). We grouped the results according to coder identity (left) or according to tandem configuration (right). A particular coding condition can be located by a cross-alignment of the left and the right plots. MOS values, by definition, are between 1 (bad) and 5 (excellent). For the coders we used, EIH-distance values are between 0 (perfect match to original) and 8, and Cepstral-distance values are between 0 (perfect match) and 6. In our database, for example, μ-PCM has an MOS of 4.2, an EIH-distance of 1.2 and a Cepstral-distance of 1.2.

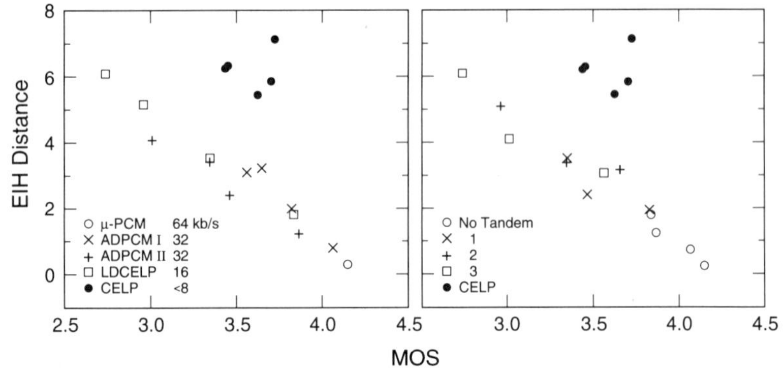

Figure 13 EIH-distance vs. MOS. The results are grouped according to coder identity (left) or according to tandem configuration (right). A particular coding condition can be located by a cross-alignment of the left and the right plots.

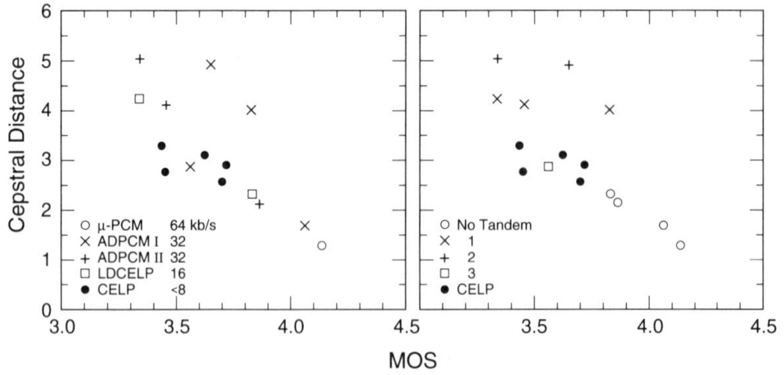

Figure 14 Cepstral-distance vs. MOS. Figure legend is as in Figure 13.

Figure 13 shows that EIH successfully predicts MOS of coders with a bit-rate of 16 kb/s and above. For this group of coders, EIH-distances and MOS are linearly related. However, MOS of the CELP coders are poorly predicted. These coders stand as a separate group in the MOS-EIH plane, with EIH-distances that are above the linear regression line that fits the other coders. Below, in Section 5.1, we shall discuss possible reasons for this behavior.

Figure 14 shows the extent to which the Cepstral-distance predicts the MOS. The scatter among the coders with a bit-rate of 16 kb/s and above is higher than the scatter observed in Figure 13. However, the group of CELP coders is properly merged with the rest of the coders.

5 DISCUSSION

5.1 Possible Reasons for Deficiency in Performance

The evaluation process of Sections 3 and 4 demonstrates that generally performance is improved by replacing a conventional speech representation by the auditory model, but is still short of achieving human performance. The shortcomings occur, in particular, in tasks related to low bit-rate coding and to speech recognition. In this section we discuss possible reasons for this deficiency. From the outset, we assume that EIH is an appropriate representation for an acoustic signal. This is so since the EIH is constructed from a detailed simulation of the human auditory-nerve firing patterns, using rules derived from general properties of observed auditory nerve activity in the cat (e.g., synchrony in low CFs, instantaneous rate in high CFs). The deficiency, therefore, may be the result of inappropriate use of the EIH in the context of speech processing.

The experiments in Section 4 demonstrate that using the L_2-norm in the EIH domain is appropriate only if the coded speech is of high quality (bit-rate of over 16 kb/s). However, this same measure fails to predict MOS of CELP coded speech. What is the reason for this behavior? Figure 15 shows the representation of the vowel [ɪ] (like in bit) produced by a male speaker. The upper plot contains the high-resolution Fourier magnitude spectrum computed by a 128 point discrete Fourier transform, using a 20-ms Hamming window. It also contains the envelope fit achieved by an 10th-order all-pole polynomial-fit. The plot uses a log-frequency/decibel scale. The lower plot is the EIH representation of the same signal. The EIH plot is in a linear-ERB/linear-EIH scale. (Note that the abscissa spans the frequency range [250,4000] Hz. In this frequency range, the ERB-rate scale overlaps with the log-frequency scale.) Since it reflects the tonotopic organization along the auditory pathway, the ERB-rate is a natural scale for the EIH, and was used during the experiments of Sections 3 and 4. The EIH spans the first 1000 Hz by approximately half the number of bins, and with fine frequency resolution. This results in a detailed representation of the first few harmonics, as shown in the figure. In turn, an L_2-norm of the difference between two such EIHs is very sensitive to the value of the first few individual frequency components. Consequently, EIH is a good predictor of MOS only if the two speech observations under comparison are similar enough (such that differences are localized in time and frequency). Indeed, this is the case for those coders in our database with a bit-rate of 16 kb/s and above. For the CELP coders, however, differences are not localized. Saving bits in low bit-rate coders relies on perceptual tolerance to changes that extend over durations of several frames. In order to achieve synthetic speech with an acceptable quality, these coding schemes are designed to spread the differences between the original speech segment and the synthesized segment smoothly over the entire segment. Consequently, a comparison between two single observation taken at the same time instant, one from the original waveform and one from the CELP coded waveform, will result in an irrelevant large errors due to the misalignment of the harmonics. Note that the Cepstral-distance is insensitive to the location of individual harmonics since the L_2-norm is measured between the spectral envelopes (via the 11th-order cepstra).

A similar observation can be made when examining the recognition experiment of Section 3. Speech recognizers rely on measuring a distance between phonemic/articulatory information that extend over durations of 50–100 ms, very much like low bit-rate coders. In the simulation of the 1I2AFC procedure we compared two speech segments of such duration, the state-template C_iV and the input segment, each represented as an ordered sequence of single observations. Even when these speech segments carry the same phonemic information, they usually differ in their length and in the way this information is manifested over the segment duration. Two sources of errors are noteworthy. First, an L_2-norm type of measure between two single EIH observations is inappropriate. For recognition tasks, we seek a measure that will reflect the **phonemic** distance between the speech observations. The exact location of individual harmonics is not relevant to the identity of a phoneme, and large errors due to misalignment of the harmonics will mask the useful error information. (This observation holds even for speaker-dependent speech recognition tasks, such as the one examined in Section 3). Figure 16 demonstrates this point. The right-upper plot (performance of a human listener) and the left-lower plot (EIH with

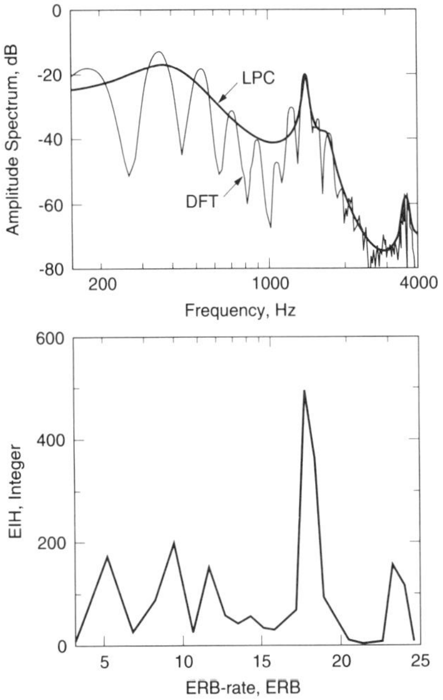

Figure 15 Amplitude spectrum and EIH representations of the vowel [ɪ] (like in bit) produced by a male speaker. The upper plot contains the high-resolution Fourier magnitude spectrum computed by a 128 point discrete Fourier transform, using a 20-ms Hamming window. It also contains the envelope fit achieved by an 10th-order all-pole polynomial-fit. The plot uses a log-frequency/decibel scale. The lower plot is the EIH representation of the same signal. The EIH plot is in a linear-ERB/linear-EIH scale. Note that the abscissa spans the frequency range [250,4000] Hz. In this frequency range, the ERB-rate scale overlaps with the log-frequency scale.

an ERB bin allocation) are from Figure 12. The right-lower plot (EIH with a linear bin allocation) is from Ghitza, [10]. In that study, the error distribution was obtained using the same method described in Section 3: The parameters of the MBPNL cochlear filters and the allocation of the level-crossing detectors were as in Section 3.3, but the interval-histogram bins were assigned differently, using 128 linearly spaced bins ranging from 0 to 4000 Hz. Mathematically, EIH was treated as a log spectrum (recall that the levels are equally distributed on a log scale) and the feature-vector was a 25th-order cepstral fit, obtained by computing the inverse DFT of the EIH and by truncating the resulting cepstral series.

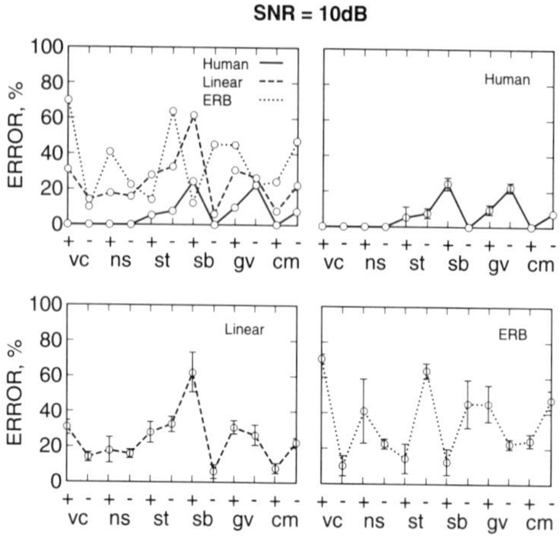

Figure 16 Distribution of errors made by the human listener, an EIH with an ERB bin allocation (from Figure 12) and an EIH with a linear bin allocation (from Ghitza, [10]) in a 10 dB signal-to-noise ratio. Figure legend is as in Figure 12.

Figure 16 shows that although the overall number of errors of the two EIH configurations is almost the same, the error distribution of the EIH with the linear bin allocation better predicts the error distribution of the human observer. Since the same recognizer have been used in both experiments, the improvement in performance is due to the difference between the feature vectors which is mainly dictated by the nature of the bin allocation (all other parameters of the EIH configurations are identical). A linear-frequency scale turns out to be more advantageous for recognition since only one quarter of the number of bins is used for the first 1000 Hz. Therefore, a cepstral fit of the EIH becomes effective in smoothing out individual harmonics and, in turn, the L_2-norm is no longer sensitive to their location. (This is illustrated in Figure 17 which replicates Figure 15, in a linear-frequency scale. The upper plot contains the same high-resolution Fourier magnitude spectrum and the same envelope fit, plotted over a linear-frequency/decibel scale. The lower plot shows the EIH representation with a linear bin allocation (128 bins over [0,4000] Hz), and the corresponding 25th-order cepstral envelope fit. Obviously, the harmonic information is no longer presented in the envelope fit of the EIH). At this point, the reader may wonder why we prefer the ERB-rate scale over the linear-frequency scale, even though the latter results in a better fit to human performance. We study the EIH with an ERB bin allocation because it represents more realistically the actual organization of the auditory periphery. Working with this scale, although leaving us with an unresolved question (namely, what is the appropriate measure between two such EIH observations), provides the appropriate framework for understanding the strategy of the human observer in performing this task.

The second noteworthy source of errors is the way we measure the distance between two speech segments that extend over durations of, say, 50–100 ms. In the simulation of the 1I2AFC we used $D(\mathbf{O}^i, \mathbf{O}^j)$ of Eq. (17.6) to measure the distance between two sequences $\mathbf{O}^i$ and $\mathbf{O}^j$, where $D(\mathbf{O}^i, \mathbf{O}^j)$ is the average distance between corresponding single observation vectors in the two sequences, after the sequence $\mathbf{O}^j$ has been optimally warped on to the sequence $\mathbf{O}^i$. But, recent study on the perceptual effects of modifications in the time-frequency domain [11] indicates that a human observer utilizes a much more complex strategy. Next, we briefly summarize that study.

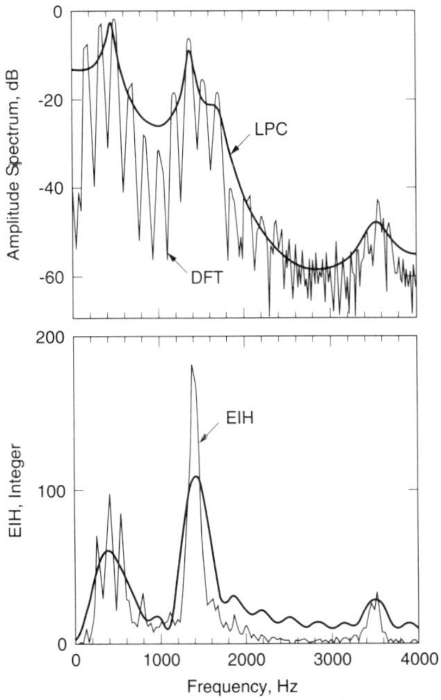

Figure 17 A replication of Figure 15, in a linear-frequency scale. The upper plot contains the same high-resolution Fourier magnitude spectrum and the same envelope fit, plotted over a linear-frequency/decibel scale. The lower plot shows the EIH representation with a linear bin allocation (128 bins over [0,4000] Hz) and the corresponding envelope fit, achieved by computing the inverse DFT of the EIH and truncating the cepstral series at c_{25}.

5.2 The Tiling study (after [11])

The study was aimed at understanding how auditory nerve activity is integrated over intervals of 50–150 ms, and over diphones in particular. From the outset, it was assumed that distinct peripheral auditory functions do exist which operate on different sections of the auditory nerve. Due to the tonotopic organization of the auditory nerve fibers, the nature of these auditory functions, as well as their accuracy of performance, can be inferred from psychophysical data on the perception of acoustic information contained in different time-frequency regions. Obtaining this psychophysical data was the subject of the study.

A database was used which covers the speech subspace associated with initial and with final diphones. It comprised two parts, Voiers' old DRT database for initial diphones (described in Section 3), and a new database, also designed by Voiers and termed DALT (for Diagnostic ALliteration Test) for final diphones [34]. Tables 1 and 2 show the lists of words in the DRT and the DALT, respectively. Like the DRT database, the DALT database also consists of 96 pairs of confusable words spoken in isolation. Each word is of the CVC type, and words in a pair differ only in their **final** consonants. The diphones are equally distributed among the same six phonemic features and the same eight vowels used in the DRT database.

Modified versions of the database were generated by introducing well-defined distortions into pre-selected time-frequency regions. Figures 18, 19 and 20 illustrate the procedure that was used to generate the modifications. Although the illustrations (examples and figures) relate to modifications of initial diphones in the DRT database, their analogous counterparts are appropriate for illustrating modifications of final diphones in the DALT database. Figure 18 shows the waveforms and the wideband spectrograms of the DRT word-pair *shock/mock* spoken by the same male speaker. Since the words differ only in their initial diphone, the main difference between the spectrograms is in the time-frequency region associated with those diphones, i.e. the region bounded by the bold lines containing the initial consonant (either [*S*] or [*m*]) and the left, coarticulated part of the vowel [*a*].

Figure 19 shows a diagram of the time-frequency region occupied by a spoken CVC word, like the word *shock* or *mock* of Figure 18. The time-frequency region of interest (i.e., the initial diphone) is subdivided into 6 "tiles". The frequency boundaries (from the bottom up) are 0 Hz, 1000 Hz, 2500 Hz and the highest frequency in the band, say 4000 Hz. The choice of the frequency boundaries was motivated by observed properties of auditory physiology and of speech

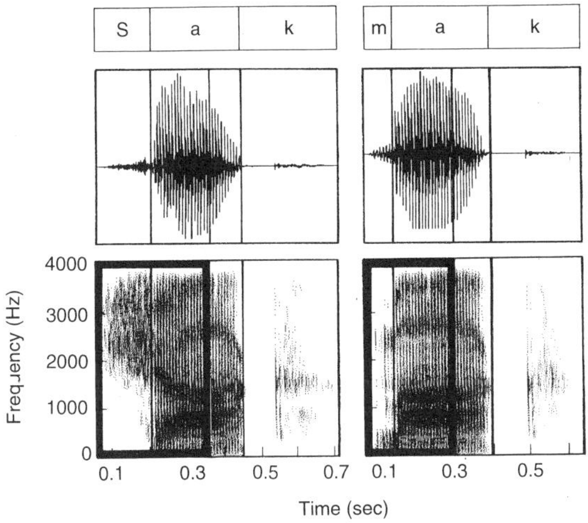

Figure 18 The waveforms and the wideband spectrograms of the DRT word-pair *shock/mock* spoken by a male speaker. Also indicated are the boundaries between the phonemes. Since the words differ only in their initial diphone, the main difference between the spectrograms is in the time-frequency region associated with that diphone, i.e. the region bounded by the bold lines containing the initial consonant (either [*S*] or [*m*]) and the left, coarticulated part of the vowel [*a*].

perception [11]. The time landmarks are (from left to right) the beginning of the word ($t = 0$), the transition from the initial consonant to the vowel, and the mid-point of the vowel.

Figure 20 shows a diagram of the time-frequency domain occupied by a prototype DRT word-pair, where the regions corresponding to the initial diphones are divided into 6 tiles each. To modify the database, an **interchange** operation was perform between the words in the pair. An interchange was defined as a "cut and paste" procedure, where a particular tile from one word was interchanged with the same tile of the opposite word. Once a particular interchange was chosen, it was applied to the entire database. The example of Figure 20 illustrates the interchange of the consonantal part of the second frequency band. Of course, one can interchange more than one tile at a time.

The modified versions were tested using Voiers' psychophysical discrimination task (see Section 3). The responses of the subjects were displayed as a distribution along Jakobson, Fant and Halle's phonemic dimensions, very much like the displays used in Figure 12. Detailed discussion of the experimental results is beyond the scope of this paper (for that, see [11]). Highlights of the results are plotted in Figures 21, 22 and 23. Figure 21 shows the average human per-

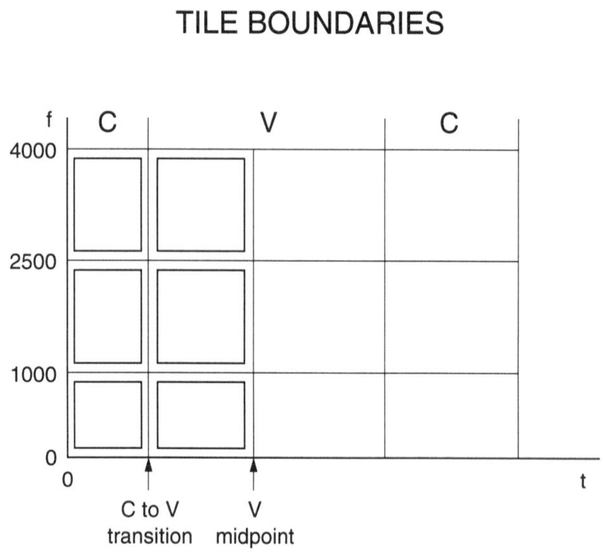

Figure 19 A diagram of the time-frequency domain occupied by a spoken CVC word. The time-frequency region of the initial diphone is subdivided into 6 "tiles". The frequency boundaries (form the bottom up) are 0 Hz, 1000 Hz, 2500 Hz and 4000 Hz. The time landmarks are (from left to right) the beginning of the word ($t = 0$), the transition from the initial consonant to the vowel, and the mid-point of the vowel.

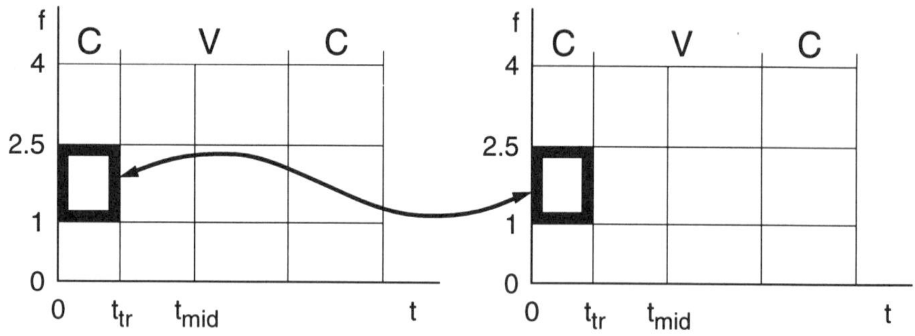

Figure 20 A diagram of the time-frequency domain occupied by a prototype DRT word-pair, where the region corresponding to the initial diphones are divided into 6 tiles each. Illustrated is an interchange of the consonantal part of the second frequency band.

formance and the associated 95% confidence interval (among the 8 listeners) for the baseline versions (i.e., no interchange) of the DRT (left) and DALT (right). The abscissa is as in Figure 12. The ordinate is termed "switch", and it represents the percentage of words in a phonemic category that, when played to the listener, were judged to be the opposite word in the word pair (i.e., the listener "switched" to the opposite category).

Figure 22 shows the human response for three selected modifications of the DRT database: an interchange of the first frequency band, an interchange of the second frequency band and an interchange of the third frequency band of the entire diphone. The abscissa is as in Figure 12. As for the ordinate, from the measured switches for the selected interchange condition we subtracted the number of switches for the baseline version (Figure 21, left). We termed the ordinate "$\Delta switch$", since it represents the *additional* number of switches, relative to the baseline version, that occurred due to the particular interchange operation. The upper right plot shows the amount of $\Delta switch$, in percent, when the first frequency band (i.e., [0,1000] Hz) of the initial diphone is interchanged. The lower left plot is for the case when the second frequency band (i.e., [1000,2500] Hz) of the initial diphone is interchanged, and the lower right plot is for the interchange of the third frequency band (i.e., [2500,4000] Hz) of the initial diphone. Several observations are noteworthy. The phonemic dimensions *Voicing* and *Nasality* are sensitive to an interchange of the first frequency band of the diphone, *Graveness* and *Compactness* to an interchange of the second frequency band of the diphone, and *Sibilation* to the interchange of the third frequency band of the diphone. *Sustention* is equally sensitive to interchanges of the diphone information in the second and the third frequency bands. (Such a behavior is expected, since the acoustic manifestation of "interrupt" (which is one of the binary attributes of *Sustention*) is an abrupt temporal change, spread over a wide frequency range.)

Figure 23 summarizes the human response for similar modifications of the DALT database. Somewhat surprisingly, the qualitative behavior is similar to that of Figure 23. However, there are quantitative differences, i.e., in the numerical values of the $\Delta switch$ in various conditions.

Summarizing the results of Ghitza's study, the following important observations have been noted: (1) the human observer uses different time-frequency tiles for different phoneme-discrimination tasks, (2) there is a direct mapping between phonemic/articulatory features and particular time-frequency tiles, and (3) tiles associated with diphones are far more important than tiles associated with the consonant part or the vowel part alone. We suggest that these observations

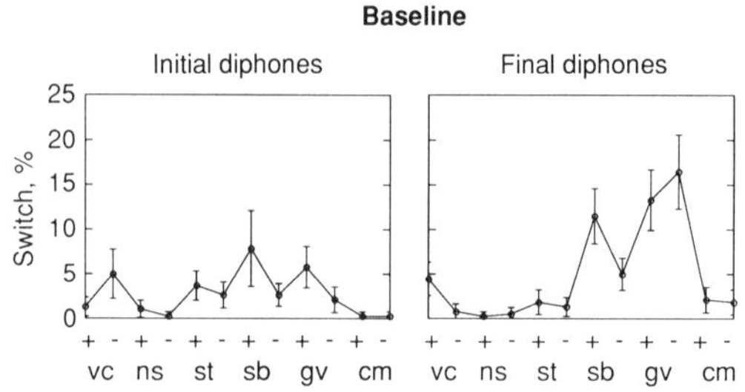

Figure 21 The average human performance and the 95% confidence interval for the baseline (i.e., no interchange) versions of the DRT (left) and DALT (right). The abscissa is as in Figure 12. The ordinate is termed "switch", and it represents the number of words in a phonemic category that, when played to the listener, were judged to be the opposite word in the word pair (i.e., the listener "switched" to the opposite category). The switch is represented in percents, relative to 16 (the total number of words per phonemic category).

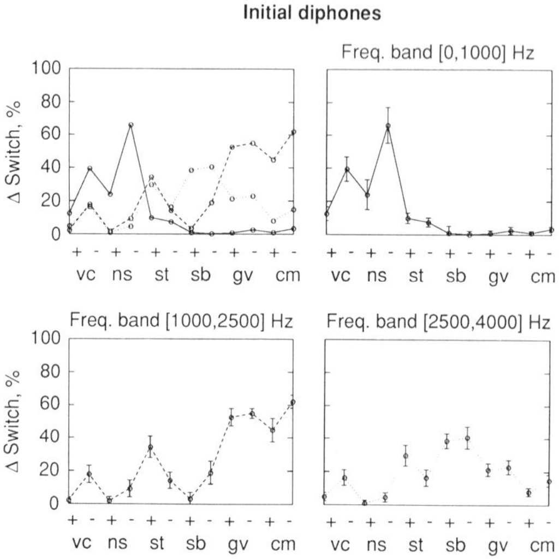

Figure 22 Human performance under the interchange of each frequency band over the entire diphone, on the DRT database. The upper left plot is a summary of the other 3 plots, with the confidence-interval bars omitted. The abscissa is as in Figure 12. The ordinate is termed "Δ*switch*", since it represents the *additional* number of switches, relative to the baseline version, that occurred due to the particular interchange operation. The upper right plot shows the amount of Δ*switch*, in percent, when the first frequency band of the initial diphone is interchanged. The lower left plot is for the interchange of the second frequency band, and the lower right plot is for the interchange of the third frequency band of the initial diphone. Notice that *Voicing* and *Nasality* are strongly correlated with the first frequency band of the diphone, *Graveness* and *Compactness* with the second frequency band of the diphone, and *Sibilation* with the third frequency band of the diphone.

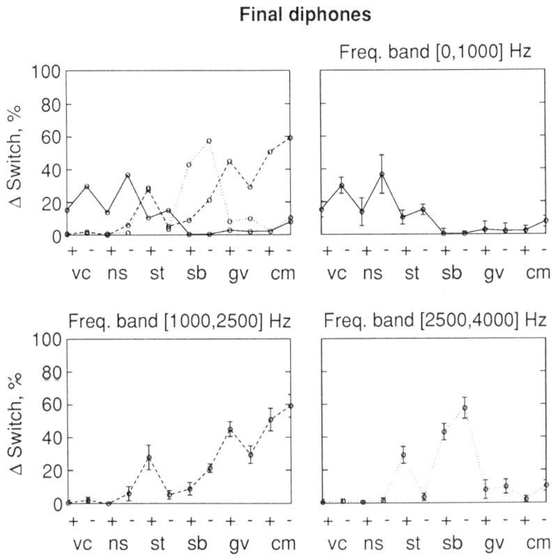

Figure 23 Same as Figure 22, for DALT.

should serve as a guideline to formulate a proper measure, to replace D of Eq. (17.6) (and d of Eq. (17.5)).

6 SUMMARY

In this paper we evaluated the adequacy of current peripheral auditory models to mimic human performance, in the context of speech recognition and speech coding. Models with such a capability are of interest since they can provide a basis for realizing effective speech processing systems. We first described our auditory model, then we outlined the methodology for a quantitative evaluation of the model accuracy, and finally we provided a critique of the current status of the model.

We focused on a particular auditory model, the EIH, that was developed using a "bottom-up" approach. Its first stage uses knowledge about the functional operation of pre auditory-nerve elements of the auditory periphery to simulate, in considerable detail, human auditory nerve firing activity. In order to achieve human performance, the simulated auditory nerve firing patterns would have to be processed according to principles derived from properties of post auditory-nerve nuclei. Unfortunately, such information is not available at present. Our approach, therefore, was to process the simulated auditory nerve firing patterns according to principles that are motivated by observed properties of the actual auditory nerve response. The resulting representation, the EIH, is a measure of the spatial (tonotopic) extent of coherent activity across the simulated auditory nerve and it signifies different physical properties of the acoustic stimulus depending on frequency, from synchrony at low frequencies to instantaneous rate at high frequencies. As such, EIH differs markedly from traditional speech representation methods and, consequently, exhibits quite different properties. The premise is that these properties accurately reflect properties of the internal human representation of speech.

The extent of this accuracy was measured in a quantitative manner by comparing the performance of a system that uses the EIH representation to that of a human listener while performing the same task. This methodology was used in tasks related to speech recognition and to speech coding. In general, performance was improved by replacing a cepstral speech representation by the auditory model, but was still short of achieving human performance. The evaluation process indicated that the shortcomings occur particularly in tasks related to low bit-rate coding and to speech recognition. Since schemes for low bit-rate coding rely on signal manipulations that spread over durations of several tens of ms, and since schemes for speech recognition rely on phonemic/articulatory information that extend over similar time intervals, it was concluded that the shortcomings are due mainly to a lack of perceptually related integration rules over durations of 50–100 ms. The results of the "tiling" experiment confirmed human usage of such integration, with different integration rules for different time-frequency regions depending on the phoneme-discrimination task. From the perspective of representation of speech, we conclude that "snap-shot" representations (i.e., representations that are local in time and frequency) provided by current auditory models are inappropriate, and that a representation based on integration over time intervals of 50–100 ms is needed. Also needed is a formulation of a proper measure of the difference between two speech segments of that duration.

Acknowledgements

I wish to thank J. L. Goldstein for providing the parameters of the MBPNL cochlear model, M. M. Sondhi for stimulating discussions throughout this work, and D. A. Berkley, A. L. Gorin, W. B. Kleijn, M. A. Randolph, J. Schroeter and M. Slaney for reviewing earlier versions of the paper.

REFERENCES

[1] Allen, J. B., "Cochlear modeling," *IEEE-ASSP Magazine*, 1985, pp. 3–29.

[2] Chen, J.-H., Cox, R. V., Lin, Y.-C., Jayant, N. and Melchner, M. J., "A low-delay CELP coder for the CCITT 16 kb/s speech coding standard," *IEEE Transactions on Selected Areas in Communications*, JASC-10, 1992, pp. 830–849.

[3] Delgutte, B. and Kiang, N. Y. S., "Speech coding in the auditory nerve: I. Vowel-like sounds," *Journal of the Acoustical Society of America*, 75, 1984, pp. 866–878.

[4] Delgutte, B. and Kiang, N. Y. S., "Speech coding in the auditory nerve: III. Voiceless fricative consonants," *Journal of the Acoustical Society of America*, 75, 1984, pp. 887–896.

[5] Delgutte, B. and Kiang, N. Y. S., "Speech coding in the auditory nerve: IV. Sounds with consonant-like dynamic characteristics," *Journal of the Acoustical Society of America*, 75, 1984, pp. 897–907.

[6] Delgutte, B. and Kiang, N. Y. S., "Speech coding in the auditory nerve: V. Vowels in background noise," *Journal of the Acoustical Society of America*, 75, 1984, pp. 908–918.

[7] Delgutte, B., "Analysis of French stop consonants using a model of the peripheral auditory system," in *Invariance and variability in speech processes* (J. S. Perkell and D. H. Klatt, editors), Lawrence-Erlbaum, Hillsdale, New Jersey, 1986, pp. 163–177.

[8] Deng, L., Geisler, C. D. and Greenberg, S., "A composite model of the auditory periphery for the processing of speech," *Journal of Phonetics*, 16, 1988, pp. 93–108.

[9] Ghitza, O., "Auditory nerve representation as a basis for speech processing," in *Advances in speech signal processing* (S. Furui and M. M. Sondhi, editors), Marcel Dekker, New York, 1992, pp. 453–485.

[10] Ghitza, O., "Adequacy of auditory models to predict internal human representation of speech sounds," *Journal of the Acoustical Society of America*, 93(4), 1993, pp. 2160–2171.

[11] Ghitza, O., "Processing of spoken CVCs in the auditory periphery: I. Psychophysics," *Journal of the Acoustical Society of America*, 94(5), 1993, pp. 2507–2516.

[12] Ghitza, O. and Sondhi, M. M., "Hidden Markov models with templates as nonstationary states: An application to speech recognition," *Computer Speech and Language*, 7(2), 1993, pp. 101–120.

[13] Goldstein, J. L., "Modeling rapid waveform compression on the basilar membrane as a multiple-bandpass-nonlinearity filtering," *Hearing Research*, 49, 1990, pp. 39–60.

[14] Goldstein, J. L. and Srulovicz, P., "Auditory nerve spike intervals as an adequate basis for aural spectrum analysis," in *Psychophysics and physiology of hearing* (E. F. Evans and J. P. Wilson, editors), Academic Press, London, 1977, pp. 337–346.

[15] Greenwood, D., "A cochlear frequency-position function for several species – 29 years later," *Journal of the Acoustical Society of America*, 87(6), 1990, pp. 2592–2605.

[16] Jakobson, R., Fant, C. G. M. and Halle, M., "Preliminaries to speech analysis: the distinctive features and their correlates," *Technical Report No. 13*, Acoustic Laboratory, Massachusetts Institute of Technology, Cambridge, Mass., 1952.

[17] Jayant, N. S. and Noll, P., *Digital coding of waveforms*, Prentice-Hall, New Jersey, 1984.

[18] Johnson, D. H., "The relationship between spike rate and synchrony in responses of auditory-nerve fibers to single tones," *Journal of the Acoustical Society of America*, 68(4), 1980, pp. 1115–1122.

[19] Kiang, N. Y. S. and Peake, W. T., "Physics and physiology of hearing," in *Stevens handbook of experimental psychology*, 2nd edition, 1, Wiley, New York, 1988, pp. 277–326.

[20] Kitawaki, N., "Quality assessment of coded speech," in *Advances in speech signal processing* (S. Furui and M. M. Sondhi, editors), Marcel Dekker, New York, 1992, pp. 453–485.

[21] Kroon, P. and Atal, B. S., "Predictive coding of speech using analysis-by-synthesis techniques," in *Advances in speech signal processing* (S. Furui and M. M. Sondhi, editors), Marcel Dekker, New York, 1992, pp. 141–164.

[22] Liberman, M. C., "Auditory-nerve response from cats raised in a low-noise chamber," *Journal of the Acoustical Society of America*, 63, 1978, pp. 442–455.

[23] Liberman, M. C., "Single-neuron labeling in the cat auditory nerve," *Science*, 216, 1982, pp. 1239–1241.

[24] Lyon, R. F., "Experiments with a computational model of the cochlea," *International Conference on Acoustics, Speech and Signal Processing – ICASSP* '86, 3, 1986, pp. 1975–1978.

[25] Moore, B. C. J., *Frequency selectivity in hearing*, Academic Press, London, 1986.

[26] Moore, B. C. J. and Glasberg, B. R., "Suggested formula for calculating auditory-filter bandwidth and excitation patterns," *Journal of Acoustical Society of America*, 74, 1983, pp. 750–753.

[27] Pickles, J. O., *An introduction to the physiology of hearing*, Academic Press, London, 1982.

[28] Quackenbush, S. R., Barnwell, T. P., III and Clements, M. A., *Objective measures of speech quality*, Prentice Hall, Englewood Cliffs, New Jersey, 1988.

[29] Sachs, M. B. and Young, E. D., "Encoding of steady-state vowels in the auditory nerve: Representation in terms of discharge rate," *Journal of the Acoustical Society of America*, 66, 1979, pp. 470–479.

[30] Sachs, M. B., Blackburn, C. C. and Young, E. D., "Rate-place and temporal-place representations of vowels in the auditory nerve and anteroventral cochlear nucleus," *Journal of Phonetics*, 16, 1988, pp. 37–54.

[31] Seneff, S., "A joint synchrony/mean-rate model of auditory speech processing," *Journal of Phonetics*, 16, 1988, pp. 55–76.

[32] Shamma, S., "The acoustic features of speech sounds in a model of auditory processing: vowels and voiceless fricatives," *Journal of Phonetics*, 16, 1988, pp. 77–92.

[33] Voiers, W. D., "Evaluating processed speech using the Diagnostic Rhyme Test," *Speech Technology*, 1(4), 1983, pp. 30–39.

[34] Voiers, W. D., "Effects of noise on the discriminability of distinctive features in normal and whispered speech," *Journal of the Acoustical Society of America*, 90(4), Pt. 2, Abst. 6SP10, October 1991.

[35] Wang, S., Sekey, A. and Gersho, A., "An objective measure for predicting subjective quality of speech coders," *IEEE Transactions on Selected Areas in Communications*, JSAC-10, 1992, pp. 819–830.

[36] Wilpon, J. G. and Rabiner, L. R., "A modified K-means clustering algorithm for use in isolated word recognition," *IEEE Transactions on Acoustics, Speech and Signal Processing*, ASSP-33, 1985, pp. 587–594.

[37] Young, E. D. and Sachs, M. B., "Representation of steady-state vowels in the temporal aspects of the discharge patterns of populations of auditory-nerve fibers," *Journal of the Acoustical Society of America*, 66, 1979, pp. 1381–1403.

18

APPLICATIONS OF WAVELETS TO SPEECH PROCESSING: A CASE STUDY OF A CELP CODER

James Ooi* and Vishu Viswanathan**

* *Massachusetts Institute of Technology*
Cambridge, Massachusetts

** *Texas Instruments, Inc.*
Dallas, Texas

ABSTRACT

Wavelets are a new family of orthogonal basis functions for representing finite energy signals. In this chapter, we provide a brief review of wavelets and their properties. We cite a number of applications of wavelets to speech processing that have been proposed recently. As a detailed case study in wavelet applications, we present our work on a wavelet-transform-based CELP coder design for high-quality speech coding at about 4.8 kbits/s. The coder quantizes the second residual using a wavelet transform approach instead of the stochastic-codebook-based vector quantization normally used in CELP coders, including the U.S. Federal Standard FS 1016 coder at 4.8 kbits/s. The wavelet coder improves the computational efficiency for encoding the second residual by requiring only 1.2 MIPS instead of 8.3 MIPS required by FS 1016. Subjective speech quality tests involving pairwise comparisons show that the wavelet coder was preferred 61% of the time over FS 1016.

1 INTRODUCTION

Wavelets are a new family of orthogonal basis functions for representing finite energy signals, and they offer three attractive properties: highly compact representation for smooth signals, nearly uncorrelated transform coefficients, and computationally efficient decomposition and reconstruction [1],[2],[3]. A number of researchers have recently exploited these and other properties of wavelets and applied them to a number of signal processing problems in areas including speech, image, video, and radar. In Section 2, we provide a brief review of wavelets. A number of references are provided to facilitate further study of wavelets for interested readers. In Section 3, we mention a number of recent investigations of wavelet applications to speech processing. In Section 4, as a case study in wavelet applications to speech processing, we present and discuss in detail our own recent work on a computationally efficient, wavelet-transform-based code excited linear prediction (CELP) speech coder at a data rate of about 4.8 kbits/s. Section 5 presents our conclusions.

2 WAVELETS BACKGROUND

A large number of papers, monographs, and books on wavelets have appeared in the last several years. Our purpose in this section is to give a brief introduction to wavelets and discuss some of their important properties. As a majority of articles on wavelets tend to be rather mathematically oriented, we suggest that readers unfamiliar with wavelets start with simpler tutorial-type articles such as [3],[18],[19] and proceed to more detailed treatments [1],[2],[12],[16]. Several special issues of technical journals have been devoted exclusively to wavelets, including the March 1992 issue (Part II) of the IEEE Transactions on Information Theory and the December 1993 issue of the IEEE Transactions on Signal Processing.

Wavelets are a family of basis functions for the space of square integrable functions or signals $L^2(\mathbf{R})$ [1]. In the well-known Fourier analysis, the complex exponentials are used as a basis of such a space. The Fourier transform is then a representation of a signal with respect to the basis of the complex exponentials. Similarly, the wavelet transform of a signal is a representation of the given signal with respect to the wavelet basis. The wavelet basis is formed by the dilations and contractions (scale changes) and the shifts (translations) of a single wavelet function, the *mother wavelet.* One popular instance of the wavelet basis is the *dyadic* wavelet basis, which contains only dilations of the mother wavelet by powers of two and integer shifts. Specifically, a function $f(t) \in L^2(\mathbf{R})$ can be represented as

$$f(t) = \sum_{j=L}^{\infty} \sum_{k=-\infty}^{\infty} \sqrt{2^j} d(j,k) \psi(2^j t - k) + \sum_{k=-\infty}^{\infty} \sqrt{2^L} a(L,k) \phi(2^L t - k). \quad (18.1)$$

The function $\psi(t)$ is known as the mother wavelet, while $\phi(t)$ is known as the scaling function. The set of functions $\{\sqrt{2^L}\phi(2^L t - k), \sqrt{2^j}\psi(2^j t - k) \mid j \geq L, j, k, L \in \mathbf{Z}\}$, where $\mathbf{Z}$ is the set of integers, is an orthonormal basis for $L^2(\mathbf{R})$. The numbers $a(L,k)$ are known as the approximation coefficients at scale L, while $d(j,k)$ are known as the detail coefficients at scale j. Together, the approximation and detail coefficients are called wavelet coefficients. Because of the orthonormality of the basis functions, $a(L,k)$ and $d(j,k)$ can be expressed easily as

$$\begin{aligned} a(L,k) &= \sqrt{2^L}\int_{-\infty}^{\infty} f(t)\phi(2^L t - k)dt \\ d(j,k) &= \sqrt{2^j}\int_{-\infty}^{\infty} f(t)\psi(2^j t - k)dt. \end{aligned}$$

There are, of course, constraints on $\psi(t)$ and $\phi(t)$ so that the basis functions are orthonormal. These constraints are described by the following equations:

$$\phi(t) = \sum_k c(k)\phi(2t-k) \text{ for some } c(k) \tag{18.2}$$

$$\psi(t) = \sum_k (-1)^k c(1-k)\phi(2t-k) \tag{18.3}$$

$$\int_{-\infty}^{\infty}\phi(t)\phi(t-k)dt = \delta_{0,k}\ \ \forall k \in \mathbf{Z}. \tag{18.4}$$

Let us give some intuition about these constraints. Let $\mathbf{V}_\ell$ be the span of the integer translates of $\phi(2^\ell t)$. Equation (18.4) says that these translates are an orthonormal basis for $\mathbf{V}_\ell$. Equations (1.2) and (1.3) break up $\mathbf{V}_\ell$ into a space spanned by integer translates of $\phi(2^{\ell-1}t)$, *i.e.*, $\mathbf{V}_{\ell-1}$, and a space spanned by integer translates of $\psi(2^{\ell-1}t)$ which can be shown to be the complement of $\mathbf{V}_{\ell-1}$ in $\mathbf{V}_\ell$. Specifically, let $f_\ell(t)$ denote the projection of $f(t)$ onto the space $\mathbf{V}_\ell$. This projection $f_\ell(t)$ provides the best approximation (in the sense of minimum error energy) to $f(t)$ at scale ℓ and can be constructed from the coefficients $a(\ell,k)$, using the equation

$$f_\ell(t) = \sum_{k=-\infty}^{\infty}\sqrt{2^\ell}a(\ell,k)\phi(2^\ell t-k).$$

As the scale ℓ increases, the approximation becomes finer, converging to $f(t)$ as $\ell \to \infty$. The difference between the approximation at scale $\ell+1$ and that at ℓ, $f_{\ell+1}(t) - f_\ell(t)$, is completely described by the coefficients $d(\ell,k)$ using the equation

$$f_{\ell+1}(t) - f_\ell(t) = \sum_{k=-\infty}^{\infty}\sqrt{2^\ell}d(\ell,k)\psi(2^\ell t-k).$$

Using these relations, given $a(L,n)$ and $\{d(j,n)|j \geq L\}$, it is clear that we can build the approximation at any scale. Hence, the wavelet transform breaks the signal up into a coarse approximation $f_L(t)$ (described by $a(L,n)$) and

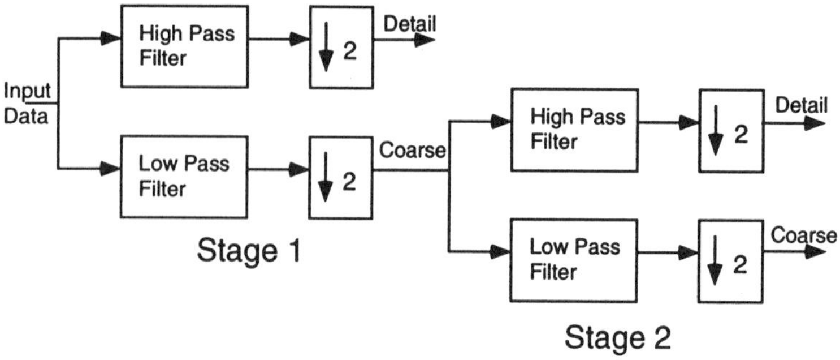

Figure 1 Two-stage discrete wavelet decomposition.

an infinite number of layers of detail $\{f_k(t) - f_{k-1}(t) | k > L\}$ (described by $\{d(k,n)|k \geq L\}$). As each layer of detail is added, the approximation at the next finer scale is achieved. Thus, equation (18.1) can also be written as $f(t) = \lim_{\ell \to \infty} f_\ell(t)$.

Because our work only concerns the processing of discrete-time signals, we must define a discrete wavelet transform (DWT). The DWT works as follows: Given a discrete-time signal $g(n)$, the DWT is found by letting $a(0,n) = g(n)$, and then finding the wavelet transform coefficients, $a(L,n)$ and $\{d(j,n) \mid L \leq j < 0\}$, corresponding to the continuous-time signal $f(t) = f_0(t) = \sum_k a(0,k)\phi(t-k)$. Together, $a(L,n)$ and $\{d(j,n) \mid L \leq j < 0\}$ are the DWT coefficients. In [2], Mallat describes an efficient, fast tree algorithm for computing these coefficients directly from $g(n)$. That is, $d(-1,n)$ and $a(-1,n)$ can be computed directly from $a(0,n)$. In turn, $d(-2,n)$, and $a(-2,n)$ can be computed from $a(-1,n)$, and so on. More precisely,

$$a(\ell,m) = \left.\frac{1}{\sqrt{2}}\left(a(\ell+1,k) * c(-k)\right)\right|_{k=2m}$$

$$d(\ell,m) = \left.\frac{1}{\sqrt{2}}\left(a(\ell+1,k) * h(-k)\right)\right|_{k=2m}$$

where $*$ denotes the convolution operation (*i.e.*, $x(k)*y(k) = \sum_n x(k-n)y(n)$); $c(k)$ is defined in (1.2), and

$$h(k) = (-1)^k c(1-k). \tag{18.5}$$

This algorithm can be described in terms of linear filtering and sampling. Figure 1 illustrates the wavelet decomposition for a two-stage example. The algorithm consists of a number of identical stages. At the first stage, the

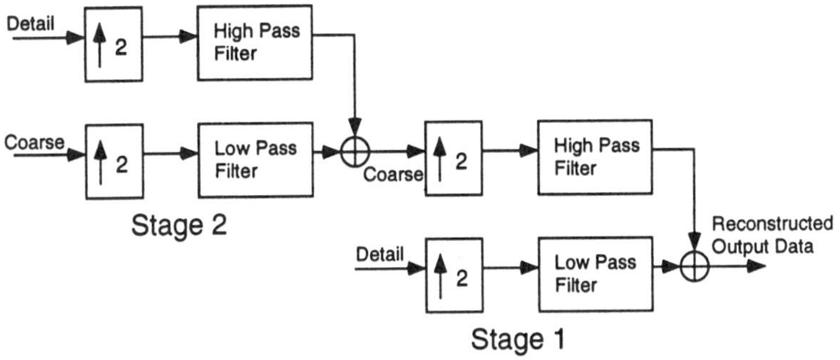

Figure 2 Two-stage inverse wavelet transform.

digitized input signal $g(n)$ is filtered through a lowpass filter $\frac{1}{\sqrt{2}}c(-n)$ and a highpass filter $\frac{1}{\sqrt{2}}h(-n)$. The filtered samples are then down-sampled by a factor of 2 to yield the approximation data $a(-1,n)$ and the detail data $d(-1,n)$. The detail data give high-frequency information about the signal and the approximation data give low-frequency information. For subsequent stages, the approximation data from the previous stage are used as input to obtain even finer low-frequency resolution. The detail data at every stage, together with the approximation data at the last stage, form the wavelet transform coefficients. Because of down-sampling at each stage, the total number of transform coefficients is the same as the total number of input data samples. Reconstruction of the signal from the DWT coefficients, or the inverse wavelet transform, is accomplished by reversing the above procedure. Figure 2 illustrates the inverse wavelet transform for a two-stage example. The approximation data and the detail data at each stage of the wavelet transform are up-sampled and passed through lowpass and highpass filters ($\frac{1}{\sqrt{2}}c(n)$ and $\frac{1}{\sqrt{2}}h(n)$, respectively) and then added to form the approximation data of the next stage.

The wavelet transform has an advantage over the Fourier transform in that the wavelet representation provides simultaneous time and frequency localization (within the limits of the uncertainty principle), thereby leading to the so-called time/frequency or time/scale representation of signals [3]. In particular, the different scaled versions of the wavelet function form a set of bandpass filters; each translated wavelet function picks up information about a certain temporal region and around a certain frequency region. This point is illustrated in Figure 3, which shows on the left three scaled versions of a wavelet function and on the right their corresponding Fourier transforms. We also see from Figure 3 that increased time resolution results as we go from (a) to (c) to (e) but at the

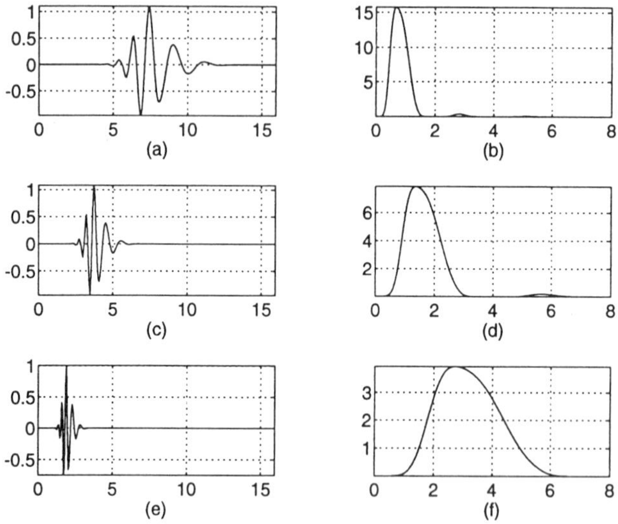

Figure 3 (a),(c),(e) Different scales of a wavelet function (time scale is in ms) and (b),(d),(f) the corresponding Fourier transforms (frequency scale is in KHz).

expense of decreased frequency resolution as seen from Fourier transform plots (b), (d), and (f).

The wavelet transform also provides a good approximation to the Karhunen-Loève transformation (KLT) of a wide class of stationary and nonstationary processes [6]. The wavelet transform thus provides us with a compact representation of signals without specific a priori knowledge of their statistics (which is required by the KLT). This can be understood intuitively as follows. Suppose an ensemble of signals is given which has the property that proximate samples are more statistically dependent than distant samples. It is plausible that the wavelet transform coefficients will be less correlated than the signal samples by the following reasoning. At the approximation level, the scale is low, so coefficients at that scale represent points in time which are distant and hence nearly uncorrelated. The detail coefficients at the next scale represent the error between the approximation at the previous scale and the approximation at the current scale. If we assume that the approximations at two adjacent scales are close, then the error between them will consist of nearly uncorrelated samples, much like quantization errors that tend to be nearly uncorrelated when the quantization is fine enough. The extent of compactness of signal representation depends on the number of vanishing moments of the wavelet function used. The number of vanishing moments indicates the smoothness of the wavelet function and hence the ease with which it can approximate smooth functions. A wavelet with p vanishing moments satisfies the following equation:

$$\int_{-\infty}^{\infty} t^m \psi(t) dt = 0 \text{ for } m = 0, \ldots, p-1,$$

or, equivalently,

$$\sum_k (-1)^k k^m c(k) = 0 \text{ for } m = 0, \ldots, p-1.$$

Thus, the number of vanishing moments indicates the number of zeros at $\omega = \pi$ of the filter $c(k)$, as well as the order of the polynomials to which $\psi(t)$ is orthogonal. Hence, if $\psi(t)$ has p vanishing moments, then $\psi(t)$ can only represent the portion of the signal which is less smooth than a $(p-1)$th order polynomial. Therefore, $\phi(t)$ can only represent the smooth portion of the signal. For representation of smooth signals, the higher the number of vanishing moments, the faster the decay of the higher scale wavelet coefficients. Thus, wavelets with a high number of vanishing moments lead to a more compact signal representation and are hence useful in coding applications. Data compression is achieved since a large number of transform coefficients will be either zero or will have negligible magnitudes.

Use of wavelets for signal coding is also motivated by connections between the wavelet transform and subband coding. First note that (18.5) implies that $(c(k), h(k))$ are conjugate quadrature filters (CQF's), as defined in [8]. (These filters are also called quadrature mirror filters (QMF's).) Next, note that (1.2) and (1.4) imply that $c(k)$ satisfies the *power complementary* property; *i.e.*,

$$|C(e^{j\omega})|^2 + |C(e^{j(\omega+\pi)})|^2 = 4,$$

where $C(e^{j\omega}) = \sum_k c(k) e^{-j\omega k}$. Together, the CQF property and the power complementary property imply that $(c(k), h(k))$ is a pair of perfect reconstruction (PR) filters, *i.e.*, filters that can be used to divide the signal into lowpass and highpass regions with no loss of information. Hence, the DWT coefficients are the output of an iterated PR-QMF filter bank, and the DWT can be interpreted as implementing an octave-band filter bank. Hence, coding with the wavelet transform is strongly tied to subband coding with QMFs, which has already been explored in [10],[9].

Although the wavelet transform is intimately related to subband coding and is, in fact, efficiently implemented using a subband algorithm as illustrated above in Figures 1 and 2, there are important differences in viewpoints that the wavelet transform offers for analysis and applications. We mention here two such differences that have been exploited in data compression applications of wavelets. We consider the data transmission application for concreteness, although the statements made below are valid for other applications including data storage. First, subband coding has been traditionally used to divide the signal into appropriate subbands and allocate bits among subbands using some meaningful (*e.g.*, perceptually based) criteria. Once the bits have been allocated, every time sample in each subband is quantized and encoded using the same number of bits. Said another way, once a given subband has been chosen for transmission, all time samples in that subband are transmitted using the same number of bits for each sample. The wavelet transform offers the perspective of an orthogonal transformation. Hence, one may decide to send only a preselected number of the highest amplitude transform coefficients from each wavelet band or a preselected number from all wavelet bands. This method of selecting a subset of transform coefficients for transmission is sometimes referred to as adaptive zonal sampling. In fact, we use this selection method in our case study on the wavelet transform CELP coder as will be clear in Section 4.

Second, in traditional subband coding, the designer uses a number of criteria and designs the filters used for dividing the signal into subbands. Once the filters have been designed for a given application, they are held fixed during data compression in that application. This method is equivalent to the use of a single, fixed mother wavelet for data compression. Because of the orthogonal transform perspective, wavelets facilitate the use of adaptive basis functions involving multiple mother wavelets, one for each set of basis functions. A large number of mother wavelets (infinite, in theory) are available to choose from. Changing the basis functions means rotation in the appropriate vector space. In an optimal transmission system, one selects that mother wavelet which leads to sending the least number of data bits under a predefined error criterion. This idea is exploited in a recent work on wavelet-based audio signal compression [4].

3 APPLICATIONS OF WAVELETS TO SPEECH PROCESSING

In this section, we cite examples of recent wavelet applications to speech processing. We hasten to point out that this is not intended to be a survey of all such applications. The reader is encouraged to look at the references cited for additional related work in this area. First, it is well known that wavelet transform coefficients in different scales exhibit extrema corresponding to locations of discontinuities in the underlying signal. This property of wavelets has been used, for example, in edge detection in image processing [2]. In speech and audio processing, we cite two cases where this property has been exploited. For event-based pitch detection, in which pitch is estimated as the duration between successive glottal closures, the wavelet transform has been used to locate instances of glottal closures [14]. In this approach, significant local maxima of the wavelet transform with respect to the translation variable are located in each scale, and a glottal closure decision is reached if such locations agree across two scales. In another case involving high-fidelity audio coding, adaptive segmentation of the audio signal into frames is accomplished using the DWT [11]. The motivation in this application is to use short analysis frames in regions of sharp signal changes as would occur in the castanet signal, for example; this strategy minimizes audio signal distortions such as pre-echoes. The adaptive segmentation is achieved by looking at sudden energy changes in wavelet transform bands and confirming their presence over multiple scales.

Second, we cite the use of adaptive wavelets for high-fidelity audio signal encoding, which we mentioned above in Section 2 [17]. The same approach may also be used for high-quality speech coding.

Third, a wavelet-based representation is used to describe acoustic signal processing in the early stages of the auditory system [15]. This auditory representation has potential applications in automatic speech recognition and low data rate speech coding.

Fourth, a pitch synchronous wavelet transform has been developed for representing speech and music signals [20]. This transform has applications in speech coding, speech and audio signal modifications involving time-scale and pitch changes, and feature extraction for speech recognition.

Fifth, the so-called AM-FM modeling has been recently investigated in speech and image processing applications (see for example [13]). This modeling approach involves solving a nonlinear set of equations for calculating the AM-FM components through some approximations. The performance of such approximations has been shown to improve if the signal being modeled is first filtered into subbands and the approximations are carried out separately in each subband. The use of wavelet-based subband filters has been shown to provide compatibility or uniformity of these approximations across bands [13].

Finally, in our own detailed case study of the application of wavelets to speech coding, we investigated the use of wavelets in each of several speech coding approaches: adaptive transform coding (ATC), baseband coding, and CELP [5]. We applied wavelet-based ATC directly on the speech signal and in conjunction with spectral and/or pitch prediction methods; perceptual error weighting of the type used in CELP was also employed. For producing speech quality comparable to that of 4.8 to 8.0 kbits/s CELP coders, wavelet-based ATC required much higher bit rates on the order of 12 kbits/s and above. In baseband coding, we investigated the use of the wavelet transform in extracting the baseband instead of using a lowpass filter and experimented with several ways of approximating the high-frequency information from the baseband. Given our objectives of low bit rate and high speech quality, the wavelet transform CELP coder produced the best results. Section 4 below describes this coder in detail.

4 WAVELET TRANSFORM CELP CODER

The U.S. Federal Standard 4.8 kbits/s CELP coder (Version 3.2, referred to as FS 1016 below) described in [7] was used as the baseline system for this work. We replaced the stochastic codebook approach for encoding the second residual resulting from the pitch prediction (or adaptive codebook encoding) of the LPC residual with our wavelet transform approach. Also, we used a frame size of 32 ms and a subframe of 8 ms, instead of 30 ms and 7.5 ms used in FS 1016.

4.1 Coder Description

Figure 4 depicts the wavelet transform CELP (WT-CELP) coder. The input speech $s(n)$, sampled at 8 kHz, is analyzed once every 32 ms using a 10th order LPC analysis. The LPC coefficients are encoded using the line spectrum pairs (LSPs) at 34 bits/frame, with the bit allocation (3, 4, 4, 4, 4, 3, 3, 3, 3, 3) among the 10 LSPs. The LPC inverse filter $A(z)$, as shown in Fig. 4, is given by

$$A(z) = 1 + \sum_{k=1}^{10} a_k z^{-k},$$

where $\{a_k\}$ are the predictor coefficients. The LSPs are interpolated once every 8 ms subframe (64 samples) and used to update $A(z)$ every subframe. For a given subframe, $A(z)$ filters the speech signal to produce the LPC residual $e(n)$,

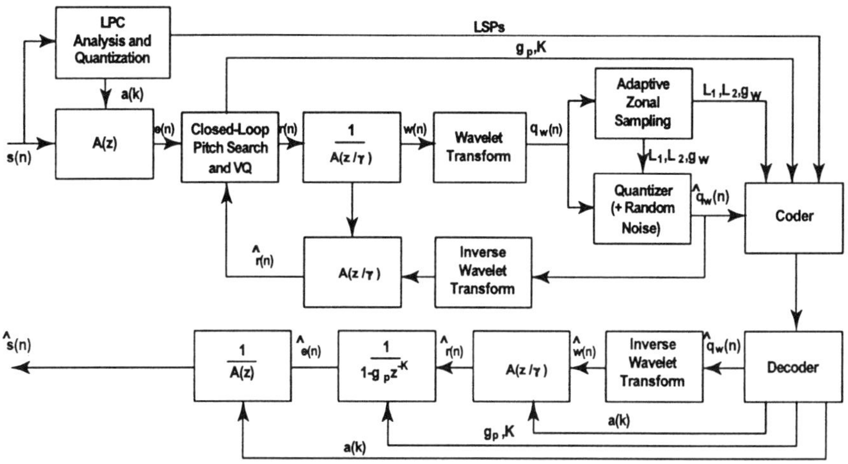

Figure 4 Block diagram of WT-CELP coder.

$0 \le n < 64$. The residual $e(n)$ is vector-quantized in a closed-loop manner from an adaptive pitch codebook, using perceptual weighting with $A(z)/A(z/\gamma)$ (we used $\gamma = 0.8$) and using 128 integer and 128 noninteger pitch delays, in exactly the same way as in [7]. Also as in [7], we used 8 bits for encoding pitch delay K for odd subframes and 6 bits for delta delay encoding for even subframes; the pitch predictor gain g_p is quantized using 5 bits. The second residual $r(n)$ is $e(n) - g_p p(n)$, where $p(n)$ is the code vector corresponding to the optimal delay K. The WT-CELP coder differs from the FS 1016 coder in the way the second residual $r(n)$ is encoded. FS 1016 searches a 9-bit stochastic codebook for the best code vector and gain, again using the perceptual error weighting $A(z)/A(z/\gamma)$. The wavelet transform coder first filters $r(n)$ with the appropriate perceptual weighting filter $1/A(z/\gamma)$, to produce $w(n)$. [This weighting filter may be viewed as a cascade of the LPC filter $1/A(z)$ and the perceptual speech error weighting filter $A(z)/A(z/\gamma)$.] The wavelet transform of the weighted signal $w(n)$, denoted as $q_w(n)$, is then quantized by the procedure described in Section 4.2. The quantized transform $\hat{q}_w(n)$ is inverse wavelet transformed to form $\hat{w}(n)$, which is inverse filtered with $A(z/\gamma)$ to produce the decoded second residual $\hat{r}(n)$. The decoded LPC residual $\hat{e}(n)$ is computed as $\hat{r}(n) + g_p\hat{p}(n)$, where $\hat{p}(n)$ is taken from the adaptive codebook. Finally, the decoded speech signal $\hat{s}(n)$ is obtained by filtering $\hat{r}(n)$ with $1/A(z)$. Table 1 gives the bit allocations used by the WT-CELP coder. The total bit rate is 4,687.5 bits/s, without including any forward error correction bits. The overall quantization noise $S(z) - \hat{S}(z)$ can be shown to be equal to $[W(z) - \hat{W}(z)]A(z/\gamma)/A(z)$[5]. Thus, the overall output speech has the desired perceptually shaped noise, pro-

Parameter	Subframe				Frame Total
	1	2	3	4	
LSP's	34				34
Pitch index	8	6	8	6	28
Pitch gain	5	5	5	5	20
DWT coeff.(2)	11	11	11	11	44
DWT coeff. magnitude(1)	4	4	4	4	16
DWT coeff. signs(2)	2	2	2	2	8
Total bits / frame					**150**

Table 1 Bit allocations used by the WT-CELP coder (32 ms frame).

vided that the quantization error $w(n) - \hat{w}(n)$ is white noise. We believe that this noise is approximately white. We note that the WT-CELP coder quantizes the second residual in an open-loop manner. In contrast, FS 1016 performs this quantization using the 9-bit stochastic codebook in a closed-loop manner. Computational savings provided by WT-CELP over FS 1016 (see Section 4.5) is a direct result of this open-loop quantization.

4.2 Quantization of Wavelet Transform Coefficients

Given our low bit rate goal, we quantize the wavelet transform coefficients $q_w(n)$ using the so-called adaptive zonal sampling. In simple terms, we transmit only the two highest amplitude DWT coefficients. We encode them with 11 bits for their locations (since there are 2,016 ways of selecting 2 out of 64, where 64 is the number of DWT coefficients), 4 bits for their average magnitude g_w, and 2 bits for their signs. The remaining 62 DWT coefficients are not transmitted (or encoded with zero bits). To maximize SNR, these untransmitted coefficients must be set to zero. However, this method yields a lowpassed, muffled output speech quality. Setting these coefficients to uniformly distributed random numbers with a maximum range that is some fraction of the transmitted average magnitude g_w yields better subjective quality. We set the maximum on the range to be 5% of g_w. The coder bit rate and speech quality can be increased by transmitting the locations and values of N ($N > 2$) highest amplitude DWT coefficients, without significantly increasing the computational complexity. This observation is important, as this is not the case with FS 1016 (see Section 4.5 for a specific complexity comparison).

4.3 Coder Optimization

We performed an extensive optimization of the WT-CELP coder using objective (segmental SNR) and subjective listening tests. The results of this optimization study are summarized below. First, we compared several different wavelets: Daubechies wavelets of length 6, 16, and 20; another 20-coefficient wavelet with 10 vanishing moments; and two cases in which a choice (based on minimum quantization error) was allowed among the Daubechies wavelets of different lengths: (6, 16), (6, 16, 20). We selected the 20-coefficient Daubechies wavelet, as it yielded good speech quality most consistently over all our test data. Second, the wavelet transform depth did not make much difference from four to six levels (stages), but speech quality was worse for fewer than four levels. We found quality to be best using the full decomposition, *i.e.*, six levels. Increasing the depth from four to six levels increases the computation only slightly. Third, when the magnitudes of the two highest DWT coefficients are set to the same value, the value that maximizes SNR is the average of the actual magnitudes of the two coefficients. However, we found that better subjective quality is achieved by setting the magnitude to less than the optimum. We use a value that is two-thirds the optimum value, as this yields consistently good speech quality. Fourth, the WT-CELP coder causes unvoiced speech sounds, such as "s" or breath noises, to sound "crackly," as if there were static. We attribute this to the fact that the pitch prediction is very poor in these sections, so the excitation signal is made up primarily of the second residual. However, this signal consists of the inverse wavelet transform of data that consist of only two large coefficients. Because the wavelet basis is highly localized in time for high frequencies, the main component of unvoiced speech, the decoded second residual consists of two pulses localized in time. Perceptually, for unvoiced sounds, it is preferable to have random noise excitation, which is more evenly distributed in time. Thus, we have added a simple voicing detector that is conservative in declaring sections unvoiced (*i.e.*, only speech that is clearly unvoiced is declared unvoiced). If a subframe of speech is declared unvoiced, then we do not perform wavelet transform coding on the second residual. Instead, we set the second residual to uniformly distributed random noise, with energy equal to 60% of the energy of the computed second residual. This method minimizes the distortions perceived during unvoiced sounds.

4.4 Speech Quality Evaluation

We performed subjective testing using pairwise comparisons of the WT-CELP coder with the FS 1016 coder. Thirteen test subjects, including two experienced listeners, were used. Each subject listened to 89 Diagnostic Acceptability Measure test sentences from 3 male and 3 female speakers, with random ordering of the WT-CELP and FS 1016 processed outputs. We found that the listeners preferred WT-CELP over FS 1016 about 61% of the time. Most listeners reported that the two coded versions were quite close in overall speech quality on many of the sentences. The two experienced listeners, however, preferred the wavelet coder 80% of the time.

4.5 Coder Complexity Comparison

In terms of number of multiplications and additions, the stochastic codebook search in FS 1016 requires about 8.3 MIPS [7]; in contrast, the WT-CELP coder requires only about 1.2 MIPS. Also, FS 1016 requires storing 512 code vectors; WT-CELP requires the storage of only 20 wavelet filter coefficients. Another advantage of the WT-CELP coder is that a higher data rate speech coder with higher speech quality can be obtained without significantly increasing the computational complexity, by simply sending a larger number of high-amplitude DWT coefficients. With FS 1016, increasing the codebook size increases the stochastic codebook search complexity rather dramatically; for example, increasing it to 10 bits nearly doubles the required MIPS.

5 CONCLUSIONS

In this chapter, we provided a brief review of wavelets and some of their important properties. We cited a number of recent applications of wavelets to speech processing. As a detailed case study in wavelet applications to speech coding, we presented a wavelet-transform-based CELP coder design at about 4.8 kbits/s. The WT-CELP coder is based on the U.S. Federal Standard FS 1016 CELP coder at 4.8 kbits/s and differs only in the way the second residual is encoded. WT-CELP uses the wavelet transform of the perceptually weighted second residual and transmits only the two highest amplitude transform coefficients, instead of the 9-bit stochastic-codebook-based vector quantization used in FS 1016. This leads to a much lower computational complexity for WT-CELP than for FS 1016 (1.2 MIPS versus 8.3 MIPS for the second residual encoding). Subjective speech quality tests involving pairwise comparisons show that WT-CELP was preferred 61% of the time over FS 1016.

6 ACKNOWLEDGMENTS

This work was performed at Texas Instruments as part of the first author's MIT 6A master's degree thesis. The authors would like to acknowledge Prof. Ahmed Tewfik and Mr. Murtaza Ali of the University of Minnesota for technical discussions on wavelets and for providing wavelet software. The authors would also like to thank the U.S. Department of Defense for providing the FS 1016 CELP coder software.

REFERENCES

[1] I. Daubechies, "Orthonormal Bases of Compactly Supported Wavelets," *Commun. Pure Appl. Math.*, pp. 909-996, November 1988.

[2] S. Mallat, "A Theory for Multiresolution Signal Decomposition: The Wavelet Representation," *IEEE Trans. Pattern Analysis and Machine Intelligence*, vol. 11, pp. 674-693, July 1989.

[3] O. Rioul and M. Vetterli, "Wavelets and Signal Processing," *IEEE Signal Proc. Magazine*, vol. 8, pp. 14-38, October 1991.

[4] D. Sinha and A. Tewfik, "Low Bit Rate Transparent Audio Compression Using a Dynamic Dictionary and Optimized Wavelets," *Proc. of Int. Conference on Acoust., Speech, Signal Proc.*, Minneapolis, MN, pp. I-197:I-200, April 1993.

[5] J. Ooi, "Application of Wavelets to Speech Coding," Master of Science Thesis, Department of Electrical Engineering and Computer Science, Massachusetts Institute of Technology, June 1993.

[6] A. Tewfik and M. Kim, "Fast Positive Definite Linear System Solvers," *IEEE Trans. Signal Proc.*, vol. 42, pp. 572-585, March 1994.

[7] J. P. Campbell Jr., T. E. Tremain, and V. C. Welch, "The DoD 4.8 KBPS Standard (Proposed Federal Standard 1016)," *Advances in Speech Coding*, Kluwer, 1991.

[8] M. J. T. Smith and T. P. Barnwell, "Exact Reconstruction Techniques for Tree-Structured Subband Coders," *IEEE Trans. Acoust., Speech, Signal Proc.*, vol. 34, pp. 434-441, June 1986.

[9] T. P. Barnwell, "Subband Coder Design Incorporating Recursive Quadrature Filters and Optimum ADPCM Coders," *IEEE Trans. Acoust., Speech, Signal Proc.*, vol. 30, pp. 751-765, October 1982.

[10] A. J. Barabell and R. E. Crochiere, "Subband Coder Design Incorporating Quadrature Filters and Pitch Prediction," *Proc. IEEE Int. Conference on Acoust., Speech, Signal Proc.*, Washington, DC, pp. 530-533, April 1979.

[11] M. Ali, A. H. Tewfik, and V. R. Viswanathan, "Real Time Implementation of Second Generation of Audio Multilevel Information Coding," *Proc. SPIE Conference*, vol. 2242, Orlando, FL, pp. 212-223, April 1994.

[12] R. K. Young, *Wavelet Theory and Its Applications*, Kluwer Academic Publishers, 1993.

[13] A. C. Bovik, P. Maragos, and T. F. Quatieri, "AM-FM Energy Detection and Separation in Noise Using Multiband Energy Operators," *IEEE Trans. Signal Proc.*, vol. 41, pp. 3245-3265, December 1993.

[14] S. Kadambe and G. F. Boudreaux-Bartels. "Application of the Wavelet Transform for Pitch Detection of Speech Signals," *IEEE Trans. Info. Theory*, vol. 38, part II, pp. 917-924, March 1992.

[15] X. Yang, K. Wang, and S. A. Shamma, "Auditory Representations of Acoustic Signals," *IEEE Trans. Info. Theory*, vol. 38, part II, pp. 824-839, March 1992.

[16] I. Daubechies, *Ten Lectures on Wavelets,* SIAM, 1992.

[17] D. Sinha and A. H. Tewfik, "Low Bit Rate Transparent Audio Compression Using Adapted Wavelets," *IEEE Trans. Signal Proc.*, vol. 41, pp. 3463-3479, December 1993.

[18] G. Strang, "Wavelets and Dilation Equations," *SIAM Review*, vol. 31, pp. 613-627, December 1989.

[19] C. E. Heil and D. F. Walnut, "Continuous and Discrete Wavelet Transforms," *SIAM Review*, vol. 31, pp. 628-666, December 1989.

[20] G. Evangelista, "Pitch-Synchronous Wavelet Representations of Speech and Music Signals," *IEEE Trans. Signal Proc.*, vol. 41, pp. 3313-3330, December 1993.

INDEX

T